THE TRAVELLER'S GUIDE TO
THE GOLDFIELDS

HISTORY & NATURAL HERITAGE TRAILS
THROUGH CENTRAL & WESTERN VICTORIA

A Warning & Request

Although every effort has been made to ensure the book is up-to-date and correct, mistakes happen, things change, and improvements can always be made. Please send any suggestions, corrections or additions for the next edition to Richard Everist via goldfields@bestshot.com.au, or BestShot! Goldfields, PO Box 850, Torquay 3228.

All contributions will be acknowledged in the next edition, and the very best will be rewarded with a free copy of BestShot!'s *The Heart of Victoria* coffee-table guidebook. Excerpts from your correspondence may appear in new editions of this guide or in website updates. Please specify if you don't want your letter published or your name acknowledged.

The writers and the publisher have tried to make the information as accurate as possible, but they accept no responsibility for any loss, injury or inconvenience sustained by any person using this book or by any business, organisation or individual featured in this book.

The Traveller's Guide to the Goldfields
History & Natural Heritage Trails Through Central Victoria

ISBN 0-9756023-3-0

Published in August 2006

Published by Best Shot! Publications Pty Ltd, ACN 100 252 926,
PO Box 850, Torquay 3228

Designed in Ballarat by Ascet Creative

Printed in Maryborough by McPherson's Printing Group

Text © Best Shot! Publications Pty Ltd 2006

Photos © as indicated: Ballarat Tourism (BaT), David Bannear (DM), Bendigo Tourism (BeT), Best Shot! (BS), Gary Chapman (GC), Richard Everist (RE), Elaine Grant (EG), Gold Museum (GM), Martin Hurley (MH), Joe Kinsela (JK), Joe Mortelliti (JM), Parks Victoria (PV), Alison Pouliot (AP), Katherine Seppings (KS), Sovereign Hill (SH), George Stawicki (GS), Pete Walsh (PW)

Front Cover Top: Bendigo Town Hall, Katherine Seppings
Front Cover Bottom Left: Castlemaine Diggings National Heritage Park,
    Parks Victoria (Alison Pouliot)
Front Cover Bottom Centre: Old Savings Bank, Castlemaine, Richard Everist
Front Cover Bottom Right: Daylesford, Pete Walsh
Back Cover Top: Castlemaine Diggings National Heritage Park, Katherine Seppings
Back Cover Centre: Central Deborah Mine, Bendigo, Central Deborah Mine
Back Cover Bottom: Mully's Café, Bendigo, Richard Everist

The Traveller's Guide to

THE GOLDFIELDS

History & Natural Heritage Trails
Through Central & Western Victoria

CONTENTS

SMALL TOWNS & VILLAGES.......... 197

PRACTICAL INFORMATION 289

INDEX.................... 304

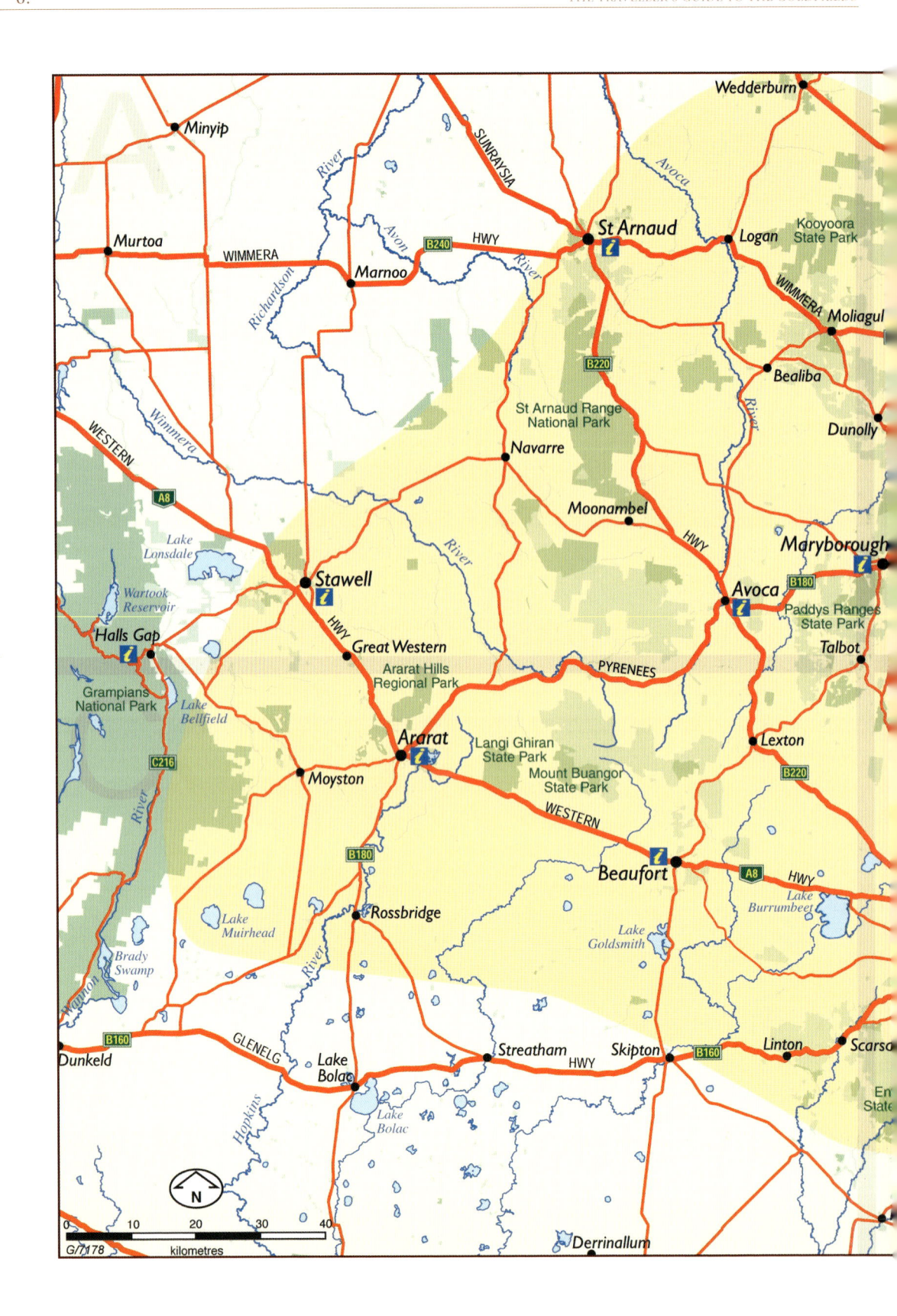

MAP A: *Towns and cities indicated in this colour are featured in this book.*

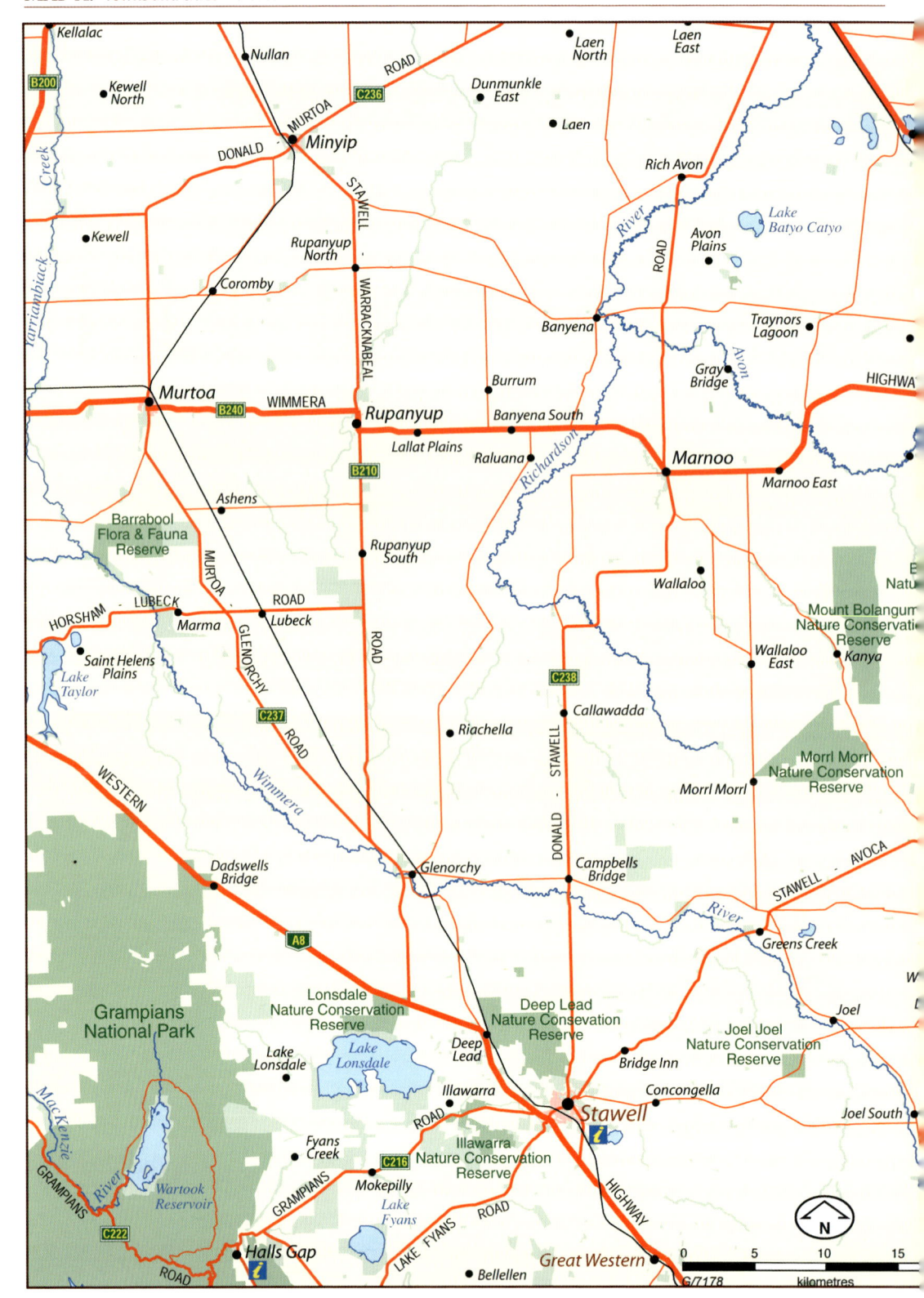

MAP B: *Towns and cities indicated in this colour are featured in this book.*

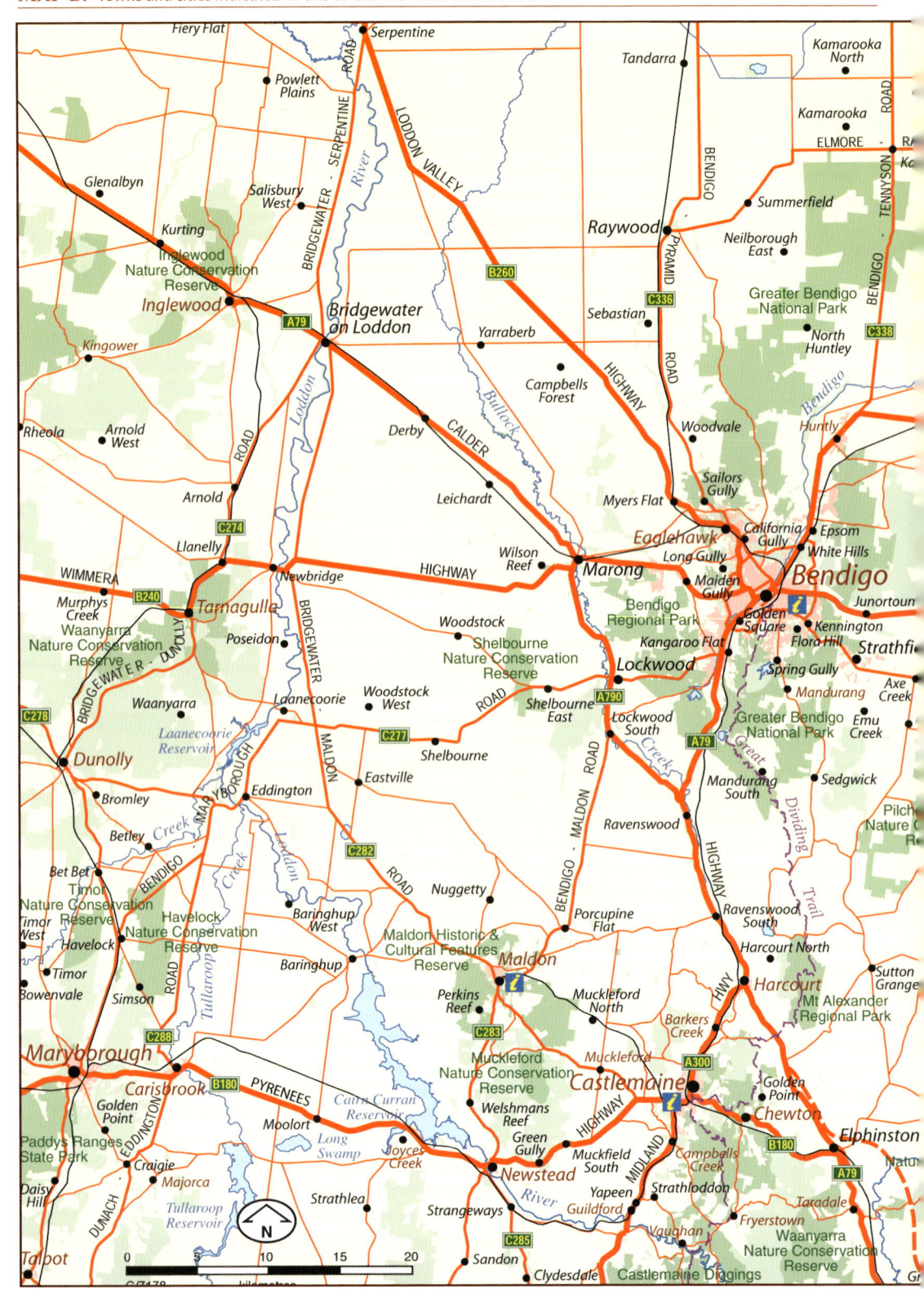

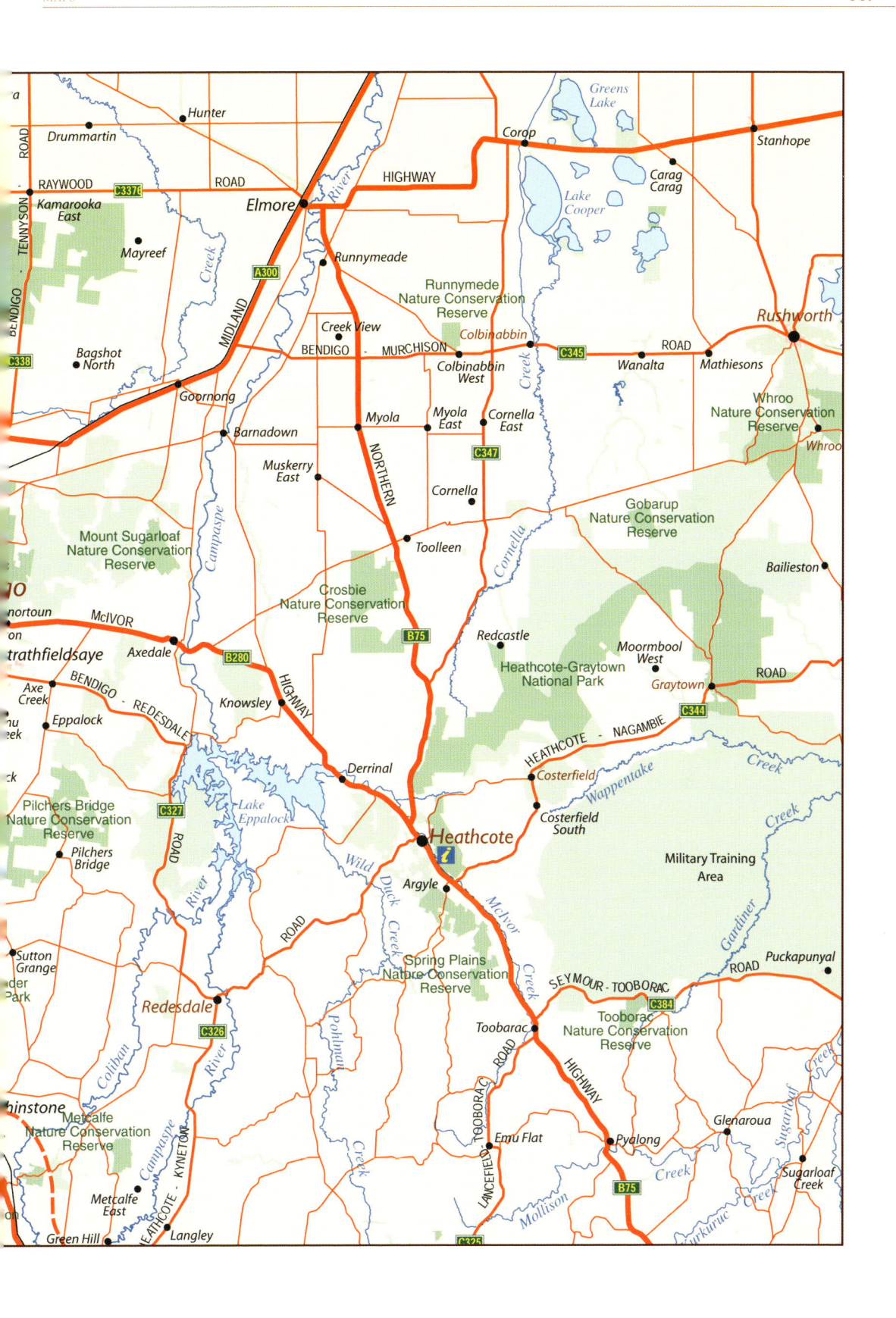

MAP C: *Towns and cities indicated in this colour are featured in this book.*

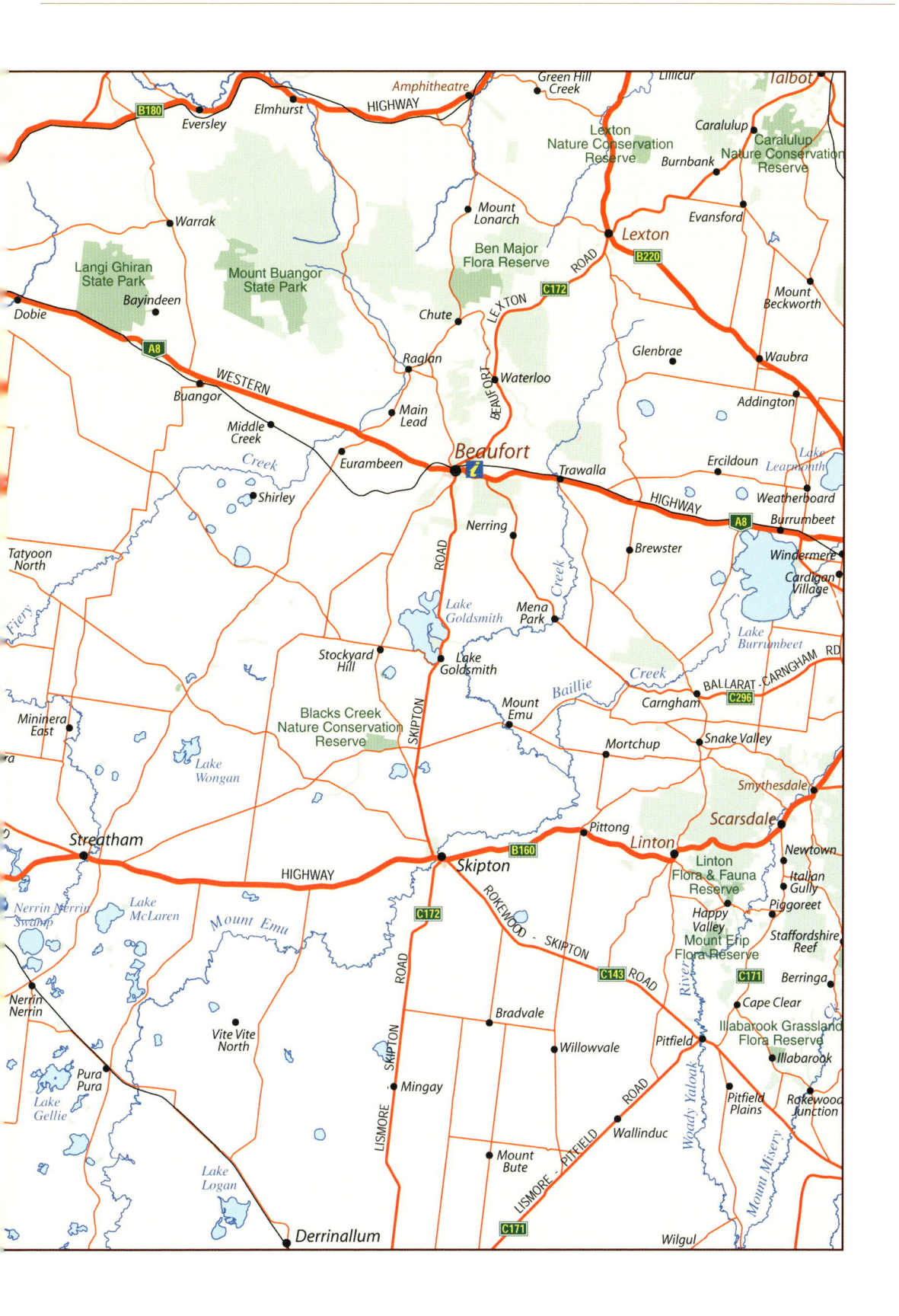

MAP D: *Towns and cities indicated in this colour are featured in this book.*

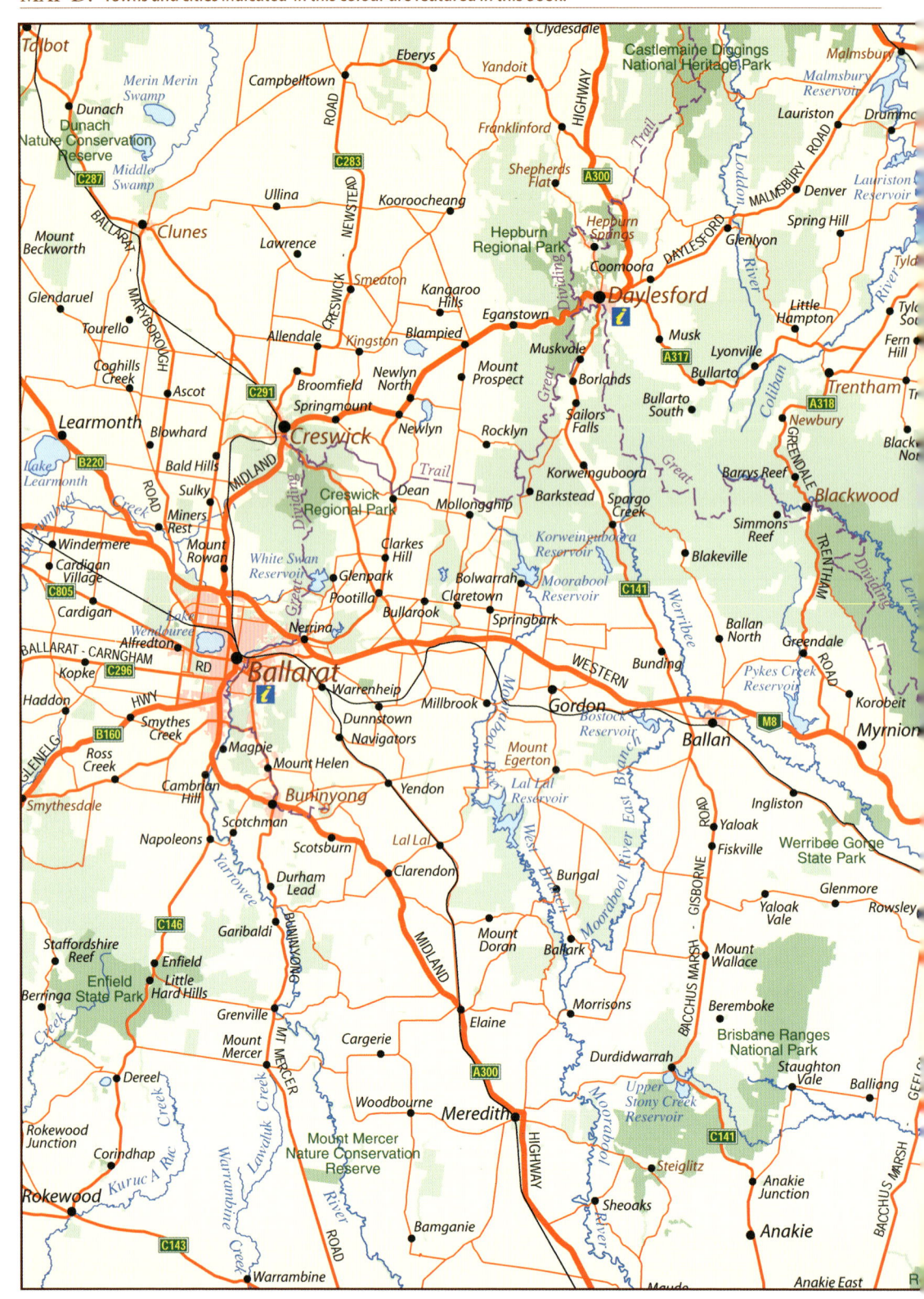

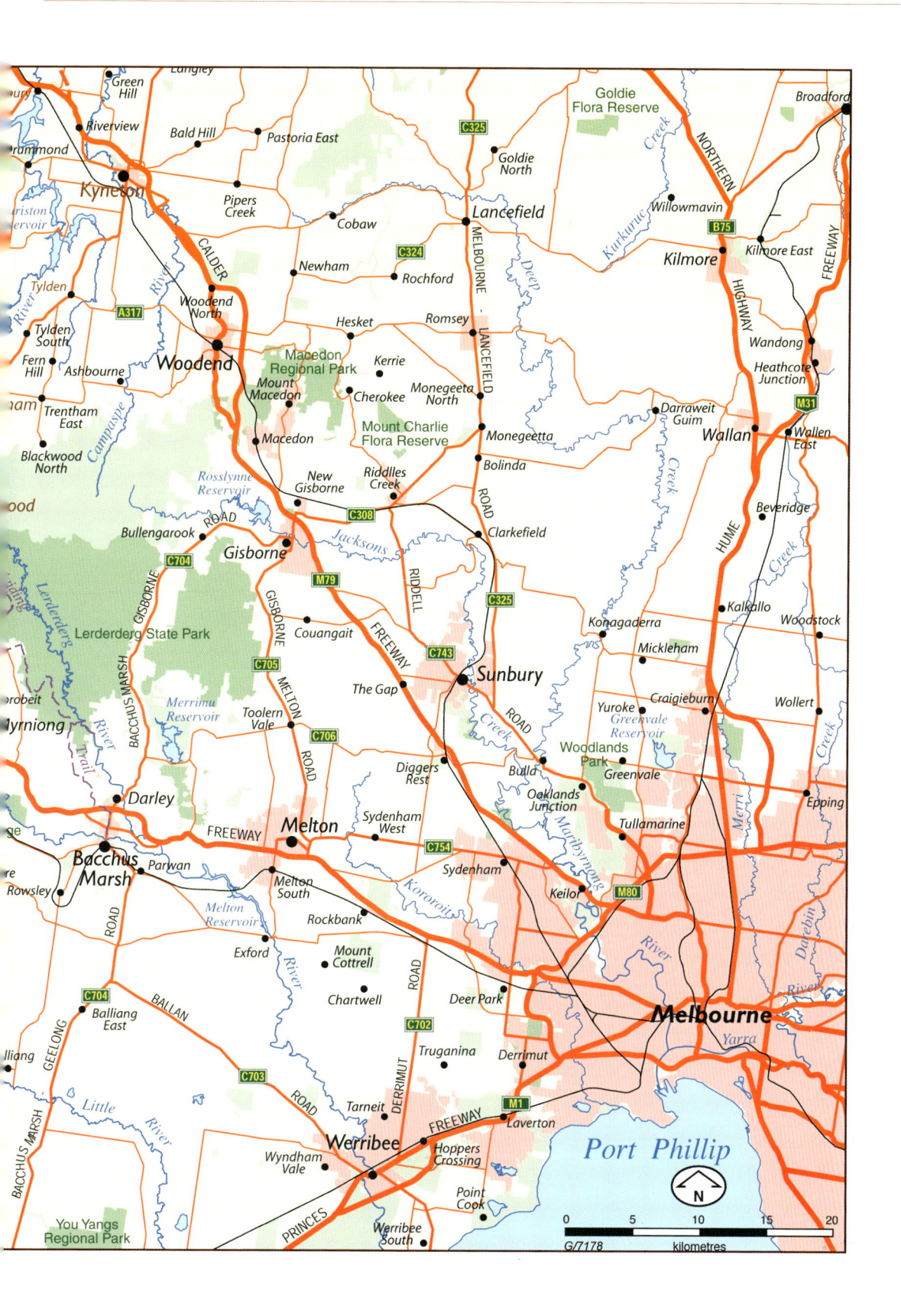

Traditional Aboriginal Landowners

At the time of the gold rush, there were many Aboriginal clans in central Victoria all speaking the dja dja wurrung language. My ancestors occupied the land for thousands and thousands of years.

The gold seekers came from societies whose existence depended on gold-based currencies. This meant it proved difficult for them to empathise with my people whose way of life was so inextricably linked to country – our mother – emotionally, spiritually and physically. The irony is that my people, traditional owners of one of the world's richest goldfields had no use for gold, too soft and heavy to be used to hunt or dig; yet it gave the newcomers, yellow fever.

I hope that readers will appreciate the Aboriginal voice as they travel through the Goldfields. I hope that hearing this voice will help them see that on this transformed country, before the times when the world came looking for gold, walked men, women and children who carried with them everything they needed for survival – tools, ceremonial implements and possum skin cloaks.

BRIEN NELSON, Jaara Elder

Foreword

The great gold rush of central and western Victoria brought energetic and optimistic people who could imagine a better life. They mined and settled in a 100 or so goldfields and their experiences transformed pastoral and bush country into Australia's most recognisable 19th century cultural landscape.

The 1850s gold boom swept the frontier away, replacing it with the affluent and architecturally sophisticated cities of Ballarat and Bendigo, and regional centres such as Castlemaine and Maryborough. Ambitious government infrastructure projects supplied railways and roads to link the towns and cities, and large-scale investment provided schools, libraries, art galleries and hospitals. Within twenty years, gold had turned Melbourne from a straggling village into 'Marvellous Melbourne' – one of the world's biggest, richest and most cosmopolitan cities.

The parks and reserves of the Goldfields further reflect the richness of a remarkable cultural heritage and offer experiences that are both rewarding and unique. Today, these sites are reserved for the significance of both their natural and cultural values and continue to contribute to the lives of travellers and residents alike. This book provides a glimpse of the wealth that still abounds in this resilient resource and is a key to help unlock many of the secrets that make the Goldfields so distinctive.

The cultural heritage of the Goldfields region is still much in evidence. Succeeding generations have done their utmost to preserve and interpret this outstanding legacy. The myriad stories it tells awaits the visitor.

The Goldfields region has a history of welcoming strangers. Enjoy your travels!

CHRIS GALLAGHER, Chair, Heritage Council Victoria   Opposite: Shepherds Flat, JM

THIS BOOK

This book was commissioned by the Goldfields Tourism, the Heritage Council of Victoria, Parks Victoria and BestShot! Publications. Ray Osborne and David Bannear were the visionaries and enthusiasts who kick-started the project. David also made an enormous contribution in terms of his knowledge, his writing, and his photography. Richard Everist, the co-founder of BestShot!, was the project manager, co-ordinating editor, writer and photographer. Paul Mah, ASCET Creative, was the designer.

Although every effort has been made to ensure the book is up-to-date and correct, mistakes happen, things change, and improvements can always be made. Please send any suggestions, corrections or additions for the next edition to Richard Everist via goldfields@bestshot.com.au, or BestShot! Goldfields, PO Box 850, Torquay 3228. All contributions will be acknowledged in the next edition, and the best will receive a free copy of BestShot!'s *The Heart of Victoria* coffee-table guidebook.

Many people who both know and love the Goldfields have generously contributed words, photos, time and energy to the project. Major contributors were:

ROBYN ANNEAR – WRITER

Robyn Annear is the author of several books about history and place, including *Nothing but Gold: The Diggers of 1852, Bearbrass: Imagining Early Melbourne* and *A City Lost & Found: Whelan the Wrecker's Melbourne.* (**Contributed *Goldrush Melbourne***)

WESTON BATE – WRITER

Weston Bate's two volumes on Ballarat, *Lucky City*, and *Life After Gold* tell the story of an astonishing human achievement which his *Victorian Goldrushes* puts into a wider context. He has an OAM for services to history and heritage, as an honorary consultant at Sovereign Hill, as Chairman of the Museums Advisory Board of Victoria and as President of the Royal Historical Society of Victoria. He has published 14 books on history and public history. (**Contributed *The Goldfields Story – An Introduction***)

DAVID BANNEAR – WRITER & PHOTOGRAPHER

David Bannear is an archaeological advisor for Heritage Victoria. He was introduced to the Goldfields region in 1989 when he began a five-year-long survey of Victoria's gold-mining heritage. David has lived in Castlemaine ever since and is passionate about the Goldfields, past and present. (**General advisor; contributed *Gold – Then & Now***)

FRED CAHIR – WRITER

Fred Cahir is completing a doctorate at the University of Ballarat on the role of Aboriginal people on the Goldfields of Victoria from 1850 to 1900. His industry partner is the Sovereign Hill Museums Association and his supervisor is Associate Professor Ian D Clark. Fred's MA dissertation and his pubications have investigated the history of contact between indigenous and non-indigenous peoples in the Central Highlands region of Victoria. (**With Dr Ian D Clark contributed *Aboriginals on the Goldfields***)

GARY CHAPMAN – PHOTOGRAPHER

Gary Chapman has lived in Central Victoria all his life and has been a professional photographer for 17 years. His day-to-day business includes weddings and commercial work, but he also specializes in large format panoramic landscapes and seascapes. Wherever he is he makes time to explore the back roads searching for his next subject and that 'magical light'.

DR IAN D CLARK – WRITER

Dr Ian D Clark is an associate professor in tourism in the School of Business at the University of Ballarat. He has been a lecturer in tourism management at Monash University, a research fellow in history at AIATSIS, manager of the Brambuk Aboriginal Cultural Centre in Halls Gap, and senior researcher in the former Victorian Tourism Commission. He holds a doctorate from Monash University in aboriginal historical geography and has published widely in Victorian aboriginal history. He collects the music and memorabilia of Ella Jane Fitzgerald. **(With Fred Cahir contributed *Aboriginals on the Goldfields*)**

RICHARD EVERIST – PROJECT MANAGER, WRITER, EDITOR & PHOTOGRAPHER

Richard Everist has written and co-written a number of guidebooks for Lonely Planet Publications, including Nepal, South Africa, Britain, Papua New Guinea and West Asia on a shoestring. He was the Co-General Manager and Global Publisher for Lonely Planet from 1995 to 2000. In 2004 he co-founded BestShot! Publications with his partner Lucrezia Migliore. BestShot! focuses on Australian regions, publishing books, photographs, calendars, cards, postcards and websites.

ELAINE GRANT – PHOTOGRAPHER

Elaine Grant loves to travel and to take photos wherever she goes. As a hoarder of photographs over the years, she started taking a more serious approach after being captivated by the wildlife and scenery during a trip to Africa in the 1990s. Now Elaine is seldom without a camera in her hand – an easier task with today's digital technology. She enjoys recording the visual image, keeping an exciting record of her travels and framing the best samples to grace the walls of home.

JOE KINSELA – WRITER & PHOTOGRAPHER

Joe Kinsela is a historian, writer, professional musician, landscape gardener, and architectural historian with a masters in heritage conservation from the University of Sydney. He was an Anglican monk for 20 years, before leaving to pursue his music with the Australian Opera and as a soloist on the United Kingdom oratorio circuit. He is fascinated by the gothic revival in Australia and is an expert on the work of Edmund Blacket. Joe was the author of The Heart of Victoria, published by BestShot! in 2004, parts of which have been adapted and reused in this book. **(Contributed *European Explorers*, plus other work from *The Heart of Victoria*)**

JOE MORTELLITI – PHOTOGRAPHER

Joe Mortelliti's passion for photography began when he was given a darkroom developing kit for his first birthday. He was hooked from the first time he saw images coming up in a developing tray…. These days Joe shoots on a digital camera, but he remains a purist and does not manipulate the images on computer. Joe and his wife are working on a book of photographs covering Australia.

ALISON POULIOT – PHOTOGRAPHER

Alison's fascination for the natural environment has grown from years of exploring the bush. As a natural history photographer she fulfils her passion for the environment as well as creatively expressing her impressions through her images. Previously a freshwater scientist, Alison's research has provided her with the ideal framework to seek and understand the subjects of her photography. Alison's images are her attempt to reflect and share her wonderment in the diversity of the natural world.

ANDREW REEVES – WRITER

Andrew Reeves is a historian, who has worked in Australian museums for many years. He has worked at the Museum of Victoria, the National Museum of Australia and most recently as Director of the Western Australian Museum. He has researched widely in Australian labour history, concentrating on mining unionism, labour migration and working class communities and has published on historical and cultural issues. He is the author (with Ann Stephen) of *Badges of Labour, Banners of Pride*, a pioneering study of Australian union banners, of *Another Day, Another Dollar: Working Lives in Australian History* and has co-edited, with Ian McCalman and Alex Cook, *Gold: Forgotten Histories and Lost Objects of Australia*. (**Contributed *Trade Unionism***)

DR KEIR REEVES – WRITER

Dr Keir Reeves' main research focus has been on the cultural history of gold mining communities, particularly the experience of Chinese during and immediately after the gold rush era. He is also interested in the Australian gold mining industry, heritage tourism, cultural landscapes and the world heritage values of the region. Keir lectures in public history at the University of Melbourne and in mid-2006 will undertake a ARC funded postdoctoral fellowship on the many layers of central Victorian history. (**Contributed *Chinese – The New Gold Mountain***)

KATHERINE SEPPINGS – WRITER & PHOTOGRAPHER

Katherine Seppings, local and international artist, writer and photographer, came to live in Chewton in 1983. A member of the National Trust Photographic Committee, Katherine was attracted to the Goldfields for visual and historical interests. For many years Katherine travelled through Australia, Asia, Europe and the Americas, living and working in London and New York. Publications include *Women of the Hills, The Complete Australian Bushfire Book, Fireplaces for a Beautiful Home* and *The Heart of Victoria*. Exhibitions include *Characters of Castlemaine, Nepal, Back Streets of the World, New York Graffiti, Castlemaine at the Crossroads* and *France*. (**Major contributor to *Small Towns & Villages***)

RAY TONKIN – WRITER

Ray is Executive Director of Heritage Victoria, a position he has held for over 10 years. He has been closely associated with the development of heritage protection legislation, policies and the administration of those matters in Victoria for many years. Ray is an architect and planner and has used these qualifications as a basis for his career in heritage conservation, even though much of his current work could be better described as public administration and issues management. He has seen substantial growth and change in the community's interest in heritage matters, and the desire of people to better understand the wide variety of places and objects that make up their cultural heritage. (**Contributed** *Architecture – Building Goldfields Towns*)

PETE WALSH – PHOTOGRAPHER

Pete Walsh's passion for photography and the outdoors crystalised during a solo mountain bike cycling journey through remote Australia in 1988. Self taught, Pete has since worked as a photographer, focussing primarily on the Australian landscape, and particularly on the spa country region.

RICK WILKINSON – WRITER & COPY EDITOR

Rick Wilkinson is a geologist turned journalist and author who has written for and edited several magazines in Britain and Australia. He was oil & gas correspondent for *The Australian*, *The Australian Financial Review* and *Australian Business* during the 1980s and 1990s and is now a freelance contributor to publications in Australia, the UK and the US. He has written eight books on petroleum history as well as five books of fiction and non-fiction for children. His most recent book, *The Bellarine…Via Rambler's Road*, is a journey around Victoria's Bellarine Peninsula. (**Major contributor to** *Major Towns & Cities*)

CLIVE WILLMAN

Clive Willman is a geologist with more than 25 years experience in Government and industry. He has extensive knowledge of Victoria's geology having mapped large areas of Victoria but is best known for his detailed maps of the Castlemaine and Bendigo goldfields. Clive has contributed to several books on Victoria's geology, has published scientific papers in Australian and international journals and was awarded the Selwyn Medal in 2003 for his contribution to Victorian geology. (**Contributed** *Geology*)

Photography Credits

Due to space limitations and the large number of photos used, photographers and image copyright owners are identified with initials as follows:

AP	Alison Pouliot	BaT	Ballarat Tourism collection
BeT	Bendigo Tourism collection	DB	David Bannear
EG	Elaine Grant	GM	Gold Museum collection
GS	George Stawicki	JM	Joe Mortelliti
KS	Katherine Seppings	MH	Martin Hurley
PV	Parks Victoria collection	PV (AP)	Parks Victoria collection (photo by Alison Pouliot)
PW	Pete Walsh	RE	Richard Everist
SH	Sovereign Hill collection		

Ruin, Moorlort, RE
Sovereign Hill, SV
Opposite: Bendigo, PW

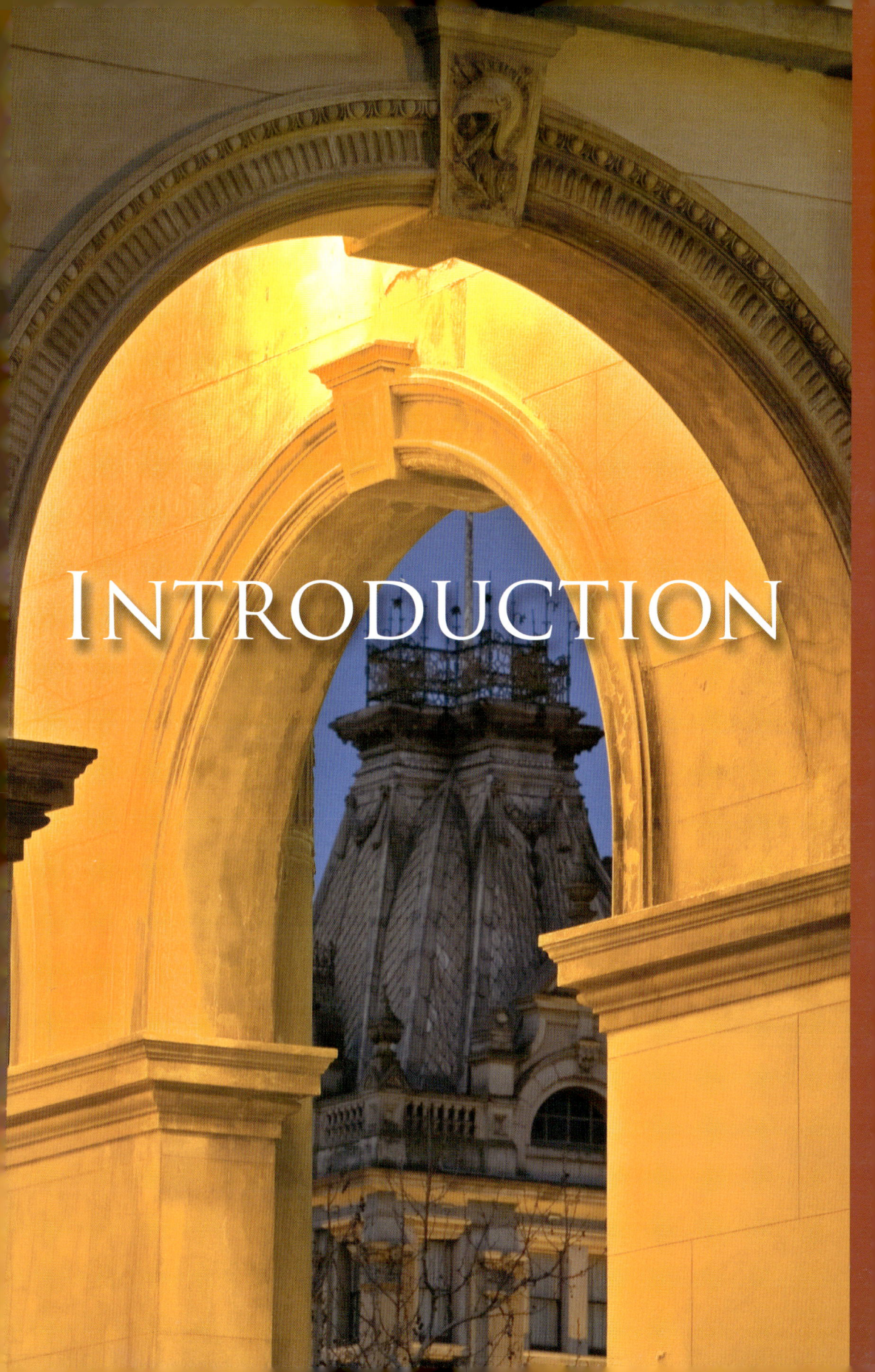

INTRODUCTION

PREFACE

Central Victoria was home to the world's richest and greatest goldrushes. Nothing to match them has ever been seen – and they have created one of the world's great travel experiences.

The impact of Victorian gold on the world was significant… and the impact on Australia was enormous. It was a substantial factor in the economic expansion enjoyed by the British Empire, and Australia changed from a penal colony and backwater into a vibrant, progressive country.

The legacy of the goldrushes can be discovered by any traveller with a spirit of curiosity and adventure. Take a chance. Explore. You can't get lost if you tour by car because every road leads to a goldfield or a town – otherwise the road would not have been built in the first place. And, yes, you can still find gold.

No other Australian region has such a strong and immediate historical presence.

The goldrushes left a ghostly but beautiful landscape that was radically changed by the miners. The vast forests were almost completely clear-felled (although miraculously they have returned) and vast areas were turned upside down by picks, shovels, sluices and dredges.

There are abandoned mines, mullock heaps and ruins wherever you look. There are ghost towns that are frozen in time, but there are also thriving regional centres with wildly beautiful – and wildly ostentatious – 19th century architecture.

The countryside is rich with signs and secrets and stories, and a visitor can sense this immediately. The physical heritage also provides signposts to the wider story of Australia.

In various ways and at various times gold empowered miners to pursue new ideas of social mobility, education and community. They challenged and changed the nature of Australia's economy, government and society.

The Goldfield region's main story keeps coming back to gold – and gold is still being discovered – but there are many other stories. There are the Aboriginal tribes, the volcanoes, the ironbark forests with their unique ecology, little towns and villages, churches and lonely cemeteries, outstanding wineries, pubs, restaurants, orchards, spas, art galleries and museums.

Today Victoria's gold is once again being seriously mined. Improved technology and high gold prices mean the remaining recoverable reserves now being targeted equal the amount that has already been taken.

Many argue, however, that the region's greatest treasure is its heritage and, increasingly, fine food and wine. You can't put a price on that!

Mt Alexander, RE

TIME TRAVEL

On a journey through the Goldfields the most important piece of luggage will be your imagination. Wherever you go you can see and feel the ghosts – whether you're scanning the plains from the rim of a volcano, soaking in the atmosphere of a ruin in the forest, or exploring the cities and towns.

The goldrushes attracted a very particular kind of person – risk-takers, thinkers and small capitalists – who have had a huge impact on the Australian character and the fabric of Australian life.

If *you* were to learn today that an extraordinary new alluvial goldfield had been discovered, somewhere remote, would you go?

Perhaps gold has been discovered in Central Asia. People with nothing more than a spade are picking up huge quantities of gold. Let's say the lucky ones are getting $2 million for a few weeks' work – which means you could dig a couple of holes and never need to work again.

It's not entirely clear from the reports where the goldfield is. You might get there only to find the gold is all gone, or the local government, although initially welcoming, has introduced punitive taxes…

Even with modern transport and communications it will take you at least a month to get there. And the whole exercise will be costly. You will certainly have to leave your current job (if you have one).

Would you do it? Would you have the courage? Would you take the chance?

In 1851 gold was discovered outside Ballarat at Buninyong, igniting the first Victorian goldrush. Over the course of the next few years, the gold at Castlemaine and Bendigo attracted hundreds of thousands of people from every corner of the globe – brave people and, perhaps, people with little to lose. They took the gamble of their lives.

They embarked on voyages that, from places like Liverpool, took 10-17 dangerous weeks. The hinterland of Victoria had only recently been explored by Europeans and settled by pastoralists. Aboriginal tribespeople still roamed their country. The gold seekers arrived in Melbourne or Geelong, which were little more than dusty villages, and as soon as possible started walking into a strange, unmapped country in search of gold. The journey to the diggings was an ordeal that could take the best part of a week.

Today, on the highway from Melbourne it takes 1-1/2 hours. But in the course of that short trip, your car can become a time capsule, and your trip can become a journey of exploration and discovery.

*If you **were to learn today that an extraordinary new alluvial goldfield had been discovered, somewhere remote, would you go?***

Central Deborah, Bendigo, PW
Historic belt buckle, HV

The joy of exploring central Victoria is that the landscape speaks. It speaks of geological time and events. It speaks of the Aboriginal people. It speaks of gold.

There are enough landmarks that remain unchanged, and enough clues (buildings, tracks, forests, ruins, mines and mullock heaps) left to enable you to recreate the extraordinary scenes of the world's greatest goldrushes. For a start, you can imagine the feelings of relief and excitement when, near Malmsbury, the new miners had their first sight of Mt Alexander…

As you continue, even if you stick to the main highways and tourist sites, you will make your own discoveries – and the most rewarding discoveries you make will be off the beaten track.

The joy of exploring central Victoria is that the landscape speaks. It speaks of geological time and events. It speaks of the Aboriginal people. It speaks of gold. It speaks of hopes, fears and stubborn courage. It speaks of the miners' optimistic faith that they were building for a long, bright future.

You'll find signs of these stories down dirt tracks, in museums and churches, in cemeteries and hotels. Along the way you'll also find memorable accommodation, wineries and restaurants.

Even today, with all the advantages of cars and mobile phones, it can take a bit of courage to cast off the city and head off down a dirt track. But the worst that can happen is that you'll end up at a goldfield diggings, or a town, or both.

There are many signs to watch for. Occasional chimneys mark the site of huts. Valley floors are often pitted with shallow holes and piles of dirt. The sides of creeks are often scarred from sluicing. Sometimes you will notice the donut-shaped circular gutter left by a puddling machine, quartz tailings, or a deeper shaft.

Initially you may find the Goldfields forest grey, unkempt and uniform. But it is not. First you must learn to look *through* it. Then you must learn to look closely *into* it. Look through it to see its light and shade, to enjoy its space and openness. Look closely into it to see the rich variety of tiny and extraordinary plants on the forest floor.

The Goldfields of Victoria cover old towns and cities and dozens of villages. Today they are the fascinating places where both the past and the present can be savoured. The towns covered in this book are shown in maroon in the map section.

Orchid, RE
Castlemaine Diggings Heritage
National Park, RE

HOW TO USE THIS BOOK

The best moments of any trip are those moments when you are given the opportunity to see the world in a fresh, new light – where for a time, you are challenged and intrigued, your surroundings cease being wall paper and you feel intensely alive. This is what the Goldfields offer.

This book is a starting resource for those who want to explore Victoria's central goldfields. Take a few minutes to familiarise yourself with the contents and then dive in. The book is not designed to be read from cover to cover, but to be dipped into according to need and interest.

The region is bursting with natural and man-made sites, so the book can only point you towards a few of the highlights. In any event, your greatest highlight is likely to be a discovery you make yourself, by chance…

To properly explore the region you would need a lifetime but, because of its excellent road network, day trips and weekend trips from Melbourne are easy. The region's rhythm and character start to make more sense if you can give yourself three days to explore – and a week is even better. After three days you can forget that Australia's big coastal cities even exist!

In this book we identify what we think (entirely subjectively) are the highlights. But depending on your own particular interests – whether they are Aboriginal culture, the Chinese diggers, architecture, geology, natural history, national parks, wineries, good food, whatever – you will no doubt create your own list.

The Visitor Information Centres throughout the region are excellent resources. They have huge amounts of printed information and they are staffed by knowledgeable locals. They can help with accommodation (including bookings) and can give detailed, up-to-date information about reaching some of the sites we talk about in this book.

Because some of the sites we recommend are reasonably well hidden and only accessible by dirt roads (the condition of which can vary from month to month) up-to-date directions are invaluable. We strongly recommend you use the Visitor Information Centres.

The book has been structured to enable you to pick and choose according to your own interests, and to put together a touring route that meets your time constraints.

We suggest you skip through the Planning Your Trip section to get a sense of some of the possibilities (Top 10 Highlights, Suggested Routes and Themes).

The Context section provides background information on the region's history, flora and fauna. The townships have been broken up into two main sections: Cities & Major Towns and Small Towns & Villages. Within these two sections the towns appear alphabetically. The towns covered in this book are shown in maroon in the map section.

The townships have been broken up into two main sections: Cities & Major Towns and Small Towns & Villages. Within these two sections the towns appear alphabetically.

Planning Your Trip

Highlights

Top 10

Ararat – Gum San Chinese Heritage Centre – p105

BaT

Ararat is an attractive town that lies between the spectacular Grampians National Park and the Goldfields – a gateway to both. Ararat played an important role in the history of the Chinese miners, and this is celebrated in the outstanding Gum San Chinese Heritage Centre.

Ballarat – Sovereign Hill, Gold Museum, Art Gallery, Eureka Stockade Centre – p115

KS

Ballarat – and Sovereign Hill in particular – is an ideal place to start your exploration of the Goldfields. Sovereign Hill is an extraordinary, vast, historically accurate recreation of the Goldfields, and a visit will give you an enjoyable first hand introduction to the region. It's as close as you can get to time travel. The Gold Museum, next door, has some outstanding exhibits and should not be missed. The Ballarat Art Gallery includes the original Eureka flag, but also has some superb artworks and the Lindsay family's sitting room rescued from their house at Creswick before it was demolished. The art gallery symbolises the civic ambitions and achievements of the town.

A visit to the Eureka Stockade Centre, where the actual battle took place, is a must.

Bendigo – Chinese Museum, Town Hall, View Street, Victoria Hill, and Central Deborah Goldmine – p138

KS

Bendigo is a proud city with a magnificent architectural heritage. Sometimes aptly called 'Vienna in the bush', the city is now surrounded by national parks. The interiors of buildings, like the outstanding town hall are, if anything, more magnificent than the exteriors. The Chinese Museum houses the oldest imperial dragon in the world and the longest Chinese dragon outside of China, while Victoria Hill and the Central Deborah Goldmine give a vivid lesson in the history and realities of the underground quartz mining that provided Bendigo's enormous riches.

Castlemaine Diggings National Heritage Park – p207

KS

The Castlemaine Diggings National Heritage Park is likely to be nominated as a World Heritage site. It surrounds the beautiful historic regional centre of Castlemaine and the straggling gold towns of Chewton, Campbell's Creek and Fryerstown. It includes some of the most beautiful forest and some of the most intriguing and impressive relics of the goldrush era. There are ruined cottages, goldfields, cemeteries, mines and Aboriginal relics. The drama of the Goldfields history is written clearly across the landscape.

CLUNES – P210

Clunes is a town where time seems to have come to a standstill. The sleepy 19th century streetscapes have attracted many film-makers.

PW

MALDON – P242

Maldon was declared Australia's first 'Notable Town' by the National Trust and is a remarkably complete and beautiful gold-mining town. Every little street has an interesting building or view – perhaps of an old chapel or an industrial chimney. It is also a tourist town, which means there are antiques, excellent cafés and accommodation options. The vintage railway, the 360° view from the top of Mt Tarrengower, Carman's Tunnel (an old mine that can be toured by candlelight) are features not to be missed.

RE

SMEATON – P263

There's not much to the town of Smeaton, but the surrounding volcanic landscape is extraordinary. Miners followed deep leads under the countryside in the golden triangle formed by Clunes, Smeaton and Creswick. Apart from impressive mullock and tailing heaps, there are the ruins of several monumental Cornish engine houses. Anderson's Flour Mill, in a sheltered valley, is an extraordinary bluestone building of massive scale. Perhaps the most impressive Avenue of Honour in the Goldfields is at nearby Kingston.

RE

TALBOT – P271

Talbot is even quieter than Clunes and feels even more like a frontier ghost town. Moving into drier country it has a very different feel to the towns further south. It's a good jumping-off point to two of the Goldfields' best kept secrets – Amherst's Big Reef and the Stoney Creek Primary School.

KS

TARADALE – P274

Taradale is a quiet little town, but it is home to one of Australia's great 19th century engineering feats – a magnificent bluestone railway viaduct. There are some pleasant short walks and it's worth tracking down the whimsical Anglican Church of the Holy Trinity.

GC

TOP 10 SECRETS

AMHERST – BIG REEF – P272

The Big Reef (also known as Quartz Mountain) is a massive quartz outcrop standing six metres above the ridgeline. It's the biggest left in Victoria, and probably the world.

JM

BALLARAT – GOLD MUSEUM – P126

Although it is just across the road from Sovereign Hill, it is often missed. Amongst other things there is an excellent multi-media presentation about the Wathaurong Aboriginal people, some fascinating original artefacts, and an amazing collection of gold nuggets.

GM

BeT

BENDIGO – TOWN HALL INTERIOR & VIEW ST BANKS – P138

The town hall is an impressive freestanding building from the outside, but nothing can prepare you for the magnificent hall – with its light, its gold and its murals.

From several viewpoints on the western side of View St, you can look over the backyards of the impressive banks that line the thoroughfare. The chimneys rising from small brick buildings belong to the banks' smelters, where the diggers' gold was smelted into ingots for transportation to Melbourne.

KS

BARKER'S CREEK – SPECIMEN GULLY ROAD – P199

Turn right into the unsealed Specimen Gully Road, about two km from Castlemaine on the Midland Highway to Harcourt. First note the magnificent culverts under the railway line, then visit the memorial to the discovery of gold, and continue to the crest of the ridge overlooking Mt Alexander and the Harcourt Valley.

KS

FRYERSTOWN – P218

On the hilly road from Chewton to Vaughan Springs, surrounded by forest, Fryerstown is a tiny, atmospheric hamlet. Mullock heaps and the remains of the Duke of Cornwall engine house lie to the north.

GS

BUNINYONG & LAL LAL – P201

Buninyong is a charming town – which is not such a secret – but the extraordinary Mt Buninyong, and its views are. Nearby the falls at Lal Lal, the Bungal Dam, and the ruins of the Lal Lal Iron Mining Company, are dramatic and fascinating by turns.

GS

SMYTHESDALE – P265

Tucked away, south-west of Ballarat, Smythesdale is a fascinating small town surrounded by a classic Goldfields landscape. There are alluvial diggings, as well as some impressive ruins and landscapes in the adjoining state forest. The nearby Nimon's bridge is the highest wooden bridge in Victoria.

RE

STONEY CREEK ELEMENTARY SCHOOL – P273

In the bush near Talbot, all that remains of the school are the evocative, remnants of rock gardens, and in particular, a large map of Australia created out of quartz stones. The modest scale of the ruins and the location are intensely moving.

KS

VAUGHAN MINERAL SPRINGS – P280

The mineral springs were first found by the alluvial miners who discovered gold in the area in 1853 and they are now part of a reserve along the Loddon River. There are walks, a swimming hole, a playground, mining relics, Chinese graves, and magnificent silver poplars.

KS

WHROO – P284

The Whroo Historic Reserve includes the site of the former Whroo township and part of the associated goldfields. The main feature of the site is the Balaclava Mine, an open cut working where gold worth millions of pounds was extracted in the 1850s from shafts and adits that are still visible today. A path and interpretive trail lead to lookouts and an exciting walk through one of the tunnels into the open cut mine itself.

SUGGESTED ROUTES

Goldfields Loop – Seven Days

Castlemaine – Maldon	20 km	15 min
Maldon - Bendigo	40 km	30 min
Bendigo - Inglewood	44 km	30 min
Inglewood - Dunolly	45 km	30 min
Dunolly - Maryborough	22 km	15 min
Maryborough - Avoca	26 km	15 min
Avoca - Moonambel	19 km	15 min
Moonambel - Stawell	62 km	40 min
Stawell - Ararat	30 km	20 min
Ararat - Avoca	63 km	45 min
Avoca - Talbot	26 km	15 min
Talbot - Clunes	18 km	10 min
Clunes - Creswick	18 km	10 min
Creswick - Ballarat	16 km	15 min
Ballarat - Daylesford	45 km	30 min
Daylesford - Castlemaine	35 km	25 min

Grand Tour From/To Melbourne – Seven Days

Melbourne - Macedon	60 km	40 min
Macedon - Taradale	50 km	35 min
Taradale - Castlemaine	16 km	15 min
Castlemaine - Bendigo	38 km	30 min
Bendigo - Maldon	40 km	30 min
Maldon - Newstead	14 km	10 min
Newstead - Maryborough	30 km	20 min
Maryborough - Dunolly	22 km	15 min
Dunolly - St Arnaud	60 km	35 min
St Arnaud - Moonambel	49 km	30 min
Moonambel - Stawell	62 km	40 min
Stawell - Ararat	30 km	20 min
Ararat - Beaufort	44 km	30 min
Beaufort - Linton	50 km	35 min
Linton - Smythesdale	13 km	10 min
Smythesdale - Ballarat	19 km	15 min
Ballarat - Daylesford	43 km	30 min
Daylesford - Trentham	28 km	15 min
Trentham - Blackwood	13 km	10 min
Blackwood - Melbourne	84 km	50 min

Goldfields Eastern Loop – Three Days

Castlemaine - Bendigo	38 km	30 min
Bendigo - Tarnagulla	45 km	30 min
Tarnagulla - Dunolly	16 km	10 min
Dunolly - Maryborough	22 km	15 min
Maryborough - Talbot	14 km	10 min
Talbot - Clunes	18 km	10 min
Clunes - Creswick	18 km	10 min
Creswick - Ballarat	16 km	15 min
Ballarat - Daylesford	43 km	30 min
Daylesford - Castlemaine	35 km	25 min

Goldfields Western Loop – Three Days

Ballarat - Creswick	16 km	15 min
Creswick - Clunes	18 km	10 min
Clunes - Talbot	18 km	10 min
Talbot - Maryborough	14 km	10 min
Maryborough - Avoca	26 km	15 min
Avoca - Stawell	81 km	55 min
Stawell - Ararat	30 km	20 min
Ararat - Beaufort	44 km	30 min
Beaufort - Linton	50 km	35 min
Linton - Smythesdale	13 km	10 min
Smythesdale - Ballarat	19 km	15 min

Sovereign Hill, Ballarat, SV

Bendigo To/From Melbourne – Two Days

Melbourne - Macedon............. 60 km 40 min
Macedon - Kyneton .... 25 km 15 min
Kyneton - Taradale .... 19 km 10 min
Taradale - Castlemaine .... 24 km 15 min
Castlemaine - Bendigo .... 38 km 30 min
Bendigo - Maldon .... 40 km 30 min
Maldon - Newstead .... 14 km 10 min
Newstead - Hepburn Springs... 26 km 20 min
Hepburn Springs - Daylesford. 4 km 5 min
Daylesford - Trentham............. 28 km 20 min
Trentham - Blackwood .... 13 km 10 min
Blackwood - Melbourne .... 84 km 50 min

Ballarat To/From Melbourne – Two Days

Melbourne - Ballarat .... 110 km 70 min
Ballarat - Creswick .... 16 km 10 min
Creswick - Clunes .... 17 km 10 min
Clunes - Daylesford................. 52 km 40 min
Daylesford - Trentham............. 28 km 20 min
Trentham - Blackwood .... 13 km 10 min
Blackwood - Melbourne .... 84 km 50 min

Melbourne to Grampians – Two Days

Melbourne - Ballarat .... 110 km 70 min
Ballarat - Creswick .... 16 km 10 min
Creswick - Clunes .... 17 km 10 min
Clunes - Talbot....................... 18 km. 10 min
Talbot - Bung Bong................. 18 km 10 min
Bung Bong - Avoca.................. 8 km 5 min
Avoca - Ararat .... 63 km 40 min
Ararat - Stawell....................... 29 km 20 min
Stawell - Halls Gap................. 25 km 20 min

Grampians to Melbourne – Two Days

Halls Gap – Stawell .... 25 km 20 min
Stawell - Ararat....................... 30 km 20 min
Ararat - Beaufort .... 44 km 30 min
Beaufort - Linton .... 50 km 35 min
Linton - Smythesdale............... 13 km 10 min
Smythesdale - Ballarat .... 19 km 15 min
Ballarat - Daylesford............... 43 km 30 min
Daylesford - Trentham............. 28 km 15 min
Trentham - Blackwood .... 13 km 10 min
Blackwood - Melbourne .... 84 km 50 min

Chewton Cottage, KS

Themes

Geology

First People: Aboriginal country and culture

Turpin's Falls, RE

Gold Mining

Alluvial Diggings, Sovereign Hill,
Ballarat, SV

CHINESE – DAI GUM SAN

CRY JOE: THE SPIRIT OF SOCIAL AND POLITICAL DEMOCRACY

PARKS & GARDENS

Golden Dragon Museum, Bendigo, BeT
Eureka Flag, BaT
Rosalind Gardens, Bendigo, BeT
Castlemaine Botanic Gardens in Autumn, KS

CLUES

The Goldfields region is rich in signs and clues that point to the ghosts of the past. Look carefully. If you sense something has happened, it probably has.

RE

Small brick chimney stacks found at the rear of former banks indicates that gold smelting was carried out during the goldrush.

GS

Miner's rights granted the holder the right to build houses and establish gardens on quarter-acre allotments – the progenitor of Australian suburbia.

PV (AP)

The sandstone ridges of the Goldfields, which were once the ocean floor, always run north-south, thanks to the forces that compressed them and squeezed them upwards.

SH

The ubiquitous miner's cottages nearly always started as tent-like structures with a pitched roof over one or two rooms – and kept on extending backwards, often with another gable, then with verandahs and lean-tos…

DB

A legacy of Chinese market gardens is wild fruit and vegetables – like the fruit trees, and spring onions that appear around Vaughan.

GS

Collections of public buildings such as police stations, lockups, court houses and post offices often mark the site of the original government camp.

KS

The most over-the-top architecture with two-storey grand facades – but *without* verandahs or balconies – was usually reserved for banks. They invariably have bluestone bases which were intended to give a look of permanence (essential when attracting savings!)

PW

Avenues of European trees (usually elms or plane trees) on the outskirts of towns were planted to commemorate local men and women who enlisted in WW1. They are moving tributes that give some idea of the scale of the communities' contribution.

KS

The windows of Goldfields shops were invariably large – benefiting from new technological advances in the 1860s in rolling glass.

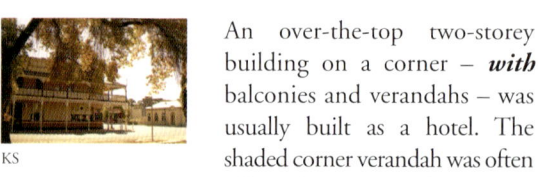
KS

An over-the-top two-storey building on a corner – *with* balconies and verandahs – was usually built as a hotel. The shaded corner verandah was often a favourite town meeting spot.

JM

The location and size of culverts, drains and bridges can indicate the size of the goldrush-era town and the scale of its inhabitants' expectations for the future.

BeT

The interiors of the Goldfields town halls are often more magnificent than their exteriors.

KS

Often, only brick and stone buildings survive in old goldfields towns. There will always be a church, sometimes a court house, and occasionally a mechanics' institute or hotel.

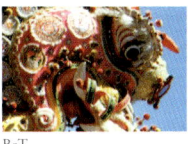
BeT

Century-old community festivals and events held at places like Bendigo, Stawell, Maldon, Ballarat, Hepburn and Maryborough demonstrate the strength of the communities built by the gold miners.

DB

Manufacturer's marks are common on mining machinery, iron fittings associated with public buildings, and iron street and park furniture. Look for them on light poles. The marks provide a great insight into the global trade established by the goldrushes.

DB

Surviving brick foundations for mining machinery will date from the 1890s when cement was first used. Foundations built prior to the 1890s are scarce because they were made with mud or weak mortar, and were easy to demolish.

DB

The clumps of slender trees common in most forests are ghosts. The old forest giants were cut down, but new shoots coppiced (grew) from the old tree's roots. Each clump is the ghostly impression of the original giant, with the ancient rootstock hidden but still alive.

KS

Public libraries, museums, heritage societies and research centres have fascinating historical collections. Local bookstores (including the many second-hand bookstores) always have local history books.

KS

The use of particular building materials – slate, sandstone, granite and bluestone – is a good clue to the surrounding geology.

RE

Cemeteries tell many stories of the goldrush: the nationalities involved, living and working conditions, and religious groupings. The earliest burials are those usually marked by weathered sandstone headstones (because sandstone was relatively easy to carve). Chinese burning towers are common.

SH

The dry environment and the miners' need for water means most mining activity required the construction of a dam. Dams in the forest are therefore good indicators of mining activity and good starting points when exploring a goldfield.

DB

Often near a dam, a doughnut-shaped earth impression marks the site of a puddling machine. Dumps of finely crushed quartz (grey or white sand) indicate a crushing battery was nearby.

DB

A collection of vats, perhaps the remnants of a boiler and a crane, and a well dome indicates you have discovered the remains of a eucalyptus (eucy) oil distillery.

RE

The sides of old buildings often carry the faint remnants of advertising signs that will signal former uses of the building.

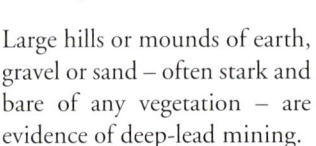
DB

Large hills or mounds of earth, gravel or sand – often stark and bare of any vegetation – are evidence of deep-lead mining.

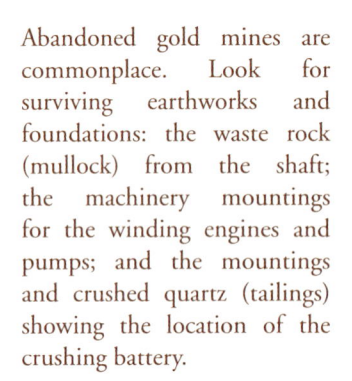
DB

Abandoned gold mines are commonplace. Look for surviving earthworks and foundations: the waste rock (mullock) from the shaft; the machinery mountings for the winding engines and pumps; and the mountings and crushed quartz (tailings) showing the location of the crushing battery.

The main landmarks left behind by a goldrush are closely spaced holes. Each hole was at the centre of an eight foot claim, the amount of land a licensed digger was able to 'own'.

SH

The Chinese miners dug round holes, the Europeans dug square holes. Some believe the Chinese had a superstition that corners harboured bad spirits – but round holes also required less timber to support them.

DB

A collection of exotic trees, especially fruit trees and peppercorn trees (pepper trees from the *schinus* species), indicate a house site.

RE

KS

Raw, exposed rock near creek or river beds often indicates the gold bearing soil on the hillside was washed away by sluicing.

RE

A raised, straight mound of rocks in or around a mine site may have been the foundations for a water channel or tramway for carrying mullock or quartz.

RE

Scars on big old trees – especially red gums – were left when Aboriginals removed bark for various purposes including shields.

GH

The power of gold is reflected in the grandeur and self-confidence of the architecture. The Castlemaine Market, for example, looks like a Grecian temple scooped up from the shores of the Aegean.

KS

Churches give an important clue to the origins and class of the population. The Church of England was the establishment church of the English upper and middle classes; the Catholic Church was Irish; the Presbyterians had a Scottish culture and ethos and were often associated with the original graziers; and the Wesleyan Methodists were working class Cornish, or northern English.

Places names are clues to the past

Miners' former homes:

Adelaide Flat, Launceston Gully, Aberdeen Hill, British American Reef, Daphne Reef, Devonshire Gully, Switzerland Reef and Italian Hill

Landscape

Bald Hill, Hard Hill, Granite Reef, Rocky Gully, Apple Tree Gully, Crabhole Gully, Pig Face Gully

Personal names

Abraham's Gully, Amman Hill, Burn's Gully, Heinriech Reef, Preshaw's Flat, Trewartha Reef, Elizabeth Gully, and Wightman's Flat

Nicknames

Black Jack, Dirty Dick's, Cranky Ned's, Granny Thomas's and Uncle Billy's, Drunken Scotchmans.

Events

Pickpocket Hill, Bung-Eye Gully, Drunkard's Gully, Chokem Flat, Murdering Flat, Burying Ground Flat, Cricketers Gully, Cut Throat Gully, Derwenters Mistake Gully, Agitation Hill, Black Monday.

Gold mining terms

Crusher's Gully, Fossicker's Point, Scramble Gully, Lucky Strike Reef, Schicer Gully, and Pennyweight Flat.

Trades

Shellback Reef, Blacksmith's Gully, Charcoal Gully, Cobbler's Gully, Shepherds Gully, Sawpit Gully, Pottery Flat, and Slaughteryard Hill.

Sailor's Gully Diggings, PV (AP)

Colles Rd near Castlemaine, KS
Near Taradale, KS
Opposite: Doorknob, View St, Bendigo, RE

CONTEXT

THE GOLDFIELDS STORY – AN INTRODUCTION

BY WESTON BATE

The remarkable heritage of Victoria's central goldfields was created by an unusual combination of great mineral deposits, rich volcanic soils, splendid forests and enterprising people.

A single generation developed a vibrant, democratic society. Their experience, unparalleled in Australian history, transformed the landscape and left structures, institutions, gardens, farms, art, literature and much other evidence of their vision and energy.

Social and political underdogs were liberated by gold. Even in its natural state it was as good as cash.

GOLD – THE DEMOCRATIC MINERAL

Before 1851, when the rushes began, Australia focused on Britain, economically through the dominant pastoral industry and socially through officials, settlers and convicts who transplanted British institutions and ideas.

Although the democratic principles of the French and American revolutions were powerful, they had been strongly resisted by the English ruling class who, in 1848, defeated the efforts of two million signatories to the Chartists' great charter for parliamentary reform. Disillusionment brought many Chartists to Victoria to be among the founders of a new democracy.

Social and political underdogs were liberated by gold. Because the world economy depended on it as the only safe means of exchange, large finds stimulated commerce generally, as well as creating a local demand for transport, food, tools, tents and other supplies. Famous clipper ships were born to carry multitudes across the world in an unprecedented movement of ordinary people, free in the new era to move where and when they wished.

Alluvial gold was a democratic mineral and a social explosive. Because it was as good as cash, no middle man or expensive processing was needed, which freed

Nugget, Gold Museum, Ballarat, GM

the miner from potential exploitation. The hardship of digging for it in frontier conditions favoured artisans and labourers, especially the Irish.

So, although middle and upper class people appeared on the Victorian fields, gold turned society upside down. It drew the young and adventurous to freedom from Old World restrictions, spear-headed by six thousand Australians who had been to California, where they learnt how to prospect and how to wash their dirt with an efficient American cradling device.

THE WORLD TURNED DOWNSIDE-UP

Discoveries at Clunes and Buninyong during July and August 1851, led to local rushes. Then, on 28 August, Ballarat ignited wider interest. The blue clay in some miners' claims yielded over 900 ounces (25.5 kg) of gold worth £3800. The equivalent, in terms of purchasing power today, would be about $2 million.

The news, when it arrived, electrified Britain. But that was only the beginning. In 1852, first Forest Creek (near Castlemaine) then Bendigo released amazing treasure. Many fortunes were made in a day from what Geoffrey Blainey has called the richest surface fields the world has known. The impact extended to Europe and the Americas, stimulating a migration that raised Victoria's population in a decade from 77,345 to 540,322, and produced Australia's first multicultural experience.

Although only 33% of 45,410 arrivals from foreign parts remained in 1861, compared with 83% of Britons, a long-term impact would come from 22,000 Chinese, 10,000 Germans, 8,000 other Europeans and 2,500 Americans.

More important than their numbers, the migrants were mostly young men, relatively well educated, entrepreneurial and skilled, ready for the opportunities created by gold. For instance, Ballarat was able to staff 10 large metal works by 1860. By 1870 its architects, builders and businessmen had constructed what the novelist Anthony Trollope described as the most surprising city in Australia.

The quality of the achievement by a single generation is probably unequalled in world history. All Goldfields communities were remarkably free to make their own way, unaffected by the policies of overseas investors. They dug their own capital and reinvested in their own future, resulting in distinctive buildings, still visible.

In a decade, Victoria's population increased from 77,345, to 540,322.

To get to that point the Goldfields had won a battle against arbitrary government. Mass protests against an unfair tax, which erupted especially at Forest Creek in December 1851 and Bendigo in August 1853, culminated at Ballarat on 3 December 1854 when government forces overran the Eureka Stockade and pulled down the rebels' Southern Cross flag.

Because of injustice, bullying and corruption the Government was discredited. Its gold-laced officials were swept away. Through election to local courts of mines in 1855 and to local and central government in 1856, control passed to the people who ran their own communities effectively. The world had turned downside-up.

Gold License, Gold Museum, Ballarat, GM

The Hinterland

The tide of mining activity has left a landscape pockmarked with shafts and spoil, the remains of puddling machines, Chilean mills and sluices. At the Welsh Village near Chewton the broken-down walls of small stone houses can be seen. Later evidence exists in the footings of quartz crushing batteries, poppet heads, engines and pumps, great mullock heaps and cyanide re-processing dams. Occasionally the bush reveals rusty boilers, or the carcasses of machines once proudly shining and hissing with steam.

Many central Victorian fields were blessed with great forests, rich volcanic soils and good rainfall, in contrast to barren hinterlands in California, South Africa and Western Australia. Until rail connections were made to the ports in 1862, they were able to develop as urban centres, protected by the expense of transport from the competition of outside manufacturers and farmers.

The pre-existing pastoral industry provided meat and, together with new farmers, produced oats and hay as fuel for horse transport. Vegetables and fruit were grown, and potatoes were planted by the Irish at forest clearings like Bungaree. Joseph Jenkins' *The Diary of a Welsh Swagman* contains important information about farming life.

Many Victorian goldfields were blessed with fertile hinterlands, unlike those in California and South Africa.

The urban impact can be seen in the Castlemaine Market as well as in banks and merchants' offices. Gold lay beneath fine soil around Creswick and Clunes where great white dumps of spoil from the mining of buried streams pattern the magnificent landscape. In the 1860s, district farmers took their wheat to Fry's large flour mill, now in ruins at Ascot, or to the Andersons' five-storey bluestone mill at Smeaton, which was powered by a large Ballarat-made water-wheel.

Modern Victorians think of a state dominated by Melbourne, which holds most of the population. Not so the goldrush generations, whose towns and countryside in 1871 held 270,428 people against Melbourne's 191,000, and whose political force led the practical march of democracy towards fairer land laws, payment of members of parliament and, later, federation.

Most of this energy was contained in Goldfields towns with over 500 people, whose total population at the 1871 census was 135,941. They contained the most visible signs of achievement. Ballarat with 47,201, was twice the size Melbourne had been in 1851, although down from over 50,000 in 1869. Bendigo, only at the outset of its quartz era, also stood out with 28,577 from Castlemaine 6935, Clunes 6068, Stawell 5166, Daylesford 4696, Creswick 3969, Maldon 3817, and Maryborough 2935.

All the Goldfields towns were significant and interestingly different in layout and building style. The dramas, successes, uncertainties and upheavals of mining were reflected in the built environments. The contrast between Ballarat and Bendigo can be used to make the point.

BALLARAT & BENDIGO

Four layers of volcanic rock had buried the western half of Ballarat's reefs and streams, and had dammed up the rest so that they lay 50 metres deep beneath wet sand, gravel and clay. Following the gold by sinking shaft after shaft from the edges of the natural basin was a dangerous and expensive task. It cemented community because storekeepers, through formal partnerships, financed miners for as long as eight months in the hope of clusters of nuggets in a 'jeweller's shop'.

Overall, the effort was so successful that the most prosperous street in Australia grew up between 1854 and 1860 along the road to Geelong through the diggings. Partly recreated at Sovereign Hill, its pubs, theatres, shops and factories made Main Street a symbol of Goldfields enterprise.

Because steam-driven pumps often needed urgent repair, those wet diggings fostered a sophisticated metal industry which was geared to help, and to become even stronger, with the more demanding phase of sinking through basalt to follow the streams westwards. New and larger mines were located on, and close to the township grid, surveyed in 1852.

As one of the first fruits of democracy, the local Court of Mines authorised the occupation by miners of quarter-acre residence areas on which distinctive tent-shaped cottages were built. This made living cheap and helped sustain the city during the 10 years it took to reach the gold. In the meantime, support came from hinterland mining, agriculture, grazing and its own industries.

Ballarat had become a regional centre, boasting some of the finest businesses in the colony. A mining bonanza in the late 1860s gave further impetus before a recession motivated the community to diversify with a woollen mill and to open a School of Mines to improve performance. The latter was to train experts for the expansion of the industry throughout Australia. Affluence and pride in the city, together with a hard-won sense of community, led to impressive street plantations, public sculptures, Mechanics Institute, Stock Exchange, Fine Art Gallery, hospital, theatres, cricket grounds, parks, Botanic Gardens and race courses. Yuille's Swamp was converted to Lake Wendouree, the home of rowing, yachting and a fleet of pleasure steamers. Each year a Welsh Esteiddford, one aspect of a strong musical tradition, was held.

By mid-1853 Bendigo's surface story was over. Many miners left to gamble in Ballarat's deep sinkings and not much happened for 10 years until the quartz – the key to Bendigo's future – was finally tackled seriously. Of 22 million ounces (624 tonnes) of gold won at Bendigo, 83% has come from quartz, whereas Ballarat's total of 13 million ounces (369 tonnes) was mainly alluvial. Economically, the gap in the Bendigo experience led to the private ownership of mines and more aggressive unionism.

Famous saddle reefs, predictable at depth, gave steady and large crushings

Bendigo was more obviously a gold town than Ballarat. Poppet heads on its lines of reef remained for decades.

Shamrock Hotel, Bendigo, RE
Ballarat Town Hall, BaT

Ballarat flourished because the road from Geelong was so bad in the 1850s, just when gold production was high, that local industry and farming was stimulated.

so that from the 1870s Bendigo blossomed. The work of its major architect, William Vahland, through buildings like the post office, town hall and Shamrock Hotel, gave the city the appearance of a German principality.

From these examples it is clear that timing was significant in shaping the Goldfields. Ballarat flourished because the road from Geelong was so bad in the 1850s, just when gold production was high, that local industry and farming was stimulated. At Bendigo, on the other hand, urban development was rudimentary before the splendid railway line was built from Melbourne in 1862. That gave Melbourne interests a large role in mines, commerce and industry.

Bendigo was unable to bite back on the capital in the way Ballarat did through Rowland's soft drinks, Suttons' music, Niven's printing and the Phoenix Foundry. The Phoenix made most of Victoria's railway locomotives between 1871 and 1905, while at the apex of agricultural implement manufacture, McKay's Sunshine Harvester Works became an interstate and overseas exporter. McKay, from a farm near Bendigo, took his inventions to Ballarat in 1885 because there were thriving implement factories there and a natural outlet to the wheat farmers of the expanding Wimmera and Mallee.

All gold towns developed industries. Thompson's Foundry at Castlemaine was notable, but there were numerous coach and wagon builders, harness and saddle makers, tinsmiths, paint factories, breweries, distilleries, gasworks, confectionary, soft drink, biscuit, clothing and furniture factories, and many others.

The largest employer in Ballarat in the early 20th century was Mrs Lucas, whose garment factory occupied the workshops of the former Phoenix Foundry, and whose 'girls' inspired the city's Arch of Victory and Memorial Avenue after the Great War of 1914-1918.

Wedderburn distilled eucalyptus and at Hepburn Springs the Gervasoni family opened a macaroni factory – known for its murals. There were links to the Swiss-Italian settlement at Yandoit, where houses still echo the distant homeland and names like Righetti resound. A treasure of mining technology, Archbold's Assay Works can be found at Chewton.

Bendigo, RE

CULTURAL CAPITAL

Gold added cultural capital. Initially it attracted artists like ST Gill, Eugene von Guerard, Thomas Woolner, and an array of photographers beginning with Antoine Fauchery. There were native-born artists, including the Lindsay family at Creswick. Notable newspapers were staffed by impressive journalists such as HR Nicholls, EJ Bateman, WB Withers and Angus Mackay.

Eyewitness accounts of the early and later experience came from William Howitt, Louisa Meredith, the photographer Fauchery, the 'Welsh swagman' Joseph Jenkins, Anthony Trollope, Charles Dickens and Mark Twain. Among actors and managers were Charles Thatcher, Lola Montez, GV Brooke and opera and minstrel companies. At Ballarat a Welsh Esteiddford was the forerunner of the South Street competitions, which in the 20th century, earned the city the title of 'Band Capital of Australia' and launched many singers on international careers.

Her Majesty's Theatre at Ballarat and the Capitol Theatre at Bendigo represent cultural maturity as do their splendid art galleries, and one at Castlemaine. Goldfields civic consciousness is visible in the Botanic Gardens at Ballarat, where a walk is flanked by the busts of Australian prime ministers and an evocative memorial commemorates prisoners taken in all Australia's wars.

SOCIAL MOVEMENT

Links to mining ventures throughout the continent and the great strength of local branches of the Australian Natives Association (ANA) helped the Goldfields to vote overwhelmingly for Australian Federation. Ballarat and Bendigo, which headed the national list, held the strongest branches of the ANA. All friendly societies flourished on the Goldfields, where their halls, especially those of the Freemasons, were great community assets.

Democracy expressed itself powerfully through the work of W.G. Spence of Creswick, whose role in the miners' and shearers' unions expanded into the truly mass Australian Workers' Union. Trades Halls at Ballarat and Bendigo were radical centres, although in the 19th century small 'l' liberalism, like Deakin's, embraced wide social goals. The Eight Hour's Day Monument at the foot of Sturt Street, Ballarat, was not only a focus for protests, but also for a major festival and procession on the day itself.

In the 20th century, influenced by WWI, community identity was cemented by Anzac, Armistice and Empire days. Memorials to servicemen and women of both world wars abound and, latterly at Ballarat, the rediscovery of the significance of Eureka has inspired events, exhibitions, plaques, the restoration of the famous flag and the sound-and-light show, *Blood on the Southern Cross*. A statue to the rebel leader, Peter Lalor, has stood for over a century in the spine of Sturt Street. Nearby are William Dunston VC, Burke and Wills, Queen Victoria and the Scots and Irish poets, Robbie Burns and Tom Moore.

Prime Ministers' busts, Botanic Gardens, Ballarat, BaT
Eureka Flag, Ballarat Fine Art Gallery, BG
Victory Arch, Ballarat, BaT

The energy, skill, vision and determination of the goldfields population is visible in the achievements highlighted by this book.

HEADSTONES & CHURCHES

Perhaps the most powerful memorials to the Goldfields story are its cemeteries. Many 19th century tombstones record places of birth so that the Welsh background of Mt Egerton miners, for instance, or the Polish origins of the Jews of Main Street, Ballarat, can be understood, just as Chinese graves and ceremonial burning towers evoke a culture often strange to Europeans.

Chinese were prominent, often comprising a quarter of the alluvial mining population. Evidence of their presence was, until recently, confined to cemeteries, a few water channels and some intricate elevator-style pumps. Now, at Bendigo's Golden Dragon Museum, Ararat's Gum San Chinese Heritage Centre and at Sovereign Hill, significant displays explain their culture and contribution.

In cemeteries, other ethnic groups, except for the Jews, are spread among the different religious denominations, each of whom had a special section. Religion was strong on the Goldfields where initially the lay ministers and tent meetings of Wesleyan Methodists gave them an impetus already assisted by the lower middle class social structure.

The buildings of the Christian churches tell the story of a powerful commitment, whose weakening in the latter half of the 20th century has led to the conversion of many small churches to homes. An exception is the fine Roman Catholic cathedral at Bendigo where major works continued until its completion in 1977.

HERITAGE

The energy, skill, vision and determination of the goldfields population is visible in the achievements highlighted by this book. Blessed and inspired by their heritage, succeeding generations should do their utmost to preserve, interpret and publicise one of Australia's great experiences, a world event that shaped our nation and put new meaning into the concept of democracy.

Pennyweight Flat Cemetery, BeT
Linton, GS

GEOLOGY

BY CLIVE WILLMAN

Imagination is the key to understanding the ancient history of Victoria. If we look carefully, the evidence for the past is all around us. Layers of sandstone in a road cutting — imagine a vast ocean floor. A mountain of granite — imagine cubic kilometres of molten rock bulging upwards. A basaltic plain — imagine red-hot lava flowing down a valley.

We have a direct connection with these past events for they have shaped the present day landscape. The shape of the land, the way we cultivate it, the building stones we use and the quality of soils are all a consequence of the geological past.

The imagination is crucial, but not enough by itself. To understand the foundations of central Victoria requires some detective work. When we look closely at the bones of the land – the sandstone layers, the granite hills, or the basalt lava flows – the evidence of their age and origins can be read as clearly as if written down.

The story that unravels shows that the sandstone layers formed first and they are the foundation stones of Victoria that extend down to great depths. The granite came next – several huge bodies of molten rock pushed upwards through the sandy layers, cooking the rocks they touched. Finally, volcanoes disgorged basalt lava that spread out over the sandstone and granite hills forming a thin cover like icing on a cake.

The shape of the land, the way we cultivate it, the building stones we use and the quality of soils are all a consequence of the geological past.

THE OCEAN FLOOR

The story of the sandstone hills began 480 million years ago in a deep ocean that once formed alongside the ancient continent of Gondwana. For about 30 million years the rocks of Gondwana were worn down by erosion — rivers carried the debris eastwards and dumped their loads of sand, silt and clay into the ocean.

Thousands of these layers, most less than a metre thick, spread across the ocean floor piling one on top of the other to eventually build a rock sandwich many kilometres thick. Amazingly we can still see the original alternating layers of sand and mud in road cuttings throughout central Victoria.

But how did the layers rise above sea-level to form land and why are they now vertical?

About 440 million years ago the ocean floor was squeezed – like a vice – between Gondwana and a tectonic plate farther east. This squashed the rock sandwich and plastered it against the eastern edge of ancient Gondwana. As the layers in the sandwich compressed they pushed upwards. The original ocean floor was four km thick and up to 1500 km wide, but by the end of the big squeeze it was a staggering 15-20 km thick and the original ocean floor had been compressed sideways to only 700 km wide. This great mass of folded sand and mud layers was welded together as hard sandstone and mudstone – which is lucky for us, because it now forms the foundations of south-eastern Australia.

Over the past 400 million years erosion has removed several kilometres of sandstone and mudstone, but there's still plenty down there. Mining in Bendigo has traced the folded layers as deep as 1.4 km and seismic surveys show they probably extend down to at least 15 km.

Sandstone Ridge, PV (AP)

NORTH-SOUTH BONES

Today the sandstone hills have a special character because the hard sandstone layers form prominent outcrops, as if the geological bones of the land are exposed. The rocks break down to form thin and poor quality soils, especially in the dry climate north of the Great Divide. While the Box-Ironbark forests have adapted to this environment, the sandstone hills are generally unsuitable for agriculture.

If you walk through the bush between Daylesford and Castlemaine you will quickly realise that the sandstone layers are all aligned north-south and therefore many of the hills are also elongated in this direction. Why? The big squeeze alongside Gondwana pushed from the west and east forcing all the layers to crumple along north-south lines. Try distorting a tablecloth on a smooth surface and you will see how it works.

Gold was formed at the same time as the big squeeze. Hot watery fluids, at about 350° C, carried gold and quartz up through the newly folded layers. The fluids quickly invaded any cracks or faults in the hardened rocks to form veins of quartz and gold.

GRANITE HILLS

The rolling and rocky hills of the granite country are a special and beautiful landscape. Unlike the sandstone hills, granite country creates better farming land and is mostly cleared of forest. Granite forms thicker and sandier soils which are much better for sheep grazing and fruit or grape growing. If you look closely at a piece of granite you will see that it is composed of large grains of quartz, feldspar and mica. The feldspar and mica eventually break down to form clays, but the quartz is very durable and gives the soil its sandy and well-drained texture. There is no layering in the granite, so the hills are more rounded than the sandstone hills.

You will quickly realise that the sandstone layers are all aligned north-south.

The Harcourt Granite formed 368 million years ago, much later than the sandstone and mudstone layers. It pushed its way upwards from deep in the earth's crust as a hot lump of molten rock, to finally cool and solidify several kilometres below the surface. Since then, millions of years of erosion have finally exposed the granite in the present day landscape. When the granite pushed the older layers aside it cooked everything it came in contact with. This baked and hardened the sandstone and mudstone layers around the granite's margin making them more resistant to erosion. This explains why there is a narrow line of hills surrounding the granite.

VOLCANIC PLAINS

The volcanic plains provide some of the best land for agriculture in central Victoria because the basaltic rocks break down to form soils rich in minerals. The lava erupted out of the volcanoes as a toffee-like fluid flowing down the valleys and sometimes spreading over large areas. The flow tops cooled to form flat surfaces, but the thickness of the basalt varies greatly, depending on whether it has covered a valley or hill over the older sandstone and granite areas.

In many places the basalt covered alluvium that had collected in the valley floors over the thousands of years before the volcanic activity. During the 19th century, miners excavated these 'deep lead' deposits for their rich alluvial gold.

The Guildford Plateau is the eroded remnant of a lava flow that originated

Sandstone Bones, RW
Dog Rocks, Mt Alexander, PV (AP)
Clarke's Hill, JM

near Daylesford and flowed northwards along the ancient Loddon River. Erosion over the last three million years has lowered the surrounding hills leaving the basalt perched relatively high in the landscape as a plateau.

Some waterfalls in the region have been created by rivers falling over the edge of lava flows that filled their old watercourses. The rivers kept running over the top of the new basalt and they are now wearing away the softer rock at the end of the lava flow. Trentham and Turpins Falls are examples.

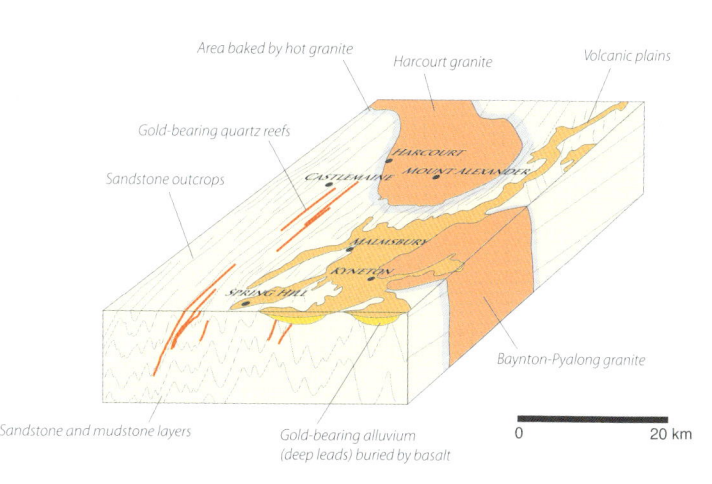

The volcanic plains are less than five million years old, which is nothing in geological time. Some of the volcanic activity in western Victoria is so recent that the Aboriginals witnessed the fireworks.

Some of the youngest volcanoes retain their volcanic shape – Mt Franklin and Mt Buninyong still have their craters. Mt Kooroocheang sits boldly above the flat plains to remind us that we still live in an active volcanic province. Between Ballarat and Daylesford 123 volcanoes have been identified, but there are hundreds of others that are not quite so easily identifiable, especially in the triangle between Lancefield, Maldon and Ballarat.

TOPOGRAPHY

Streams from the southern side of the Great Dividing Range flow to the shores of Port Phillip Bay. Over the Great Dividing Range the flow is northward across the great plain to join the Murray River.

The road from Trentham to Blackwood passes through Newbury, which is the summit of the Great Dividing Range. Within a radius of 10 km around Newbury six major rivers flow from natural springs.

The Campaspe and Little Coliban Rivers rise near East Trentham, the Coliban just west of Newbury, and the Loddon a little further toward Bullarto. These four streams, which flow north or north-west, provided natural highways for the aboriginal clans and formed the basis for the water reticulation schemes that were developed in the 19th century.

The Werribee and Lerderderg Rivers rise immediately to the south-west of Newbury. The Werribee River runs south-west before turning towards Ballan on its way to Bacchus Marsh via Werribee Gorge. The Lerderderg River flows south-east in a deep little valley past Blackwood and on through the forested slopes of Lerderderg Gorge before joining the Werribee River on the alluvial flats of Bacchus Marsh.

The western boundary of the region is defined by two mountain ranges – the Pyrenees and the Grampians. Mt Ararat is a residual peak of hard metamorphic rock, called hornfels, that has been baked by contact with granite intrusion. Other isolated peaks, like Mt Langi Ghiran, Mt Buangor and Ben Nevis, are made of the granite itself.

Wombat State Forest, PW

HISTORY

For tens of thousands of years the Goldfields region was home to various Aboriginal tribes and clans whose descendants today form part of the world's oldest living culture. Nearly two centuries ago 'new Australians' from the British Isles first came to the region as pastoral settlers. Before long, however, gold was found and the rest of the world discovered the region.

Goldrushes were a 19th century phenomenon. Their scope and influence dwarfed everything that had previously happened in the history of gold.

The arrival at London's port, in the first three weeks of April 1852, of eight tons (8.13 tonnes) of Victorian gold could not fail – and did not fail – to excite the press and the public. *The Times* newspaper predicted: 'When the spring begins, there will be a gathering from every quarter of the compass, such as has probably never been seen before on the face of the earth'.

Ballarat, Mt Alexander and Bendigo were the goldfields whose names reverberated around the globe. By the end of 1853, 200,000 newcomers had flocked to Victoria in search of gold and within a decade the population of Australia had trebled.

Immigration brought new ideas and new skills to Victoria. After trying their luck at mining many people moved back to their old trades and professions. In fact, more people got rich supplying services and supplies to the gold miners than got rich by actually digging for gold.

In the space of a single generation significant provincial cities – Ballarat, Bendigo, Maryborough and Castlemaine – grew up, and a large-scale and sophisticated infrastructure was established.

On the strength of the goldrushes Melbourne quickly grew from a small town to Australia's financial capital and most glamorous city. For a short time it even held the world record for urban growth.

By 1861 Melbourne had a population of 125,000 people, paved roads and quality housing, and it was on its way to being known internationally as 'Marvellous Melbourne'.

Nowhere on earth has a single generation created so much. But nowhere on earth has there been such a special mix of ingredients – a gathering of people from virtually every existing nation, deeply held democratic aspirations, stunningly rich surface gold deposits, and a fertile hinterland.

480 million years – The Goldfields region is a deep ocean

440 million years – The ocean floor is compressed and lifted; gold deposits are formed

1775–1783 American Revolutionary War, also known as the American War of Independence

1788 First Fleet arrives at Botany Bay

1788-1799 French Revolution

1832 British middle class males gain voting rights (First Reform Act)

1835 Melbourne founded

1836 Major Mitchell is first European to view the grassy valley where Castlemaine stands today

1837 Eighteen-year-old Queen Victoria takes the throne

1838 First recorded massacre of the Dja Dja Wurrung

1840 'Penny Postage' introduced throughout British Isles

1843 Last recorded big corroboree held by Dja Dja Wurrung

1846-1851 One million people die in the Irish Famines

1848 Californian goldrush commences

1851 (July) – the Colony of Victoria is created and gold is officially discovered at places like Clunes, Warrandyte and Buninyong

1851 Victoria's goldrushes commence with the discovery of Ballarat, Castlemaine (Mt Alexander) and Bendigo

1851 (Dec) – Monster Meeting protesting against Miners' Licenses at Forest Creek (Chewton)

1853 Red Ribbon protest against Miners' Licenses at Bendigo

1853 T.S. Barnes, local confectionery maker, produces first Castlemaine Rock

1853 Avoca, Heathcote, Maldon, Maryborough, Rushworth, Stawell goldrushes

1853/54 Deep lead mining commences at Ballarat and quartz mining commences at Bendigo and Castlemaine

1854 Charge of the Light Brigade, Crimean War

1854 Horse-powered puddling machines are used

1854 First Cobb & Co coaches run from Melbourne to the Goldfields

1854 The Museum of Victoria is founded

1854 First telegraph in Australia

1854 Chinese begin arriving in the Goldfields in large numbers

1854 Beaufort, St Arnaud, and Talbot goldrushes

1854 (Dec) Eureka Rebellion

1854 Castlemaine's Christ Church founded

1855 First fossil of Neanderthal man found

1855 All Victorian males get the right to vote

1855 Ararat goldrush

1856 Lola Montez performs her 'Spider Dance' at various Goldfields theatres

1856 Dunolly goldrush

1857 All Victorian men eligible to vote for Legislative Assembly (Lower House), but not the Legislative Council (Upper House) where property qualification remained

1857	Bush's Store built, one of the Goldfields earliest surviving shops
1857	First Maryborough Highland Gathering, Australia's oldest continuous sporting event
1858	Bendigo Pottery commences, now Australia's oldest operating pottery
1858	Population of Australia passes one million
1858	Bendigo Bank has its origins in the Bendigo Goldfield
1859	Nicholas and Edwin Fitzgerald establish the Castlemaine Brewery (Castlemaine XXXX)
1859	Charles Darwin's 'On the Origin of Species' published
1859	Macaroni factory built at Hepburn
1860	Bendigo's Gas Works, one of the first in Victoria
1861	First Melbourne Cup horse race
1861	Explorers Burke & Wills die after completing the first south-north crossing of the continent
1861	American Civil War starts
1861	New Zealand goldrushes commence; tens of thousands of diggers leave Victoria
1865	End of American Civil War; slaves freed
1867	Most British males in towns get the right to vote (Second British Reform Act)
1867	Henry Lawson born
1867	First compressed air rock drill in Australia used to excavate a tunnel into Mt Tarrengower, Maldon
1871	The Victorian countryside had a population of 270,428; Melbourne had 191,000
1871	Bendigo's Easter Festival commences
1872	Discovery of the Berry Deep Lead (Smeaton)
1872	British novelist Anthony Trollope visits Bendigo and Ballarat
1874	Amalgamated Miners' Association is created, Australia's first national union
1874	Committee is formed to organise Maldon's first Easter Fair
1876	Custer dies at the Battle of Little Big Horn
1878	Stawell Gift commences, Australia's most famous foot race
1879	Stone from Heatherlie Quarry in the Grampians is used for Parliament House in Melbourne
1880	Ned Kelly is hanged
1880	Ballarat's Welsh Esteiddford (now South Street) Competition begins
1884	All British males (including rural voters) gain the right to vote (Third Reform Act)
1887	Ballarat Fine Art Gallery is opened, the first provincial gallery in Australia
1889	The Eiffel Tower is built
1890	Maryborough Railway Station is built
1892	Bendigo's Chinese imperial dragon, Loong, makes first appearance
1893	New Zealand is the first country to grant women the right to vote
1895	Mark Twain visits Maryborough
1901	Federation; Victoria's last shallow alluvial goldrush at Mafeking, Grampians

Aboriginal History

Before Europeans

Archaeologists continue to make significant discoveries that impact on our understanding of the human settlement of Australia, but most agree Aboriginals arrived at least 60,000 years ago. In Victoria, stone tools and bones in the Keilor area of the Maribyrnong valley date back 30,000 years.

Over that time Aboriginals have seen volcanic eruptions come and go in several places, at least two 'ice ages', and rising seawater form the inland sea we now call Port Phillip Bay. Geologists have found a stone axe head trapped in a basalt rock from a lava flow, and descriptions of volcanic activity are enshrined in tribal stories. One in particular talks of 'young' Mt Franklin (or Lalgambook) throwing stones at 'old man' Mt Tarrengower (which is still called by its Aboriginal name).

Five different communities made up the Kulin nation (perhaps best understood as a cultural/linguistic group) that surrounded Port Phillip and Westernport Bays. Three of these communities are relevant to the Goldfields region. The **Wathaurong** occupied from the Bellarine Peninsula to the Pyrenees, including the region now encompassed by Geelong, the Surf Coast, Golden Plains, Ballarat and part of the Pyrenees. The **Woi Wurrung** were centred on Melbourne and the Yarra catchment, but also occupied the Macedon Ranges. The group most affected by the goldrushes, the **Dja Dja Wurrung**, were centred on the Loddon River, occupying the region now encompassed by Hepburn Springs, Mt Alexander, and Bendigo.

Further to the west were the **Djab Wurrung** (also known as the Djap Wurrung) and **Dhauwurd Wurrung** (also known as the Gundidjmara) groups, among others. Based around the Lake Condah region, the Dhauwurd Wurrung people constructed channels, weirs and stone fish traps. This allowed them to harvest large quantities of eels and fish, which in turn resulted in high population densities and the construction of permanent stone huts clustered into villages.

The Kulin people of central Victoria shared a belief in an all-powerful creator, Bundjil, the eagle. They had complex trading networks stretching as far as South Australia and New South Wales, including an important trade in greenstone (used for axes and spearheads) which was quarried at Mt William, to the east of Lancefield, and at Green Hill, north-east of Maryborough. They built sophisticated fish traps and rock wells and, among other finely crafted artefacts, were notable for the elaborate possum-skin capes they made.

It is still possible to find scar trees (scarred where bark was removed), stone drinking wells, stone arrangements and rock art. Many signs of their occupation have survived, despite the devastation wrought by the gold diggers. Many of the most important aboriginal sites were close to water where, unfortunately, the banks of creeks and rivers were turned upside down by alluvial mining and sluicing.

Prior to European farming and mining, the Loddon River was a series of long, deep waterholes, stocked with native fish. In effect, the rivers and creeks were Aboriginal highways – complete with take-away food and drink – that linked different parts of a tribe's territory, as well as one group to the next.

The Aboriginal people used managed burning to prevent catastrophic wildfires, and to allow the regeneration of fresh grass shoots that nourished the kangaroo

Loddon River, PV (AP)
Scar Tree, Joyces Creek, KS

Introduced diseases had a greater numerical impact than violence. Nevertheless there certainly was significant violence.

population, the basis of their diet. Firestick agriculture, as it is sometimes called, played a crucial part in creating the open woodland that was so characteristic of the Goldfields region.

A huge range of plant food was used by the Aboriginals, including yam and daisy roots, orchid tubers, lily bulbs and nectar-rich ironbark flowers (which were used to sweeten water). It is thought that a 15-hour working week would have been sufficient to hunt and gather a plentiful supply of food.

Reliable population figures do not exist, but what is certain is that the moment Europeans arrived on the continent Aboriginal numbers began to decline rapidly. A small-pox epidemic reached central Victoria 40 years ahead of the first explorers who therefore witnessed Aboriginal nations that had already been reduced in number by as much as two thirds.

Introduced diseases had a greater numerical impact than violence. Nevertheless there certainly was significant violence. On the lawless frontier, Aboriginal deaths were deliberately kept secret.

Possibly because of the relative population density, large-scale Aboriginal resistance to the European invasion of Victoria was centred on Lake Condah and the Stony Rises, and became known as the Eumeralla War (after the Eumeralla River). However, across the state, including in the Goldfields region, there are records of numerous attacks, skirmishes, mass poisonings and shootings.

There was clear resistance by the Dja Dja Wurrung and some squatters were driven from their land, but the Europeans' guns gave them an overwhelming advantage. By some estimates one quarter of the surviving Dja Dja Wurrung – 150 people – were killed between 1838 and 1842 as the squatters seized land for their farms.

Although the Europeans were clearly taking Aboriginal land by force, the issue was complicated by the mutual failure of both groups to understand their very different notions of land and ownership.

Attack on Dray Store, ST Gill, Courtesy
the Sovereign Hill Museums Association

ABORIGINALS IN THE GOLDFIELDS REGION
BY PETER MILLS

Only a few years before the goldrushes, much of land north-west and north-east of Melbourne was a frontier openly in dispute between the Dja Dja Wurrung and squatters.

The Europeans rated the Dja Dja Wurrung as relatively peaceable compared to the neighbouring Goulburn River Daung Wurrung, but a number of bloody clashes were recorded in the Mt Alexander area. Troopers were permanently stationed near Kyneton to protect squatters' interests in the late 1830s, but by the mid-1840s the squatters of the Mt Alexander area were no longer concerned about armed resistance or excessive interference by Aboriginals in the squatter economy.

The numbers of Aboriginal occupants of central Victoria had been greatly reduced, but they continued to inhabit their land and to make use of the resources not yet denied them by the squatters. Still, the fact that the squatters employed hut keepers on their outstations to protect supplies is a good indication of this continuing competition.

The Victorian Government pursued a policy of concentrating Aboriginals onto reserves, ostensibly for their protection. To this end a system of Aboriginal Protectorate districts was set up in the Port Phillip District in 1839. The protector for the Loddon District, which covered Bendigo and Mt Alexander region, was Edward Stone Parker.

In 1839, on the Campaspe Plains Run (25 km from Bendigo) two shepherds were speared and flocks were driven away. In reprisal troopers attacked the first Aboriginals they came across, killing six and wounding many more. Parker accused those implicated of cold-blooded massacre.

Parker may have had some success in protecting the Aboriginals from abuses by the whites, but not from the other forces which were leading to their population decline. He was involved extensively with the Dja Dja Wurrung, many of whom were present at a station that he established for a period at Neeriman on the Loddon River near Tarrengower (1840-41), before settling on a reserve at Mt Franklin in 1841.

ABORIGINALS ON THE GOLDFIELDS
BY IAN D. CLARK AND DAVID (FRED) CAHIR

The goldfields provided Aboriginal peoples with many economic and cultural opportunities. Owing to the exodus of European labourers to the diggings, many pastoral stations survived only by employing Aboriginal labourers. Aboriginal culture persisted despite the influx of 300,000 new immigrants. Elizabeth Laye's 1861 account of a meeting with the Avoca tribe in the 1850s confirms that despite the goldrushes they were living by 'traditional' means.

Aboriginal culture persisted despite the influx of 300,000 new immigrants.

Aboriginal Rock Wells, Castlemaine Diggings Heritage National Park, RE

Aboriginal people were attracted to the goldfields for a range of reasons. For one thing, the fields were on traditional Aboriginal lands, and they were keen to continue their association with their clan estates. Dr George Wakefield, in a letter to his father dated 1 May 1856, discussed displays of indigenous weapons, corroborees and multiculturalism at Black Hill, Ballarat. '…here we have representatives of all the nations of the face of the globe not the least wonderful of which is the aboriginal nation. I have frequently been present at their corroborees, and their skill in throwing the spear, boomerang etc., is wonderful.'

It is also possible to find examples of Aboriginal people succumbing to gold fever on the Victorian goldfields, and immigrant Aboriginal people joining the rush to Victoria. They moved quickly to grasp the economic opportunities presented to them by the miners, trading and selling possum skin cloaks, fish, and game such as possum. JF Hughes, a Castlemaine pioneer, recalled that possum skin and kangaroo skin rugs were 'sold to settlers and lucky gold-diggers at £5 a-piece'. Dja Dja Wurrung farmers at Mt Franklin capitalised on the nearby goldfields by selling excess produce from their farms.

When asked to show their licenses [the Aboriginals] replied to the police that the gold and the land were theirs.

In many miners' accounts it is also possible to find examples where Aboriginal people were willing to share their resources and bush craft with miners. For example, Lord Robert Cecil who made a visit to the Kyneton diggings in 1852 recalled how the diggers at Specimen Gully 'showed me what the natives call blackfellows sugar. It is a species of manna falling plentifully from the white-gum. It tastes very much like the second layer in a wedding cake'. For their part, many miners welcomed the opportunity to embrace Aboriginal ways, such as clothing, building bark shelters, and eating local foods.

Mining was not a foreign and incomprehensible spectacle for Aboriginal people, as they had been quarrying ochre, greenstone, kaolin, and in some regions, crystals, for thousands of years. There is even some cultural evidence that Aboriginal people regarded gold as a precious stone. Other observers confirm that there were many Aboriginal people prospecting for gold on the central Victorian goldfields.

One of the most significant and best-documented impacts Aboriginal people had on the goldfields was through the role of the Native Police Corps in establishing order. Miners noted that some Aboriginal people on the goldfields asserted their rights, such as the group of Aboriginal diggers at Forest Creek in 1852 who when asked to show their licenses replied to the police that the gold and the land were theirs – so why should they pay money to the Queen?

Native Police, ST Gill, Courtesy the
Sovereign Hill Museums Association

European Explorers
By Joseph Kinsela

Australia's great explorers were the astronauts of the 19th century, and their paths across the Goldfields can be traced through memorial statuary, stone cairns, buildings, and placenames that pepper the region.

The excitement and hero-worship they generated can only be explained in the context of their own time. It is hard to comprehend, from our position in the 21st century, how utterly frustrated the early settlers were by their inability to find answers to questions about the geography, resources and potential of Australia. For example, it took a full 25 years after European settlement at Sydney to pick out a route over the Blue Mountains, which began a mere 60 km west of the town. The continent did not give up its secrets easily and the great inland regions continued to defy party after party of explorers.

Major Mitchell

Major Sir Thomas Livingstone Mitchell, 'Major Mitchell', Surveyor General to His Majesty's Colony of New South Wales, was one of the few explorers who did have success, and he is consequently one of the most famous. He explored the fertile pastures of western and northern Victoria in 1836.

Born in 1792, Mitchell entered the army at 16 and was made a major in 1826 after being an aide to the Duke of Wellington. In 1827 he came to Sydney to take the position of Assistant-Surveyor to John Oxley. The following year he succeeded to Oxley's position and made several expeditions, chiefly tracing the course of the western rivers of NSW. Governor Bourke, disappointed at failure to resolve the riddle of the rivers that flowed inland (it was assumed they flowed into an inland sea) and the fact that no fertile inland region had been discovered, sent Mitchell on a quest to explore the course of the Darling River and its connection with the Murray River.

Mitchell's 1836 Expedition

Major Mitchell set out in March 1836 with a party of 24 armed men and an ox-wagon carrying a boat. Included in the party were two Aboriginal guides, known to the Europeans as John Piper and Tommy Came Last. On 3 June they found the junction of the Darling and the Murray Rivers. The party turned upstream along the Murray, reaching Swan Hill at the end of the month. It was a wet winter and Mitchell was excited by the quality of the soils and agricultural potential of the land.

Hanging Rock and Mt Macedon, BS Harcourt and Mt Alexander, BS

Leaving the Murray near Kerang, the party went south-west and encountered the dramatic wall of mountains which Mitchell named the Grampians after the Scottish ranges they do indeed resemble. Mitchell continued south-west via the Wannon and Glenelg Rivers toward Portland where he encountered the Henty family, who had settled the area from Launceston the previous year.

On their return journey, the party crossed ranges which Mitchell named The Pyrenees. The name must have resulted from the delusionary effect of having been so long on the plains – the border ranges of France and Spain reach 3000 metres; Victoria's Pyrenees may just reach 1000 metres.

There is a story that settlers found the imprint of wheel tracks more than a decade later.

Mitchell continued to be impressed by the regular streams and extensive grassland. When the party reached the Loddon River the bullock cart bogged and it took more than a day to get free of the muddy flats. This was near where the town of Newstead is today and there is a story that settlers found the imprint of wheel tracks more than a decade later.

Travelling along the valley near Chewton, Mitchell saw the heights of Mt Alexander. Passing through a ravine they reached a spot near Harcourt where a wheel of the ox-wagon broke, so they made camp. Mitchell took the opportunity to climb Mt Alexander and from there saw a high range to the south. Next morning he left with a small party hoping to make a sighting of Port Phillip Bay.

They crossed the Coliban near Taradale and reached a second stream which Mitchell named Campaspe. He crossed it at the point where Mollison Street, Kyneton, crosses it today. At Carlsruhe the river was crossed again and Mitchell headed for the wooded slopes of the mountain he was to name Mt Macedon.

MITCHELL'S EXPEDITION REACHES MACEDON

Major Mitchell approached Mt Macedon from the north. The slopes were so thickly timbered that it took a long time for him to find a viewing point. He finally achieved this late in the afternoon as the light was beginning to go, but he saw the sails of a boat on Port Phillip Bay. By complete coincidence, on 30 September 1836 a ship from Sydney did sail across the bay bringing the new administrative party for the settlement at Melbourne.

Many towns sprang up along the route of Mitchell's exploration – some by chance, others because Mitchell crossed streams and ranges at points where others later found it convenient to settle. There are monuments in many of these towns, and indeed throughout western and northern Victoria, illustrating how thoroughly Mitchell covered the territory during the course of nine months.

Grampians, near Stawell, RE

EUROPEAN SETTLEMENT BEFORE GOLD

BY PETER MILLS

Wool was the ideal export commodity for the new European colonists and they lost little time in establishing a significant industry. Wool was an easily transportable commodity that was 20 times more valuable than the equivalent weight of wheat. By 1840 the Australian colonies were providing half the needs of Britain's woollen mills and for the first time it could be said that Australia's prosperity rode on a sheep's back.

The pastoral land grab in the Goldfields region began when expanding stock numbers strained the capacity of pasture in New South Wales and Van Dieman's Land. Squatters moved towards the central highlands of Victoria, both from the north and from Port Phillip, from as early as the mid-1830s, disregarding government restrictions. The Government yielded to pressure and legalised squatting in 1836.

To allow for future expansion the squatters occupied runs far larger than their flocks initially required. Their homesteads were sited in rich areas (usually alluvial flats) where crops and vegetables for their own consumption could be grown and where water was relatively reliable.

Flocks of up to a thousand sheep were looked after by shepherds with dogs, with the only fences forming a few holding pens. The flocks were fanned out into outstations in the less fertile hilly and forested country to make use of seasonal feed. These wooded hill areas were more sparingly used and less altered by squatters than the fertile grasslands, and Aboriginals still hunted and gathered there.

The squatters' workers also included hut keepers who protected outstation huts, and bullock drivers who brought in supplies and took away the wool. The squatters copied the Aboriginal practice of burning off to provide green feed for their flocks and to maintain the open country.

There weren't many self-made squatters. Even with the peppercorn leases and minimal input into improvements and labour, a good deal of capital was needed to purchase costly stock to increase the flocks, and to see out the long wait between sending off the wool and receiving payment.

Squatter numbers were low and the early homesteads were small and modest in construction. Consequently, there are few tangible remains of this early pastoral era. Those structures constructed in stone have survived best and there are still buildings, or parts of buildings, in the region dating from before the goldrushes.

The superficial appearance of the landscape was little altered by the squatters' occupation. But grazing set in train lasting changes to the various ecosystems of the area. The hooves of sheep and cattle compressed the soft surface layers of soils previously trodden only by the soft pads of marsupial grazing animals, and native grasses were overgrazed by the introduced sheep and cattle.

There weren't many self-made squatters. Even with the peppercorn leases a good deal of capital was needed.

Clydesdale, JM

This resulted in increased run-off and flooding, causing scouring of watercourses, increased erosion of topsoil and silting up of waterholes. Water became scarce. With faster run-off, creeks ran for shorter periods of the year and the silted waterholes were reduced in volume. Introduced grasses and weeds gradually took over from the disturbed native vegetation.

Wool was an inefficient industry that supported only a tiny population. If the success of an economy is judged by the size of population it successfully supports, then the economy of Victoria had declined steadily since European settlement. The discovery of gold would alter this equation with cataclysmic speed.

Goldrush Melbourne

By Robyn Annear

As the sailing ship Kent was piloted into Port Phillip Bay in 1852, near the end of its long voyage, a crewman on a passing vessel called out, 'Come along, we'll show you the way to the diggings!'

'Hurrah!' cried the gold-seekers aboard the Kent. 'It's no hoax, then! There are diggings!'

Sure enough, the diggings were real; but they weren't as close-by as the new arrivals supposed. Though landfall was in sight, their destination – the diggings – was still quite a way off. And there was Melbourne to contend with first.

Sovereign Hill, SV

For most gold-seekers from across the seas, Melbourne was the gateway – or stumbling block – to the Victorian diggings. At least one man spent his first day ashore expecting to find gold on the city footpaths. 'I was surprised to see so few signs of it in Melbourne,' he wrote home. 'I had read that most people had a patch in their back yard.' In fact, he need only have raised his eyes to see gold at every step. 'Gold was as common in Melbourne as gingerbread at a country fair,' reported another newcomer. 'In some of the shops in Great Collins-street, I saw nuggets as large as my hand.'

A display of gold in a shop window announced that the proprietor would pay cash or exchange goods for gold. These weren't just the windows of jewellers or pawnbrokers, but of drapers, gunsmiths, wine merchants, and confectioners. Almost every Melbourne shopkeeper was a gold-buyer as well. The window display of Hood the chemist, in Collins St, featured 'an ingenious miniature representation of hut, cradle, diggers, and washers engaged in their gold hunting avocation'.

His first day in Melbourne convinced one new chum that 'almost every tenth person you met in the streets had gold dust or nuggets in his pocket'. The streets swarmed with people who were either down from the diggings or heading for them. 'Crowded with rude-looking diggers and hosts of immigrants, with their wives, their bundles, and their dogs' - that's how William Howitt, writing for The

Times of London, described the town at the end of 1852.

Before the discovery of gold in 1851, Melbourne's population had been just 23,000. Back then, there'd been only a trickle of immigrants and, excepting for Race Week, visitors to the town were few. By the spring of 1852, however, 15,000 new chums were arriving in Melbourne every month. Most of them were passing through, en route to the diggings. But even the keenest gold-seeker needed at least a day or two in Melbourne to outfit himself and find his bearings. Where was he to rest his head?

One new arrival found lodgings in the sitting-room of a house in Little Collins St. 'There were myself and another gentleman on the table, two under the table, two before the fire-place, and another on the sofa; we slept very comfortably,' he wrote. In a makeshift boarding house in Flinders Lane, 'Many of the beds held two huddled together, and here and there a complicated bundle with feet sticking out, looking like three.' Publicans even let out their stables as lodgings, fitting three men (or a family) in each horse stall.

The alternative was Canvas Town. Sprawling on the south side of the Yarra, the vast encampment housed as many as 7,000 immigrants at a time. Some arrived with ready-made tents, others fashioned shelters out of blankets and bark and cast-off clothing. What began as a campground quickly came to resemble a town, with its own restaurants, bakers, blacksmiths, and doctors – all housed under canvas. The rows of tents were even formed into streets, named after London thoroughfares: Holborn, the Strand, Regent, Bond and Oxford Sts, a world away from their namesakes.

'Hard up! Compulsory sales!' That was the cry of the Rag Fair. All too often the luggage which gold-seekers had carried across the seas ended up here: spread out along the river margin of Flinders St.

Just a day or two ashore was usually long enough for immigrants to realise that they had brought too much baggage and of the wrong sort. Those who had bought mining equipment in London, for instance, found that they'd been cruelly misled. Their cumbersome 'California-patent' gold cradles bore little resemblance to those on sale for half the price in Melbourne – let alone the rough-hewn models in use at the diggings. And their clothes they'd brought (boots 'expressly for the diggings' and El Dorado shirts) turned out to be all wrong and far too numerous. Many families had arrived almost penniless and discovered that the cost of transporting their luggage to the diggings was beyond them. Hence, the Rag Fair.

Clothes and personal effects, never unpacked, were sold from open suitcases and trunks, or spread out on rugs along the waterfront. Underwear, waistcoats, dress coats, dark-coloured trousers, footwear that wouldn't tolerate mud, all headgear other than the plainest bonnet or wide-awake hat, jewellery, umbrellas, books, musical instruments, watches, even a canary – anything surplus to the basic goldfields kit had to go.

Meanwhile, those who hadn't overpacked had to decide which of the countless 'essentials' on offer in Melbourne he ought to stock up with before leaving for the diggings. Outfitting for the diggings was a tricky business. To arrive under-equipped could be fatal; but overburdening meant a slow journey with a good chance of finding, too late, that you'd overpaid for the wrong gear. In fact, all the essentials could be bought – secondhand, most of them – at the diggings. But the shopkeepers of Melbourne made it their business to convince new chums otherwise.

'Many of the beds held two huddled together, and here and there a complicated bundle with feet sticking out, looking like three.'

Sovereign Hill, SV

Eureka: the spirit of social and political democracy

Given the rigidly hierarchical world of the mid-19th century, it is easy to imagine how the discovery of gold could have been seen as a prelude to anarchy. Gold gave everyone the luxury of imagining – and achieving – a radical change in their circumstances.

After the precious metal was dug up in Victoria, most was almost immediately buried again – in bank vaults in London.

Gold License, GM

Alluvial gold was a remarkably democratic mineral. You didn't need capital or great expertise, to find and extract it. If you did find it, you didn't need a middleman to realise its value. You could exchange it for goods at the nearest shop, or for cash at the nearest bank.

In the 19th century, in most countries, a small elite sat at the pinnacle of society, and controlled most wealth, power and status. Democratic change was in the Victorian air however, not least because of the French and American Revolutions. The Goldfields attracted a volatile mixture of people – including ex-convicts, Irish, Americans and veterans of revolutionary conflicts in Europe – many of whom passionately believed in the right of each man to vote, the right of education for all, the right to freedom of expression and the right to unhindered public assembly.

To put these radical beliefs in context, in Britain at the time of the first gold rushes only the upper classes and property-owning middle classes could vote – a situation that was naturally reproduced in the Australian colonies. It was not until 1884 that all British men were finally given the right to vote, and not until 1918 that British women were eligible…

Separated from New South Wales, the new colony of Victoria was created in 1851. The first governor, Charles LaTrobe, ruled with the help of an executive council comprising four senior bureaucrats and five private citizens, all chosen by LaTrobe. The other twenty members of Legislative Council were elected by the small number of adult males in the colony who owned or occupied property worth £10 or more per annum in rent, or held a pastoral license.

Gold was officially discovered immediately after the new colony was proclaimed in July and in August LaTrobe announced a license fee of 30 shillings a month for all gold seekers – partly to meet government expenses and partly to deter people from leaving their real 'positions' and rushing to the goldfields. Ships lay idle in the harbour for lack of crew, and there was real concern that famine would result from a farm labour shortage and a consequent failure to bring in the summer harvest.

Amongst the diggers, however, the license was seen as onerous and unfair from the start and within five days of its proclamation diggers at Buninyong staged the first protest against the law.

Thirty shillings was a considerable sum; it was payable in advance and there was no refund if you didn't find gold. It was clearly a form of taxation, but it did not confer any voting rights. A squatter with a vote and 20 square miles of land (5000 ha) paid annual tax of just £10, yet a digger with no vote, no land – and potentially no gold – paid £18 for a year's worth of licenses.

Despite the initial disquiet at Buninyong, the goldfields proved so rich that 30 shillings deterred no-one and initially created little hardship. Opposition to the license fees subsided. On 1 December, however, the Executive Council issued a proclamation doubling the fee to £3 and extending the license to every adult male working on the goldfields – not just the miners, but to cooks, shop-keepers, doctors and teachers.

These changes sparked the next major protest, the Monster Meeting at Forest Creek, Chewton, on the fabulously rich Mt Alexander diggings. Fourteen thousand diggers gathered to protest against the new fee. Within two days the increase was revoked, but the government stuck by insistence that everyone on the goldfields must pay.

The iniquity of the license fee was compounded by the methods employed in collecting it. A motley collection of police and troopers hounded the diggers to show their licenses, sometimes several times a day. Men who failed to produce their licenses were roped or handcuffed and driven to the Government camp where heavy penalties were imposed. Those who could not pay were put to work on road gangs.

Police or troopers officially took a cut of the fees and the fines, and the the system was open to corruption and abuse. After two years service on the goldfields, the hated Assistant Commissioner David Armstrong boasted he had cleared £15,000, although his salary was only £300.

The police proved remarkably effective at gathering license fees, but their tendency to mete out summary justice (and injustice) did little to prevent robbery and control crime in general. The diggers' expectation that in exchange for their fees real crime would be addressed and that they would receive police protection was not met.

Opposition to the license fee continued to grow as alluvial gold became harder to find. In June 1853 a deputation from Bendigo took a petition to LaTrobe asking for the fee to be reduced and condemning the harsh treatment of those unable to pay. LaTrobe refused the diggers' requests.

Just north of the Midland Hwy on the Ballarat side of Buninyong, a cairn marks the site of the first protest against the Miner's License.

A huge protest meeting in Bendigo on 21August prompted LaTrobe to offer to appoint a diggers' representative to the Legislative Council. The offer was rejected by the diggers and it was agreed that red ribbons would be worn on hats as a pledge not to pay license fees. In the following days it was said that red ribbons were worn by 90% of the diggers.

And red ribbons were only part of it… Men talked openly of armed resistance. Soldiers moved into barracks on Camp Hill and LaTrobe prepared to defend government offices in Melbourne. After a tense stand-off between the diggers and the government camp at Bendigo LaTrobe compromised by reducing the license fee to £1. Once again the immediate sting was taken out of the protests, but the problem did not go away.

Pistol from Eureka Dig, HV
Digger License Forest Creek 1852,
 ST Gill, Courtesy the Sovereign Hill
 Museums Association

Initially, the colony's new governor, Charles Hotham, was greeted optimistically by the miners, especially after he had asserted, 'all power proceeds from the people'. But the rhetoric was not matched with real change – the hated license hunts continued – and old resentments at the corruption and heavy-handedness of the authorities soon welled up again.

In particular, the administration at Ballarat seemed intent on provoking the diggers, and a series of incidents in 1854 exposed a systematic failure of justice: an Irish priest's servant was arrested on a dubious charge, and an American was

framed for selling sly-grog. The final spark was provided when a rich local publican, James Bentley, was acquitted of killing James Scobie. One of the local magistrates who heard the case and dismissed the charges was a friend of Bentley and a shareholder in the hotel. This miscarriage of justice led, unsurprisingly, to a protest. Eight thousand diggers converged on Bentley's hotel and it was burnt to the ground.

As has happened many times in history, the authorities responded with their own show of force, which further inflamed the situation. They arrested three men for arson, and brought in extra troops from Melbourne. Protest meetings gathered fury and demanded the release of the three arsonists. Hotham bridled at the 'demands' of the mob and encouraged the local commissioner, Robert Rede to mount a provocative license hunt.

The events are re-enacted at Sovereign Hill in a spectacular sound-and-light show, 'Blood on the Southern Cross'.

On 29 November 1854 twelve thousand people – roughly one third of Ballarat's population – assembled at Bakery Hill to listen to speeches by Frederick Vern (a German republican), Raffaello Carboni (an Italian revolutionary) and Peter Lalor (an Irishman) – all of whom were local diggers. They demanded the release of the gaoled arsonists, that gold licenses be abolished and that political representation be granted.

The Southern Cross, or Eureka flag was unfurled for the first time and the following day Peter Lalor and several hundred diggers made an oath of allegiance, 'We swear by the Southern Cross to stand truly by each other and fight to defend our rights and liberties'. They then marched off to the Eureka diggings in the centre of the goldfields and began construction of a rough timber stockade that blocked the Melbourne road.

Throughout Friday 1 December the diggers continued to build the stockade, but their numbers dwindled. The remaining men gathered as many firearms as possible, and began forging pike heads under Peter Lalor's direction. By Saturday night there were only 120 diggers in the stockade.

Well-armed troops and police made a pre-dawn attack – without warning – with fixed bayonets on Sunday 3 December 1854. The fight lasted only 15 minutes, but there was considerable loss of life and outrageous behaviour by the

Blood on the Southern Cross,
Sovereign Hill, SH

police and troopers. Five British troops (including a captain) and 22 diggers were killed or later died of their wounds; scores were wounded. Many diggers were taken prisoner, but Peter Lalor escaped to Geelong (although minus an arm that was shattered in the battle).

The authorities had won, but it was a pyrrhic victory. Public opinion was firmly on the side of the diggers; after a brief trial none of the captured rebels were gaoled. The government was discredited and a Royal Commission condemned the administration of the Goldfields. The license was abolished and the Goldfields were given eight members in the Legislative Council. By the end of 1855 Ballarat had a municipal council, and Peter Lalor (who proved to be surprisingly conservative) was elected to parliament. Lalor represented Ballarat for another 34 years.

GOLD – THEN & NOW

BY DAVID BANNEAR

Goldrushes as we understand them are a mid-19th century phenomenon. Two goldrushes – California (1848-1853) and Victoria (1851-1855) – were pivotal to making the world gold-minded by the sheer volume of gold they produced. The role of this gold as bullion, hoarded in central banks in places like America and England, provided the basis for the world's economic expansion.

The world discovered the goldfields of California and Victoria through the expanding role of newspapers and, within a year of the initial discoveries, hundreds of thousands of immigrants left Europe for the goldfields. Passage to America was relatively cheap and the country was already a favoured destination. In comparison, immigration to Australia was expensive and perilous, and the country was still viewed as a penal settlement.

The original Eureka Flag is on display at the Ballarat Fine Art Gallery. The Eureka Centre in East Ballarat is built on the site of the stockade and includes a museum and interpretive displays of the rebellion.

But the power of Australian gold, particularly Victorian gold, swept away these concerns and brought on prodigious immigration, largely from Britain. Australia was transformed into a gleaming gold beacon – it was the bank till 'free to all' where ordinary men and women could find riches beyond their dreams.

Today it's easy to regard the rolling wave of discovery, digging and development that followed as the goldrush—like one big earthquake. In fact, it was a long series of tremors and aftershocks – more than 200 of them.

Some of the rushes to new fields – Heathcote, Alma, Ararat, Dunolly, and Landsborough – were to be equal or even greater in size to Mt Alexander and Bendigo. Others were much smaller, like the Berlin (or Rheola) rushes in 1868, which were characterised by the discovery of a large number of nuggets. By 1910 the region had produced 1,300 nuggets over 20 ounces (0.5 kg), the bulk of the world's known total.

Family Digging for Gold 1852, ST Gill, Courtesy the Sovereign Hill Museums Association

Bendigo – Victoria Hill;
Ballerstedt's Open Cut
and Lansell's 180 Gold
Mine – p155

Castlemaine Diggings
National Heritage Park
– p207

Eureka Reef – p209

Clunes – Port Phillip
Company – p212

Whroo Historic Reserve
– Main Gully; puddling
machine – p284

Cradle, RE
Mine ruins, Eureka Reef, Castlemaine
   Diggings Heritage National Park, RE

MINING IN THE 1850S

The very earliest gold processing methods used by the gold diggers were the cradle and gold pan. Cradles were rocked from side to side, agitating gravel mixed with water that had been ladled in. The gold was trapped by riffles within the cradle.

When shallow alluvial ground was deemed 'worked-out', mining legislation permitted larger claims to be taken up for puddling purposes. Miners usually removed and processed all the dirt down to bedrock, stripping (known as surfacing) both gullies and hill slopes.

Puddling machines, powered by horses, could treat several tonnes of earth a day and comprised a circular wood-lined trough. On the central mound formed by the trough was a wooden pivot post. One end of a long, horizontal wooden pole was attached to this post by an iron pin and a horse was harnessed to the pole's other end. As the horse trudged in a circle, iron rakes hanging from the pole were dragged around the trench to break up the clay and free the gold. Water was obtained from a dam (the right to construct one came with a puddling lease) or from a water race.

Puddling operations led to considerable changes in a goldfield's landscape. 'Sludge' – the residue of puddled wash dirt – choked watercourses and covered auriferous ground and roads. In some places special measures were introduced, such as the construction of sludge channels and the employment of workers (paid for by the proprietors of puddling machines) to keep the channels running free. The peak period for puddling was 1854 to the mid-1860s. After that, puddling declined in importance when suitable deposits were worked out or became too deeply buried by sludge to be economically treated. Puddling continued on some goldfields, such as at Dunolly, Rushworth and Beaufort, well into the 20th century.

Quartz reef mining proved to be the most widespread industry in the region. It was undertaken on all goldfields and still continues today with success. The quartz reefs were the primary sources of gold in most goldfields. Generally steeply-dipping, they were mined in sections, leaving large slots (called stopes) in the ground where the reef had been. The quartz had to be crushed to fine sand to extract the gold. The reefs were first quarried or open-cut to reap the benefit of their rich surface exposures. Then shafts were sunk, or tunnels were driven from the side of hills, to trace the gold at greater depths. Initially, hammers, crushing mills and stamping batteries were powered by hand, but horse and eventually steam-powered machines, were installed. Large companies, such as the Port Phillip Company at Clunes, helped develop the industry through continual experimentation to improve the economy of gold recovery.

Deep lead mining was less widespread, being confined to country where gold nuggets lay in the ancient stream beds that had been buried under lava flows. Underground deep lead mining was fraught with danger from cave-ins or flooding. Where possible, the main shafts were sunk in solid ground and tunnels were dug from them out underneath the old, buried river beds (called leads). Vertical connections to the leads were made at intervals and the ancient gold-bearing gravels were excavated from the sides of a main connecting tunnel. This form of mining was pioneered on the basalt plateau around Ballarat by skilled tin and coal miners from Cornwall, Wales and Scotland. By the mid-1850s, Ballarat shafts were being sunk to depths of 30 metres through waterlogged ground to reach the rich deposits of gold. Many men became suddenly wealthy.

MINING IN THE 1860S

During the 1860s Ballarat continued to be the focal point for deep lead mining with phenomenal gold yields at a depth of 90 metres and more. The sustainability of deep lead mining had a pronounced affect on Ballarat. In less than 20 years it had become a big, progressive, bustling city with over 40,000 inhabitants, 56 churches, three town halls, 477 hotels, 135 km of made streets, 264 km of footpaths, 24 km of stone channelling, a reticulated water supply with 60 main water pipes and a gas works with 80 km of gas mains.

Water was a crucial ingredient, not only for supporting the growth of the region's new towns, but also for extracting gold from clayey soil or crushed quartz, and for generating the steam that powered mining and manufacturing plant. Throughout the region, miners established extensive water supply schemes by constructing channels called water races. In the Talbot district, Messrs Stewart and Farnsworth constructed 400 km of water races and one large reservoir. The goldfields of Bendigo and Castlemaine were supplied with water through a government scheme known as the Coliban Water System. This vast initiative included more than 20 reservoirs and 500 km of races, tunnels and syphons. It still operates today.

From the 1860s various forms of sluicing took place using the water supplied from the races and dams. Sluicing used running water to break down gold-bearing earth, and a sluice box recovered the gold. This simple, long, open-ended wooden box had transverse cleats, or riffles, tacked onto the base, and usually coarse matting placed between the riffles. When the earth and gravel was washed through the box, the heavier gold stuck in the matting, or behind the riffles.

There were two forms of early sluicing. Some miners worked stream beds, getting rid of the running water by diverting it through tunnels or cuttings, or by constructing embankments or wing-dams. Others extracted gold from the higher terraces (remains of old stream beds) by directing water from a high elevation down the faces of excavated soil to separate the gold from the soil and rocks. This form of mining was called ground sluicing and was undertaken on most goldfields.

Maldon – Beehive Mine Site – p244

Whroo Historic Reserve – Balaclava Hill – p284

Old Mines near Moliagul, KS
Whim, Sovereign Hill, Ballarat, SH
Chilean Mill, Sovereign Hill, Ballarat, SH

Gold mine near Talbot, JM

MINING IN THE 1870s-1880s

Quartz reef mining transformed Bendigo into the region's second provincial city. The boom was sustained by mining that followed quartz further and further from the surface. Geology experts in the past had predicted that Australian gold would disappear at quite shallow depths, but several Bendigo companies were passing through the 150 metre level without any drop in the yield of gold.

Another factor that proved a boon was a growing reliance on a Cornish system of working mines called tributing. Tributing involved a party of working miners (tributers) being contracted by a mining company to work a particular section of its mine. In return, the tributers paid the mining company a percentage of the gold they obtained, usually about 25%. Tributing quickly became widely used and helped sustain the quartz reefing industry.

The 1870s saw the demise of Ballarat's deep lead industry, but this was matched by a rise in prominence of new deep lead fields. Leading the way was the Maryborough goldfield and its flagship, the Duke & Timor Company. At its height, the company's proprietors boasted that nearly all the inhabitants of the surrounding towns of Timor and Bowenvale depended upon the mine.

Elsewhere in the Maryborough district – at Alma, Majorca and Carisbrook – other significant deep lead centres developed. Perhaps the most important event, however, was the discovery of gold at Spring Hill, near Creswick. A familiar scene of small company mining unfolded during the mid-1870s as the gold was traced down the hill and below the basalt plain. The main deep lead was traced northwards to Smeaton where it was given the name the Berry Lead, which became synonymous with many of Victoria's richest deep lead mines. However, the Berry Deep Lead also holds a less enviable record. On 12 December 1882 a drive from the New Australasian Company's No. 2 shaft was flooded and 22 miners perished in Australia's worst gold mining disaster.

Quartz reef mining was given a boost during the early 1880s with the introduction of the rock drill. Before its introduction, all holes required for blasting out horizontal mining tunnels (called cross-cuts) were driven into the rock face by the tap-and-hammer technique where one man held the iron drill while another hit it with a hammer. Using this method, a team of two men took 10 hours to punch a horizontal blasting hole, 9 ft 3 in long. Imported rock drills could cut the same hole in one hour which represented great savings in time and labour costs. The introduction of this drill was due, at least in part, to a visit by Bendigo's 'Quartz King', George Lansell, to the Nevada and California goldfields.

The rock drill had a profound influence. Shafts were sunk deeper to enable further cross-cutting. By the end of the 1880s, nineteen shafts on the Bendigo field were down to 600 metres and Hansel's 180 Mine was approaching a depth of 900 metres. Elsewhere other companies prospered using the new technology. The Magdala Company commenced driving towards a large ore body and dominated

Stawell's gold production until 1917. At St Arnaud, the Lord Nelson Company became a mighty gold producer, as did the Maxwell Company (Inglewood), North Cornish (Daylesford) and North British (Maldon).

Gold miners also began to extract gold for the first time from ore that was heavily laced with iron pyrites and other sulphides. In this enterprise they used a chlorination process that involved using a combination of heat and chemicals. Gold is readily attacked by chlorine gas and the chloride formed is soluble in water. Chlorine gas was generally used for the treatment of pyritical concentrates obtained from tailings. Physical evidence of the chlorination technology is now scarce in Victoria because most sites have been cleaned to prevent damage from the spread of toxic residues produced by the process.

MINING IN THE 1890S

During the early 1890s the world's economy went into recession. On the Victorian goldfields the economic squeeze forced hundreds of farmers off their farms as banks foreclosed. Many men returned to mining, finding and opening a number of new rich mines. Old mining towns came back to life and new ones were created. To support the industry, the Victorian Government provided assistance to quartz gold prospectors through the installation and operation of small quartz crushing facilities (known as government or state batteries) in localities where no privately-owned battery was available for public use.

The wave of prospecting resulted in the discovery of new goldfields. In 1895 a quartz reefing field was opened at Fosterville, 24km east from Bendigo. There was a large rush in 1901 to the Grampians area, south of Mt William, to what became known as the Mafeking Rush.

In 1896 Victoria's total gold output was 805,087 ounces (22.8 tonnes) the biggest annual yield since 1882. The Bendigo quartz mines contributed well over a quarter of this total, but new deep lead mines also played a part. One near Beaufort, on the Raglan Lead, was profitably worked for about two kilometres of its length by the Sons of Freedom group of companies. Ararat district also experienced a deep lead mining boom.

The 1890s mining boom coincided with the introduction of a new form of alluvial gold mining, called pump or jet elevator sluicing. This involved hosing the face of a gully away, then puddling up the resulting wash dirt and elevating it by pressurised water into raised sluice boxes, from whence it passed over a series of ripples where the gold was captured. Gold mining companies also commenced using bucket dredges to mine alluvial deposits.

Dredging was a large-scale means of extracting gold from extensive river-flat deposits. Dredges were like huge floating factories. Operating in a pond, they excavated the gravels in front of them with large buckets, processed the gravels on-board and dumped the treated material behind. By far the most successful use of pump sluicing and bucket dredging was in the Castlemaine district, where an estimated 18 sluicing and dredging plants were operating by 1905.

Contributing to Victoria's high gold production was the introduction of a new chemical-based gold treatment process. By 1897 there were major cyanide works at Tarnagulla, St Arnaud, Stawell, Ballarat, and Maldon. One of the most stunning successful introductions of the cyanide process in Victoria was at Costerfield. After 50 years of unsuccessful experiments to unlock the gold

Victoria Hill, Bendigo, PW

from the field's antimony-rich ore, the application of the cyanide process brought immediate success.

From 1905 to WW1 the State's gold production declined. For far too long, problems inherent in deep sinking had been largely ignored. At Bendigo, the New Chum Railway and Victoria Reef Quartz Mining companies were engaged in sinking main shafts to 4,300 ft (1310 m). The following year, Dr Walter Summons released his findings on ventilation in the Bendigo mines, presenting a picture of young men sickening and dying from working in ill-ventilated shafts. In June 1910, deep mining at Bendigo was abandoned and all operations confined to the shallow levels of mines. With the onset of WW1 the industry stalled across the region.

THE 1930S MINING REVIVAL

Gold fetched a good price during the depression of the 1930s and this led to resurgence in gold mining. Local men with mining backgrounds were joined by unemployed men who hoped to eke a living out of the creeks. The Government's Sustenance Department issued each with a gold pan, a rail ticket and a prospecting guide, and left them to it. With the help of their prospecting guides, many newcomers revived the art of cradling. Some panned in the creeks while others worked at ground sluicing on the slopes.

Maldon – Porcupine Flat Dredge – p245

Castlemaine Diggings National Heritage Park – Forest Creek Gold Workings p209

Chewton – Wattle Gully – p207

Bendigo – Central Deborah Tourist Mine – p146

One of the most staggering successes of the 1930s' prospecting revival led to a relatively unknown operation at Chewton, Wattle Gully, becoming one of the state's leading 20th century mines. Cyaniding and dredging companies also carried out successful post-1930s mining. Cyaniding was dominated by Gold Dumps Ltd, which operated on a large scale. The company began its work on the slime dumps near Carisbrook. After processing these dumps, the company moved on to Bendigo where it embarked on an even more grand and lucrative scheme of cyaniding.

Dredging barges once more cranked their way up creeks and major gullies, gobbling away swampy flats, and jet elevators tore down the last remaining faces of the old alluvial workings. The most successful alluvial mining company to work during this time was at Newstead. The Newstead dredge operated from July 1938 to March 1948. Despite many difficulties, including a time when the company's Caterpillar tractors were appropriated for the war effort (they were substituted by teams of horses), the dredge worked continuously through this period. During its 10 years of operation the dredge handled 15 million cubic metres of soil.

From 1941, Bendigo's mining industry began to feel the effects of the war. The shortage of manpower and equipment seriously hampered further development and made full production impossible. One by one the remaining mines closed and at the end of 1954, after 103 years of continuous production, gold mining at Bendigo was halted.

Victoria Hill, Bendigo, RE

TODAY

The region is still producing large amounts of gold. Nuggets are still being found by fossickers using metal detectors. Some of the recent nuggets have been large, such as the 875 ounce (24.8 kg) *Hand of Faith* which was recovered from Kingower in 1980. This nugget is now on display in the Golden Nugget Casino, Las Vegas.

Mining companies, large and small, are operating at Stawell, Fosterville, Bendigo, Ballarat, Castlemaine, Costerfield and Maldon. Improved technology and high gold prices mean the remaining recoverable reserves now being targeted equal the amount that has already been taken. In other words, today's miners are chasing another 80 million ounces (2268 tonnes) of gold which, assuming a price of US$550 an ounce and an A$ worth US$0.75, will equal a A$58 billion bonanza.

Two of the biggest operators are targeting Bendigo and Ballarat in the belief that the historical mines left much behind. Bendigo Mining and Ballarat Goldfields have used historical records and old newspapers to make up three-dimensional maps of the old workings to locate the reefs and what has been left behind.

Mines at Stawell and Fosterville are already producing more than 150,000 ounces (4.3 tonnes) a year. Ballarat Goldfields predicts it will extract 200,000 ounces (5.7 tonnes) a year. Bendigo Mining has constructed a 5.5 km tunnel under the city and when it begins production the company expects to produce 700,000 ounces (19.8 tonnes) a year – making it the third largest gold mine in Australia (after two in Western Australia).

Porcupine Flat Dredge, near Maldon, PV

ARCHITECTURE – BUILDING GOLDFIELDS TOWNS

BY RAY TONKIN

The towns and cities that we see in the region today are either grand gestures of provincial wealth such as Ballarat, Bendigo, Castlemaine, Stawell and Maryborough, or remnants of much more extensive but modest mining settlements, such as Chewton, Campbell's Creek, Creswick and Fryerstown.

The goldfields were great cultural melting pots and the settlements and towns ultimately reflected that.

Whilst the mining activities themselves wrought unprecedented change to the landscape, it was the establishment and growth of towns and cities which inflicted the greatest and most permanent change.

The miners came from many corners of the world and brought with them a great array of social backgrounds, philosophical attitudes and religious beliefs. The goldfields were great cultural melting pots and the settlements and towns ultimately reflected that. Certainly the towns and cities of the Goldfields region did not solely reflect traditional and conservative English approaches to urbanisation and architecture.

The British colonial administration had a substantial influence (the work of government surveyors was alone a significant factor in the form of these towns and cities) as did English settlers. But no one can reasonably ignore the German influences on the Bendigo architects lead by Vahland and Getzschmann, the inventive background of the naval architect Henry Casselli in Ballarat, the views of Bendigo priest Henry Backhouse and the editor of the *Ballarat Star*, the expectations of the Irish, or the non-conformist settlers from places like Cornwall and Wales. Let alone the Chinese.

The development of the Goldfields towns and cities was not left solely to the aesthetic tastes and training of a few 'architects'. Whilst they were often the people who put pen to paper and designed the buildings, they did so in the face of a range of external pressures. Not only were their clients interested in establishing a certain image and presence for their institution or business, but also they were clearly responsive to community views on what was appropriate, and applied pressure for places of substance and permanence.

Anthony Trollope wrote about the presentation of these places and, after visiting Sandhurst (Bendigo) in 1872, he wrote: '…here and there (is) an attempt at architecture, made invariably by some bank company eager to push itself into large operations.'

In all the main centres the banks competed with one another to reflect permanence solidity and wealth. In Ballarat the Melbourne based architect, Leonard Terry designed a series of neo-renaissance palazzos along Lydiard Street

Opposite: Capital Theatre, Bendigo, RE
Capital Theatre detail, RE

The buildings reflected a desire to proclaim the progressiveness of this new world.

whilst the Bank of New South Wales and the National Bank in Bendigo followed suit. However, by the later years of the century when the development of the centre of Bendigo hit its stride, the Colonial Bank, the Commercial Bank and the Union Bank used their architects to give them extravagant, mannered and eclectic interpretations of traditional designs.

There is also strong evidence that town councils were often a powerhouse of ideas which generated townscapes where the buildings reflected a desire to proclaim the progressiveness of this new world. These same organisations effusively welcomed outside publicity and comment, and co-operated with the local press to provide these commentators with room to express their wonderment at the development achievements of these new settlements. Anthony Trollope also visited Ballarat and he described that town in wondrous terms. After all, it had gone from being a tent city in 1851 to a substantial and handsome metropolis by the mid-1870s.

The Goldfields region was also well serviced by men practising as architects. Aside from the Melbourne and Geelong-based practitioners, who made regular forays into the area, there were no less than 19 local architects advertising their practices in 1875 (Source: *Balliere's Victorian Directory*). Whilst Ballarat and Bendigo accounted for many of them, there were also architects in Castlemaine, Maryborough, Daylesford and Ararat. The buildings in these places were not simply being thrown up by any old builder without professional direction.

PUBLIC BUILDINGS

The role of the Victorian Government's Public Works Department cannot be underestimated. In every town and settlement the Government was pressed to provide facilities for the delivery of government administration – be that police stations, post offices, court houses or schools. The level of investment in these facilities was extensive and the professional skills made available by the Public Works Department to these rural centres was no less than was made available to projects in Melbourne.

As a consequence, Castlemaine got a post office designed by JJ Clarke, the designer of the Melbourne Treasury Building; Eaglehawk got a court house designed by Peter Kerr, the designer of the Victorian Parliament House; and Ballarat a post office designed by William Wardell, the architect of Government House and St Patrick's Cathedral. Alongside these more prominent individuals sits the work of highly talented architects, such as George Watson and J.H. Marsden.

The designs of the Public Works Department also led to the establishment of fine civic complexes. The civic precinct in Maryborough, or the grand run of public buildings in Bendigo's Pall Mall, or the East Ballarat Civic area are all good examples, and the civic ambitions of the residents of these places were well met.

Post Office, Castlemaine, KS

DOMESTIC ARCHITECTURE

Early illustrations show us tent cities – temporary arrangements reflecting the urgency and uncertainty of the rushes. However it didn't take long for these temporary structures to become more permanent, whether it be by the construction of timber frames to support the canvas cladding, or more significantly the cladding of those frames with more permanent materials – be that split boards, bark slabs, rough stone, or even brick.

Weston Bate writing in his 1978 history of Ballarat, *Lucky City*, describes how the introduction of the deep shafts led to tradesmen being available to construct more substantial buildings. The greater permanence of this form of mining also ensured a willingness to seek more permanent forms of accommodation, and for commercial entrepreneurs to erect more substantial buildings.

It is not unreasonable to view the simple gable-roofed houses of the Goldfields as direct successors to the miners' tents. Usually two rooms wide with a central door on the long elevation, they became the residential norm and examples can still be found around the Goldfields region. Usually added to by a further one, or even two, gable-roofed structures, they were clad with weatherboards or brick. After the arrival of corrugated, galvanised iron in the 1860s, they were roofed in this evocative Australian material (often applied directly over the original hardwood shingles).

As settlements became more established, industry more prevalent, skilled tradesmen more available (along with professional architects and engineers) and many members of the community more affluent, the houses became more elaborate, often based on overseas models and decorated. It isn't difficult to draw links between the proliferation of decorative cast iron in Ballarat and the foundries established to serve the mining and other industries of that town.

The buildings of the Goldfields region were not simply thrown up to serve a practical need. They are often the conscious efforts of architects, engineers or professional builders strongly influenced by community pressure to create substantial and permanent settlements. In this they succeeded.

Sovereign Hill, Ballarat, SH
Sovereign Hill, Ballarat, SH
Cottage near Castlemaine, RE
Sovereign Hill, Ballarat, SH
Sovereign Hill, Ballarat, SH
Wesley Hill, KS

CHURCHES & MECHANICS INSTITUTES

The Goldfields towns were well-served by hotels, sly-grog shops and brothels. But, either because of this, or perhaps in spite of this, there were also strong religious beliefs and a commitment to self improvement. The Goldfields towns were dominated by religious buildings and, very often because they were built of local brick and stone, they are the only buildings that survive. Mechanics Institutes were also ubiquitous.

The Goldfields towns were well-served by hotels, sly-grog shops and brothels.

MECHANICS INSTITUTES

Over 1000 Mechanics Institutes were established throughout Victoria as part of an optimistic belief in self-improvement and the value of education. 'Mechanic' simply meant worker, tradesman or artisan, and the institutes were to encourage 'moral and mental improvement' for working people. The institutes were generally built by the community from locally raised funds and small government grants.

They were used as libraries, museums, art galleries, theatres, meeting rooms, and lecture halls. They sometimes failed in their aim of educating the working class and became middle class salons instead. Lectures vied with theatre performances as important forms of entertainment and ways of gathering information. It was not unusual for lectures on obscure topics – like ancient architecture, or fossils – to go for hours.

Mechanics Institute, Heathcote, KS

CHURCHES

There were strong links between ethnicity, class and religious affiliation. To generalise, the Church of England was the establishment church of the Government Camp and the English upper and middle classes; the Catholic Church was Irish; the Presbyterians were Scottish, with a link to the pioneer Scottish graziers; and the Wesleyan Methodists were Cornish, or northern English working class.

The Goldfields region has a rare and diverse collection of 19th century church buildings – all built with local donations. Unlike in Britain, the different denominations' churches can be easily compared because they are spread uniformly across the region and, although many are now used as houses, most have survived. In Britain the denominations tended to cluster regionally, and economic pressures within towns and cities mean that many 19th century churches have disappeared.

Despite diversity and significant regional variation, by 1861 there were more Wesleyan Methodists in Victoria than all the other denominations combined. This straightforward faith-based and emotional religion resonated with the Goldfields population and the Wesleyan's organisation suited the local conditions. Other churches were hindered by a lack of clergy and an unwillingness to use laymen, whereas the Wesleyan tradition depended heavily on both laymen and laywomen.

The various types of Methodism also provided an opportunity for upward social mobility. A working class Primitive Methodist (a democratic and revivalist offshoot of Methodism) could, upon growing more affluent, transfer membership to a more middle class and conservative Wesleyan Methodist Church without being lost to the Methodist movement.

Many of the region's earliest churches were, by necessity, modest, simple structures, not far removed from the canvas and wood chapels they replaced. Some denominations – the Baptists and Primitive Methodists – deliberately built simple, classical structures by choice, to avoid the papist (Catholic) associations with the Gothic style. A fine example is the Talbot Primitive Methodist Chapel.

On the other hand, the Roman Catholics and Anglicans (High Church) favoured a Gothic-revival style that employed a strict symbolic code. Ideally Gothic churches are built east-west, with the entry at the west end (representing sunset and death) and the altar at the east end (representing the rising sun and rebirth). If complete they take the form of a crucifix with a central nave and transepts. A double belfry symbolises Saints Peter and Paul and triple lancets symbolise the Trinity.

A third style, Romantic Gothic, was more free-flowing and experimental, and was adopted by evangelical Anglicans, Presbyterians and Methodists. The striking Anglican Holy Trinity Church at Taradale is an interesting example.

The Church of England was the establishment church; the Catholic Church was Irish; the Presbyterians were Scottish; and the Wesleyan Methodists were Cornish, or northern English working class.

St John's, Heathcote, KS
Moonlight Flat, KS

BOX-IRONBARK FORESTS

BY DAVID BANNEAR, WITH ROB PRICE

Box-Ironbark forests and woodlands are located on the dry inland slopes of the Great Divide and lie between the wetter forests to the south and the more open plains to the north. Box-Ironbark forests refer to the distinctive box and ironbark trees that are found throughout the Goldfields region.

The Box-Ironbark forests that surround cities like Bendigo, Castlemaine, Rushworth, Heathcote, Maryborough and St Arnaud are the signature forests of the Goldfields region. The vegetation is typically dominated by eucalypts with a diverse understorey of shrubs, heaths and scattered grasses. The ancient, weathered soils are nutrient poor and porous, making survival for both plants and animals challenging. Plants have made some amazing adaptations, which include deep root systems, underground dormant buds or bulbs, and leaves with reduced surface area to reduce transpiration.

Because the Victorian Government decided to retain possession of its gold bearing land, the surviving forests now mark many of the goldfields.

Because the Victorian Government decided to retain possession of its gold bearing land, and to ensure supplies of timber for mining needs, the surviving forests now mark many of the goldfields of the 19th century. The state forests of the 19th century are the basis of today's parks and reserves system.

Although they can seem grey and monotone, in spring a Box-Ironbark forest bursts into flower with wattles, a myriad of tiny orchids, lilies and other wild flowers, and the air is alive with the sound of birds.

FOREST DESTRUCTION & MANAGEMENT

Since the 1830s, land clearance for agriculture and gold mining has significantly reduced, modified and fragmented the Box-Ironbark country, with only 15% of the original forests remaining today.

Prior to the arrival of Europeans, the Goldfields region was dominated by old-growth forests – with large, wide-crowned, hollow-rich, and widely spaced trees. A history of intensive timber harvesting has resulted in a crop of younger, smaller, same-age trees (see the Coppicing section following).

The forests provided Aboriginal people with an abundance of wood, plants, animals and minerals. The limbs of ironbarks, for example, were commonly used for making boomerangs, called *wonguim*. The bark of box trees was used for constructing *willams* or bark huts. Box bark was also used for ceremonial purposes.

From the time the Europeans arrived, and especially to meet the insatiable

Box-Ironbark forest, Castlemaine Diggings Heritage National Park, PV (AP)

demands of the miners, the old-growth forests were massacred. A Polish gold digger remarked on the wanton destruction by the English who he observed: '…often cut trees without actual need. They do it simply for fun, because they can indulge their liking for fuel, which is as rare as saffron in their own land'.

Gold mines had an almost insatiable demand for timber, for supporting and lining shafts, and for firewood, which was needed to power steam engines and boilers. An average-sized quartz mine at Bendigo would consume 100 tonnes of green wood every 12 days and 100 tonnes of dry wood every 16 days. As a result, by the 1880s mining timber was being carted to Bendigo from hundreds of kilometres away.

In addition there was large-scale charcoal production (for blacksmithing purposes), a huge demand for railway sleepers, and an enormous wattle-bark industry (for leather tanning).

Wattle bark contains high levels of tannin and was considered one of the world's best barks for use in tanning leather. It was used locally and exported. The bark was simply stripped from the trunk of a tree and the tree usually died. An 1878 inquiry found that years of unregulated stripping had brought wattle to the edge of extinction in many places, and a licensing system and cultivation schemes were established.

Relatively early, the problems of deforestation were acknowledged to be acute. From 1878 to 1884, five forestry nurseries were established in Victoria, including two in the Goldfields region, at Macedon and Creswick. The nurseries raised eucalypts as well as exotic conifers.

The Macedon nursery provided trees for public distribution and Creswick was used for public plantations. Plantations were established to meet future needs, but also to rehabilitate devastated mining land. State Schools were encouraged to plant trees in their school grounds as well as in small plantations on Crown Land – largely to educate and encourage children to care for trees.

In 1907 the Forest Act led to the establishment of Creswick School of Forestry – the first in Australia to formally train cadets in forest management. The Forests Commission, now subsumed within the Department of Sustainability & Environment, was established in 1919.

The English 'often cut trees without actual need. They do it simply for fun'.

Ironbark, Castlemaine Diggings Heritage National Park, PV (AP)

Each clump is the ghostly impression of its original self, with the ancient rootstock hidden, but alive.

COPPICING

In pre-goldrush times the forests were dominated by large, wide-crowned, hollow-rich, and widely spaced trees. Today it is difficult to imagine trees that measured two metres across the base – because most of the surviving forest is made up of younger, smaller trees, often in clumps.

But appearances are deceiving. Cut off near the base, some of the forest giants have since coppiced – that is, grown new shoots from their underground rootstock. Sometimes the old tree stump has completely rotted away, so the new growth appears to be a clump of separate trees (there are sometimes up to 20 trunks). But they may all come from the single surviving underground base of a giant tree, which could have started life hundreds of years ago. Each clump is the ghostly impression of its original self, with the ancient rootstock hidden, but alive.

The coppice regrowth trees today tell a remarkable story of survival, transformation and regeneration. In gold country they form part of a cultural landscape that also contains the evidence of abandoned gold mines, earthworks, tracks and hut sites.

Conservation and enhancement measures are now being implemented in the hope of returning large trees, like those that used to exist, to the landscape. The significance of large old trees to conservation is that they can – because of bark density, increased flowering capacity and multiple nesting hollows – support higher numbers and a richer variety of faunal species.

The problem of low-quality regrowth has long been recognised, with major thinning operations commencing in the late 1880s. Thinning has provided relief work for unemployed men and youths in the economic depressions of the 1890s, after WW1, in the 1930s, and after WWII.

In the 1930s the Forestry Commission provided its workers with a tent, three pieces of corrugated iron to make a chimney, two blankets, a billy, a frying pan, a wash dish, floor boards, a lamp and some fencing wire!

From 1939 to 1945, so-called enemy aliens were interned at Graytown and used as workers in the forests. After the war, immigrants (known as displaced persons) were housed in ex-Army buildings and were also put to work.

TODAY

The number and range of plant and animal species has declined considerably in Box-Ironbark forests because of past exploitation. However, many of the native birds and animals show a remarkable resilience that matches that of the coppiced trees. Approximately 1500 species of plants and 250 vertebrate animals have been recorded. Furthermore, many plants and animals that occur in the Box-Ironbark forests occur nowhere else in the country – or the world.

Rare and endangered species such as the Eastern Quoll, Magnificent Spider Orchid, Malleefowl, and Regent Honeyeater are all residents.

The dominant tree species of Box-Ironbark forests and woodlands are Grey Box, Yellow Box, White Box, Yellow Gum, Red Stringybark and the distinctive, dark-fissured Red and 'Mugga' Ironbarks. Box-Ironbark forests are renowned for their winter flowering, with Grey Box, Yellow Box and the Ironbarks flowing from late autumn through winter.

The heavy and prolific flowering of trees in Box-Ironbark forests is a magnet

Coppiced Tree, RE
Koala, Wombat State Forest, PW

for many birds and animals. Nearly 200 species of native birds are found in Victoria's Box-Ironbark forests and woodlands, with the nectar-feeding honeyeaters, lorikeets and wattle birds being most abundant. For the many bird watchers that visit Box-Ironbark forests, the sighting of the endangered Swift Parrot, which is a regular winter visitor from Tasmania, is a valued experience.

Larger mammals such as kangaroos and wallabies are common throughout Box-Ironbark forests and woodlands, while the echidna, and the evidence of its diggings, may be observed in the forest during spring. Sightings of nocturnal animals, such as the Brush-tailed Phascogale, Sugar Glider and Brush-tailed Possum are an added bonus for campers spotlighting at night.

The Box-Ironbark forests put on their best wildflower performance in spring. The Golden Wattles and Gold Dust Wattles set much of the forest awash with a sea of yellow, and many other plants and wild flowers are in bloom.

The forests are a challenge for today's land managers – Parks Victoria and the Department of Sustainability and Environment – because of the need to balance environmental, social and economic needs.

Fortunately, in the past 20 years Victorians have recognised the value and importance of this environment. In addition to their natural values, the forests harbour tales of a golden past, with burial grounds, the crumbling stone walls of huts and pubs, and the gold mines and gullies that yielded up fortunes.

Maps of national parks, conservation reserves and state forests can be obtained from Parks Victoria and Department of Sustainability and Environment offices. If you would like to know more about the flora and fauna of the Box-Ironbark area, obtain a copy of *Wildlife of the Box-Ironbark Country* by Chris Tzaros or *The Forgotten Forests* by Victorian National Parks Association. Both of these books can be obtained from local bookshops in the Goldfields region.

Box-Ironbark forest, JM

CHINESE – THE NEW GOLD MOUNTAIN

BY KEIR REEVES

During the middle of the 19th century the discovery of major gold reserves transformed the infant European settler societies of the Pacific Rim. These regions included northern California, British Columbia, Canada, and Central Otago in New Zealand and, of course, the greatest of them all, Victoria, Australia. Chinese gold seekers, mostly from southern China, were key members of all these rushes.

Shafts dug by Chinese miners are often side-by-side with European shafts. You can distinguish the two because the Chinese shafts are round and those of the Europeans are square!

Today the Chinese legacy can be seen at the Gum San Chinese Heritage Centre in Ararat, the Golden Dragon Museum in Bendigo (including Sun Loong, the oldest imperial dragon in the world), and at numerous Chinese market gardens and former camp sites scattered through the Castlemaine Diggings National Heritage Park. At Sovereign Hill you can also visit a faithful recreation of a Chinese village.

Following the introduction in Victoria of a poll-tax on the Chinese (this was a head tax that was paid upon landing) many of them landed at Guischen Bay, near Robe in South Australia. It is still possible to retrace the overland route of the Chinese and follow their footsteps as they made their way towards Dai Gum San (or the New Gold Mountain as central Victoria was more commonly known).

Despite the difficulties they faced, the Chinese gold seekers were the largest group of non-English speaking diggers on the Australian goldfields.

The story of the Chinese miners begins in the Pearl River Delta region of southern China. With the outbreak of the first Opium War in 1840, this area was in the frontline of a direct confrontation between Britain and China. People from the region moved throughout the south-west Pacific seeking peace and economic opportunities.

During the 1850s, the Chinese – who accounted for one in 10 Victorians – settled in the key Goldfields centres of Bendigo, Ballarat and Castlemaine. They brought with them their distinctive way of life and specialised mining techniques.

Shafts dug by Chinese miners are often side-by-side with European shafts. You can distinguish the two because the Chinese shafts are round and those of the Europeans are square! Some say the Chinese were dug round to eliminate corners where evil spirits might gather. But since they were content to build square houses, it seems more likely there was a more prosaic reason. In fact, a round shaft is more efficient because it needs less timber to prop up the walls.

The Chinese encountered hostility and racist attitudes, but they were also renowned for their industry. Although best known for their role in gold mining,

Easter Procession, Bendigo, KS

they were involved in many other pursuits on the Goldfields. Many worked as herbalists, merchants, and restaurateurs.

Others played an important role in the development of the region by working as market gardeners – and continued to do so well into the 20th century. Today the remnants of their market gardens can be seen in the southern reaches of the Castlemaine Diggings National Heritage Park situated in and around Vaughan and Glenluce. One legacy of their market gardens is the wild spring onions that appear along the Loddon River!

As did many other diggers, a number of Chinese came to Australia via California. Lee Heng Jacjung, for example, was a 'forty-niner' in California before he headed to Victoria when news of the gold finds in Ballarat, and soon after at Mt Alexander, became known. He was equipped with knowledge and experience in gold seeking, had good connections with the Chinese fraternal organisations that regulated and coordinated much of Chinese life, and possessed excellent bilingual skills.

Soon after his arrival in Victoria, Lee Heng settled on the Fryers Creek diggings, five km south of present day Castlemaine, where he acted as an official interpreter at the local court. He met and married a young woman, Katherine Hornick, and they made a home together at the tiny hamlet of Vaughan in the southern region of the Mt Alexander diggings. Here the family became integral members of the local community.

Even today new evidence of the Chinese presence on the Victorian goldfields continues to be uncovered. A recent archaeological dig at Bendigo has unearthed the only 19th century Chinese brick kiln to be found outside China. The artefacts that have been recovered, like the mining landscapes of the Goldfields region, offer glimpses into the past that reveal the history of the Chinese in central Victoria from the goldrush era until the 20th century.

Chinese Grave Stone, Talbot, RE

TRADE UNIONISM

BY ANDREW REEVES

When a young Scotsman from Orkney, William Guthrie Spence, arrived on the Victorian goldfields in 1853, he can hardly have imagined the effect that his work as a trade unionist, writer and Labor politician would have on the lives of thousands of those who came to call Australia home.

Like the mass of immigrants from those years, Spence tried his hand at mining. Like so many others he became a jack of all trades – a bit of prospecting, work in the deep lead alluvial mines north of Ballarat, labouring, or working for a butcher when mining was slack.

Gold mining was harsh, unforgiving work. Hours were long and accidents common.

Gold mining was harsh, unforgiving work. Hours were long and accidents common. For miners working in the mines that dotted the fabulously rich deep leads between Creswick and Clunes, the danger was water. Deep leads were old, buried river beds rich in alluvial gold. To breach an active underground stream or break into old, flooded workings was an ever-present danger and the risks were high. In 1882, 22 Creswick miners working the New Australasian Mine drowned in just such a disaster. A more insidious danger occurred in the deep quartz reefs of the Bendigo field. Here the introduction of pneumatic drills filled the shafts and drives with gritty dust – dust that lodged in miners' lungs and eventually killed them.

Safety was a constant pre-occupation for miners. Victorian miners sought innovative solutions. For example, in the early years of the 20th century, with union support, doctors at the Bendigo Hospital undertook some of the world's first research into the diagnosis of miners' respiratory diseases – the dreaded 'dusted lung' – that struck fear into even the bravest miner.

Maldon Miners' Banner, DB

Such dangers shaped William Spence's outlook on life, as it did for many others. In 1874 he attempted to form a miners' union at Clunes. In 1878 he was the secretary of the Creswick Miners' Union and later, having been blacklisted by mine owners for his union activities, he spent time travelling the back blocks organising for the Amalgamated Miners' Association (AMA). A description of his philosophy could stand for the AMA as a whole: 'moderate and conciliatory, but firm on fundamentals.'

His reputation as a superb negotiator led to him being appointed foundation president of the shearers' union in 1886, while also general secretary of the AMA. W. G. Spence is now remembered as a founder of the shearers' union – later the Australian Workers' Union – and also recalled by a contemporary as: 'the mildest mannered man that ever ran a strike'.

Trade unionism came early to the Victorian goldfields, although the earliest unionists were not miners, but skilled tradesmen. By 1861, within a decade of the first gold discoveries, small but growing unions of printers, masons, bricklayers and engineers were well established in the larger gold towns – particularly in Ballarat, but in Bendigo and Castlemaine as well.

It was not until 1870 that the first miners' union was established. The Ballarat and Sebastapol Miners' Mutual Protection Association is regarded as Australia's first gold miners' union. This union was short-lived, but others followed its lead. The time was right for the ready acceptance on unionism among gold miners. A similar union was established in Bendigo in 1872.

This Bendigo union was established in response to growing resistance by miners to falling wages, increasingly dangerous working conditions and the 10-hour working day in the mines. Bendigo miners were the first in the world to win an eight-hour working day. This stands as an achievement of international significance. The attraction of the eight-hour day was the carrot that drew delegates of small, newly established unions from a dozen central Victorian gold towns to Bendigo in June 1874 to establish the AMA.

That June meeting formed what became Australia's first national (inter-colonial) union. From 12 branches concentrated in central Victoria, the AMA was soon a presence on all Victorian mining fields. Within a decade it was organising miners from Tasmania to Far North Queensland.

National it might have been, but its heartland remained the central Victorian goldfields. By the late 19th century the AMA could boast branches in more than 30 Victorian mining communities. The bulk of the branches were concentrated in a wide arc of central Victoria, from Ballarat and Buninyong in the south, west to Ararat and Stawell, north to the nuggetty fields of Wedderburn, and further east to the massive deep reefs of Bendigo and the outlying fields at Heathcote and Graytown. Within that perimeter lie many of the famous names of Victorian gold – Mt Alexander, Castlemaine, Creswick and Clunes, Dunolly, Maldon and Daylesford – union centres all.

The unique character of these gold towns remains a feature of central Victoria today. The vernacular architecture of the ubiquitous miners' cottages, with their 'up-down, up-down' roof, finished by a skillion kitchen or wash house, is one such defining feature.

So too are the mechanics institutes, schools of art, temperance rooms and friendly society halls, markets and trade union buildings that seemingly appear

Castlemaine statue, KS
Stawell Town Hall, KS

We benefit even today from its insistence on issues, such as working hours and health and safety.

at every corner in such numbers in such communities (or those corners not claimed by pubs). Hotels were also important union institutions. In the 1860s and 1870s, Ballarat engineers used the British Tavern Hotel in Bridge Street as their 'club rooms', while in Bendigo at the same time, unionists would meet at the Belvedere Hotel, later moving to Simpson's Hotel in Dowling Street.

Throughout the 19th century the breadth of unionism on the Goldfields widened. This can be traced in growing numbers of unions affiliated to trades and labour councils throughout the goldfields, as well as in the impressive nature of the trades' halls themselves. The best example is to be seen in Ballarat, in Camp Street close to the centre of town. Constructed in the 1880s, at the height of Victorian prosperity, this classically-designed building was the setting for the 1891 inter-colonial trade union congress, when Ballarat unionists representing more than forty trades and industries hosted delegates from all Australian States as well as New Zealand.

Mining unions around the world are great banner carriers, and the unions of the central Victorian goldfields were no exception. In Australia, the eight-hour day was marked by popular celebrations and processions in scores of Australian cities and towns. In these processions, union members paraded behind their union's banner on which were painted the symbols and emblems of their trade or industry, together with slogans and mottos arguing for a better world.

Today we know of two surviving Victorian miners' banners from among the 30 or 40 that would have been commissioned. Significantly, both come from the central Victorian goldfields – from Maldon and Stawell. They now provide us with an insight into the character of those miners of the 19th century – willing to collaborate with mine owners, but equally determined to defend their rights; passionate in the cause of an eight-hour day and a safer industry; concerned to show to the world at large that, as unionists, they too had a stake in society and would work for its improvement.

The heyday of mining unionism in Victoria passed with the decline of the industry around the time of WW1. But we benefit even today from its insistence on issues, such as working hours and health and safety. And when admiring the marvellous surviving banners at Maldon and Stawell, we can still hear the echoes of their practical ideals that contributed in so many ways to the creation of that unique region that we call the central Victorian goldfields.

Forest Creek, KS

PUBLIC WORKS & ENGINEERING FEATS

In the early years of the goldrushes, transport, communications, water supplies and sanitation were poor, and food and supplies were extremely expensive. Schools, hospitals, courts, jails, asylums, post offices, and local government simply did not exist. Often within a space of a few short years, however, this crucial infrastructure was created. Often the largest projects were funded by government (which depended heavily on overseas borrowings).

In 1857 Cobb & Co could do the 100 mile (161 km) trip from Melbourne to Bendigo in less than 10 hours. Carrying 14 passengers the coaches were pulled by four horses that were changed every 10 miles (16 km).

ROADS

For the first few years of the goldrushes, poor roads caused great discontent. In winter, tracks turned to quagmires deep enough to trap a horse, and attempts to find firm ground meant that roads extended wider and wider – in some cases they were more than a mile (1.6 km) wide, and littered with abandoned carts and the bodies of bullocks and horses.

In 1853 the Government began the work of road building, with most of the initial effort going into the Melbourne to Murray River road via the central goldfields. The projects had their fair share of financial difficulties, and toll bars and houses were established throughout Victoria. They were unpopular and proved, ultimately, to be uneconomical, so the toll system was discontinued in 1877.

The Maldon toll house was relocated and survives as a lodge in the Muckleford Cemetery on the Muckleford-Walmer Road just south of its intersection with the Maldon-Castlemaine Road. The Campbell's Creek toll house was in use in 1864 and was later used as a shire hall. It's signposted at 118 Castlemaine-Guildford Road.

RAILWAYS

The first railway in Melbourne was completed in 1854, and the government was soon convinced of the value of a northern railway to Echuca on the Murray River. The aim was to link Melbourne with the Goldfields and then to the river trade and the Riverina. The line was built in stages between 1859 and 1864.

After much lobbying from goldfields interests, and at great additional expense, the route was modified to pass through Kyneton and Castlemaine, rather than by an easier eastern route. The Castlemaine route required a series of major bridges, tunnels, cuttings and embankments – and the labour of 10,000 men. Castlemaine and Bendigo were reached 1862, and Echuca in 1864 – but not before there were a number of bitter pay-related strikes.

The railway was the largest public work in the colony in the 19th century, with a set of bridges and stations consistently designed with grand style and purpose.

Stations and ancillary sheds, bridges and viaducts are all beautifully, and very solidly, built in brick and stone. The construction material used changes depending on the geology of the surrounding country – bluestone in the south (around Kyneton and Malmsbury), granite where it is available (especially near Mt Alexander), and brick everywhere else (notably around Castlemaine). Most

Cart, Moonlight Flat, KS
Railway Culvert, Specimen Gully Rd,
   near Castlemaine, RE
Railway Bridge, Chewton, KS

*The railways'
specifications
were adopted
from British
India where the
monsoon and
climatic extremes
demanded
rigorous
engineering
standards. As
a result some
of the bridges
and viaducts
seem ludicrously
over-engineered
– but they have
certainly stood the
test of time.*

of the bridges are masonry-arched and some road bridges over the line are of riveted steel beams – the latest thing in the 1860s. Two in particular are worthy of mention as great railway structures – the Malmsbury Viaduct and the Taradale Bridge.

The Malmsbury Viaduct is a massive structure of bluestone arches built across the deep valley of the Coliban River, just before Malmsbury Station. It is easily seen from the Botanic Gardens and the Calder Freeway, and can be approached by turning off the freeway at Malmsbury Town Hall.

The magnificent bridge at Taradale consists of alternate stone and steel piers supporting a steel trestle deck. The bridge is 40 metres above Back Creek, and was for many years the highest on the Victorian railway system. The road that leads beside the creek should be followed as far as the base of these great piers, so that its true scale can be appreciated.

The Elphinstone Railway Tunnel was another extraordinary achievement. It is 382 metres long and 23,000 cubic metres of rock were removed during its construction.

WATER

The main element required for the recovery of alluvial gold is water. Large amounts are needed to separate the heavier flakes of gold from mud and gravel. Because of the reliable flow of the Coliban and Little Coliban Rivers, a project was developed to channel water to the valleys around Castlemaine and to Bendigo. The water was to be used for sluicing the hillsides and old mine workings for residual alluvial gold.

The Victorian Government took on the expense of this huge scheme in the late 1860s. A large dam was built on the Coliban River above Malmsbury to bring water by a system of gravity-fed channels to the gullies of Mt Alexander Diggings. The main channel continued for a distance of 70 km to the Bendigo Goldfield where the Sandhurst Reservoir was opened in 1874, filling up by 1877.

Malmsbury Viaduct, BS
Coliban Reservoir, Malmsbury, BS

The Malmsbury scheme was the largest water conservation system in Australia in the 19th century. The dam wall is directly up-river from the railway viaduct and can be seen from trains approaching the station. The reservoir can be reached by road, taking the Daylesford road from the Calder Freeway in Malmsbury, turning left after the railway bridge and proceeding to the gate into the picnic ground.

Malmsbury Dam has a long earthen embankment faced with basalt blocks, with concrete spillways at either end. From there a channel runs beside the river, underneath the great stone railway viaduct. By the time it flows under the Calder Freeway, the channel moving on its contour is already quite a distance from the river, by then descending into a deepening valley. Towards Elphinstone the water-channel passes under the road, and can be seen winding around the hillsides on its contour. Subsidiary channels are encountered in the old workings around Chewton, Fryerstown and Castlemaine.

In the 1960s Lake Eppalock was established 20 km north where the Coliban joins the Campaspe River. Both rivers have by then descended about 500 metres from their source on the Dividing Range near Trentham, and from the high tableland country near Kyneton and Malmsbury. The lake itself was dammed to provide water for the growing towns of the Bendigo region, and for irrigation on the plains to the north.

The first telegraphic message from Geelong to Melbourne is believed to have been the news about the Eureka Stockade. It was relayed from the first Customs House at Geelong, a small pre-fabricated building now in the Geelong Botanical Garden.

HOSPITALS & ASYLUMS

The diggers faced many health problems and often lacked family support to help them. Poor diets, polluted water, mining accidents, alcohol abuse, primitive housing, poverty and disease led to physical illnesses. Sunstroke, drinking and disappointment were often seen as the cause of psychiatric conditions.

In response, a number of district hospitals were established in the first decades of the goldrush. Lunatic asylums were built in the 1860s at Kew, Ararat and Beechworth. Their grand and imposing exteriors belied the grim and overcrowded conditions within.

SCHOOLS

The first schools on the goldfields were tents but, very early on, private schools were set up in houses and were operated by various religious denominations (especially by the Wesleyan Methodists and other Non-conformists).

In 1872 education became free, compulsory and secular in Victoria and, over the following years, many new State Schools were opened. In bigger centres they were large buildings that fulfilled the need for generous, easily managed playgrounds and well-lit, well-ventilated classrooms. The Public Works Department produced some fine buildings. Often Gothic in emphasis, they retained a secular flavour with their prominent bell-towers helping to distinguish them from churches. There are notable examples in Bendigo (the Camp Hill School of 1877 overlooking Rosalind Park), at Castlemaine (Castlemaine North, 1875) and at Eaglehawk.

Water Channel, DB
Camp Hill Primary School (now Bendigo
Senior Secondary College), PW

Uniting Church, Paradise, near St Arnaud, RE
Gazebo, Buninyong Botanic Gardens, RE
Opposite: Pyrenees, JM

PARKS & RESERVES

NATURAL HERITAGE

Very early in the history of the colony, the Victorian Government began proclaiming Crown Land Reserves to control the scope and intensity of settlement, and to manage the physical resources essential for the longer term development of Victoria.

Pre-goldrush, small areas of land were set aside for Aboriginal protectorates. These reservations were similar in purpose to those that were put aside for the American Indians – to protect, control and reduce conflict.

As goldrush followed goldrush, and permanent townships were established, the Government began to protect the region's important physical resources – gold, timber, stone and water – with state forests.

State forests protected the known goldfields from being purchased and cleared for agricultural uses. So, paradoxically, the gold that had originally seen the forests plundered also contributed to their protection and regeneration. It is no coincidence that the remaining Box-Ironbark forests align with the richest deposits of gold unearthed in Victoria. The decision of the Government to retain gold-bearing land has also produced a unique aspect of the Goldfields region today – towns set in forests.

Smaller areas of land during the late 19th century were also protected in the belief that certain places were of such fundamental value to a community that they should be held in trust for the benefit of society. For example, a small informal burial ground in Castlemaine which was closed in 1857, was described in 1862 as possessing '…an ancient history as real as any other world has produced'. The perception of Pennyweight as hallowed heritage stuck and in 1874 the cemetery was officially gazetted as a 'sepulchral reserve'.

The management and use of certain forest areas have changed dramatically since the 1970s as governments have focussed on conservation. This has resulted in a range of parks and reserves being created to be managed for the protection of natural and cultural values.

View south from Pioneer Lookout, KS
Mt Cole, KS
Stawell State Forest, KS

There are four relatively new national parks (Greater Bendigo National Park, Heathcote-Graytown National Park, St Arnaud Range National Park, Chiltern-Mt.Pilot National Park), one national heritage park (Castlemaine Diggings National Heritage Park), five state parks (Kooyoora State Park, Paddy's Ranges State Park, Warby Range State Park, Broken Boosey State Park, Reef Hills State Park) and numerous other reserves and parks across north-central Victoria representing the Box-Ironbark forests and their diverse natural and cultural values. Parks Victoria now manages the majority of these reserves to protect these values and to provide for visitors.

The following parks and reserves within the Goldfields region are easily accessible, and information relating to the discoveries to be found today is readily available through Parks Victoria's website (www.parkweb.vic.gov.au) or information line (tel 13 19 63).

WESTERN GOLDFIELDS

AMHERST REEF NATURAL FEATURES RESERVE

Amherst Reef Natural Features Reserve is noted for its existing geological and geomorphological features, particularly the remnant above-ground quartz reef which is one of the world's largest. Although it is only a small geological reserve near Talbot, the site has recorded historic, natural, aesthetic and social community heritage values. See the Talbot & Amherst section for more details.

8 HA
NEAR TALBOT
200 KM FROM MELBOURNE

ARARAT REGIONAL PARK

The ancient hills in this regional park offer beautiful panoramic views of the surrounding area. To the west is Grampians National Park and to the east Ararat, Green Hills Lake and Mt Langi Ghiran. See the Ararat section for more details.

3670 HA
1 KM FROM ARARAT
210 KM FROM MELBOURNE

CRESWICK REGIONAL PARK

Centrally located between Ballarat and Daylesford, this park embraces the hills once roamed by the artistic Lindsay family. Today you can explore the park on a variety of short and longer walks, picnic by tranquil St George's Lake or relax in the park's peaceful surroundings. See the Creswick section for more details.

2500 HA
1 KM FROM CRESWICK
128 KM FROM MELBOURNE

HEPBURN REGIONAL PARK

Hepburn Regional Park is nestled around the famous mineral spring townships of Daylesford, Hepburn and Hepburn Springs. The park contains natural mineral springs and significant relics of the gold mining era, all set in delightful bush surroundings. Each season provides a different experience with spring being a time of activity, growth and colour.

The Tipperary Walking Track runs from Lake Daylesford to the Hepburn Springs Mineral Reserve. It's an easy 16 km walk following Sailors and Spring Creeks through foothill forest. The complete walk would take five or six hours, but it can be broken into a number of shorter sections. See the Daylesford section for more details.

2900 HA
5 KM FROM DAYLESFORD
110 KM FROM MELBOURNE

LANGI GHIRAN STATE PARK

The Langi Ghiran State Park protects the granite peaks of Mt Langi Ghiran which reach a height of 949 m.

In the west, a gently sloping plain supports an open woodland, rich in wildflowers. There are two reservoirs in the park built from locally hewn granite blocks in 1880 to supply water to Ararat.

Four rock art sites, numerous shelters, scar trees and other artefacts have been recorded in the park. In the south-east of the park, near the Western Highway, the Lar-ne-Jeering Walking Track leads to an aboriginal shelter displaying art unique to the area. See the Ararat section for more details.

2695 HA
14 KM FROM ARARAT
195 KM FROM MELBOURNE

Amherst Reef, RE
Waterlily, Creswick Regional Park, PW
Bridge, Hepburn Regional Park, PW
Mt Langi Ghiran, KS
Ararat Regional Park, RE

Mt Buangor State Park

The park takes in varied eucalypt forests, creek flats, granite peaks and steep escarpments. There are magnificent tree-ferns – and even snow gums on the higher peaks.

A 21 km moderate two-day forest walk (the Beeripmo Walk) takes in the views from a number of scenic lookouts. Settlers moved into this area following Major Mitchell's expedition and, in the 1840s, small timber mills set up, booming in the gold era. See the Ararat section for more details.

2400 HA
30 KM FROM ARARAT
180 KM FROM MELBOURNE

Mt Franklin Reserve

Mt Franklin, or Lalgambook as it is known by the Dja Dja Wurrung people, offers fine panoramic views from the summit via a short walk from the picnic area. Located eight km north of Daylesford, the sheltered crater of Mt Franklin is an ideal setting for a picnic or short-term basic camping. See the Franklinford section for more details.

50 HA
8 KM FROM DAYLESFORD
115 KM FROM MELBOURNE

St Arnaud Range National Park

The park has one of the largest intact areas of Box-Ironbark vegetation in Victoria and is an ideal place to view forests as they were before the goldrushes. The Upper Teddington Reservoir is a peaceful place for picnics, camping and fishing. See the St Arnaud section for more details.

13,900 HA
35 KM FROM STARNAUD
200 KM FROM MELBOURNE

St Arnaud Regional Park

Rich with natural and cultural features, this park west of St Arnaud offers great opportunities for bushwalking, picnics, scenic drives and horse riding in Box-Ironbark country, with magnificent views from Bell Rock and View Point. It provides habitat for many threatened species.

957 HA
1 KM FROM ST ARNAUD
230 KM FROM MELBOURNE

Orchid, St Arnaud Range National
    Park, RE
Mt Franklin, RE
Mt Buangor, KS
View from Bell Rock, St Arnaud Regional
    Park, RE

CENTRAL GOLDFIELDS

CASTLEMAINE DIGGINGS NATIONAL HERITAGE PARK

The first national heritage park in Australia, this park harbours fascinating tales of a golden past and retains much of its goldrush character. Spring wildflowers, scenic drives, bushwalking, cycling, guided tours and gold fossicking are some of the highlights. See the Castlemaine section for more details.

7500 HA
5 KM FROM CASTLEMAINE
120 KM FROM MELBOURNE

KOOYOORA STATE PARK

With magnificent views and a rich variety of plants and animals, Kooyoora protects some of north-central Victoria's outstanding natural features, including Melville Caves. Kooyoora contains scarred trees, rock holes, shelters, ochre, quartz and mica quarries, tool manufacturing sites and three stone arrangements. There are more Aboriginal sites here than anywhere else in Victoria. See the Inglewood section for more details.

11,350 HA
20 KM FROM INGLEWOOD
220 KM FROM MELBOURNE

MALDON HISTORIC RESERVE

Maldon Historic Reserve comprises about 2500 hectares of public land and forest surrounding the historic township of Maldon, Victoria's first notable town. The Reserve was established to protect relics from the gold mining era that gave birth to the township. The reserve contains

large areas of regrowth Box-Ironbark forests and grassy woodland which were extensively harvested for the gold mining and farming industries.

2500 HA
1 KM FROM MALDON
135 KM FROM MELBOURNE

MARYBOROUGH REGIONAL PARK

Situated between the Paddy's Ranges State Park and Maryborough, this park is noted for its impressive wildflowers and avifauna and has developed tracks for local informal recreation.

524 HA
5 KM FROM MARYBOROUGH
170 KM FROM MELBOURNE

MT ALEXANDER REGIONAL PARK

Rising 350 m above the surrounding valleys and plains, Mt Alexander east of Castlemaine is a large granite mass with steep slopes, tall trees and rocky outcrops. The park offers magnificent views and a natural forest setting for picnics and bushwalks, and is an important habitat for rare and threatened species. See the Harcourt section for more details.

1240 HA
20 KM FROM CASTLEMAINE
140 KM FROM MELBOURNE

Dog Rocks, Mt Alexander, PV (AP)
Castlemaine Diggings National Heritage Park, RE
Lookout, Kooyoora State Park, KS
Near Spring Gully, Castlemaine Diggings Heritage National Park, RE
Mine near Moliagul, KS

MT BECKWORTH SCENIC RESERVE

A lone pine tree stands as a sentinel crowning this picturesque granite outcrop, a sanctuary for native plants and animals. Magnificent views and a rich variety of spring wildflowers are two of the reserve's many features. See the Clunes section for more details.

200 HA
8 KM FROM CLUNES
140 KM FROM MELBOURNE

PADDY'S RANGES STATE PARK

Paddy's Ranges adjoins the Maryborough Regional Park and is a fine example of Western Goldfields Box-Ironbark vegetation with wonderful spring wildflower displays. Walking tracks radiate from the picnic area and camping ground and remnants of past activities are dotted throughout the park.

2010 HA
3 KM FROM MARYBOROUGH
170 KM FROM MELBOURNE

ANDERSONS MILL, SMEATON

In a postcard setting on the banks of Birch's Creek at Smeaton, Anderson's Mill stands as a powerful reminder of an industry that flourished after the goldrush of the 1850s. Standing today much like it was over 100 years ago, the five-storey bluestone building and its magnificent iron water-wheel are still in place. See the Smeaton section for more details.

12 HA
15 KM FROM CRESWICK
130 KM FROM MELBOURNE

NORTHERN GOLDFIELDS

BENDIGO REGIONAL PARK

Surrounding the city of Bendigo, this regional park links the Greater Bendigo National Park to the urban open space within the city. Diamond Hill within the park displays evidence of a range of gold mining activities from the 19th century. Mullock heaps, mud, dust, tailings and water races form a realistic picture of past disturbances to the landscape, but the resilience and durability of the forest is clearly evident from the regrowth of native Box-Ironbark vegetation that now harbours a plethora of unique flora and fauna close to a large urban centre.

8745 HA
5 KM FROM BENDIGO
170 KM FROM MELBOURNE

GREATER BENDIGO NATIONAL PARK

The park protects some of the highest quality Box-Ironbark forests in north-central Victoria, along with mallee and grassy woodlands. Visit between August and November when the wildflowers are most abundant and colourful. Ideal for nature study,

Paddy's Ranges National Park, RE
Blossom, PV (AP)
Anderson's Mill (DB)
Greater Bendigo National Park, RE

bird watching, walking, picnics, horse riding and camping. The 60 km Bendigo Bushland Trail goes through part of the park and the Great Dividing Trail has a track head in this park linking Bendigo, Castlemaine, Ballarat and Daylesford.

17,007 HA
8 KM FROM BENDIGO
180 KM FROM MELBOURNE

HEATHCOTE-GRAYTOWN NATIONAL PARK

The park protects some of the most significant environmental, cultural and recreational values in the largest remaining Box-Ironbark forest in Victoria. Explore goldrush and war era historic features at Graytown, as well as highly accessible areas that offer solitude in a bush setting. See the Heathcote section for more information.

12,700 HA
5 KM FROM HEATHCOTE
150 KM FROM MELBOURNE

HEATHCOTE PINK CLIFFS RESERVE

This interesting and colourful phenomenon was brought to light by early gold-mining activities. Sluicing work was carried on until the early 1880s and it was in this period that work in the Pink Cliffs area brought to light these colourful 'hills'. Tracks and interpretation lead around the unusual formations providing an insight into the activities of miners that caused this strange attraction.

10 HA
1 KM FROM HEATHCOTE
150 KM FROM MELBOURNE

WHROO HISTORIC RESERVE

The reserve includes the site of the former Whroo township and part of the associated Whroo goldfields. The main feature of the site is the Balaclava Mine, an open cut working where gold worth millions of pounds was extracted in the 1850s from shafts and adits that are still visible today. A path and interpretive trail lead to lookouts and an exciting walk through one of the tunnels into the open cut mine itself. See the Rushworth section for more information.

490 HA
7 KM FROM RUSHWORTH
190 KM FROM MELBOURNE

Pink Cliffs, Heathcote, KS
Whroo Historic Reserve, KS
Heathcote-Graytown National Park, KS

Castlemaine Diggings Heritage National Park, PV (AP)
Wildflower, Castlemaine Diggings Heritage National Park, PV (AP)
Opposite: Daylesford, PW

CITIES & MAJOR TOWNS

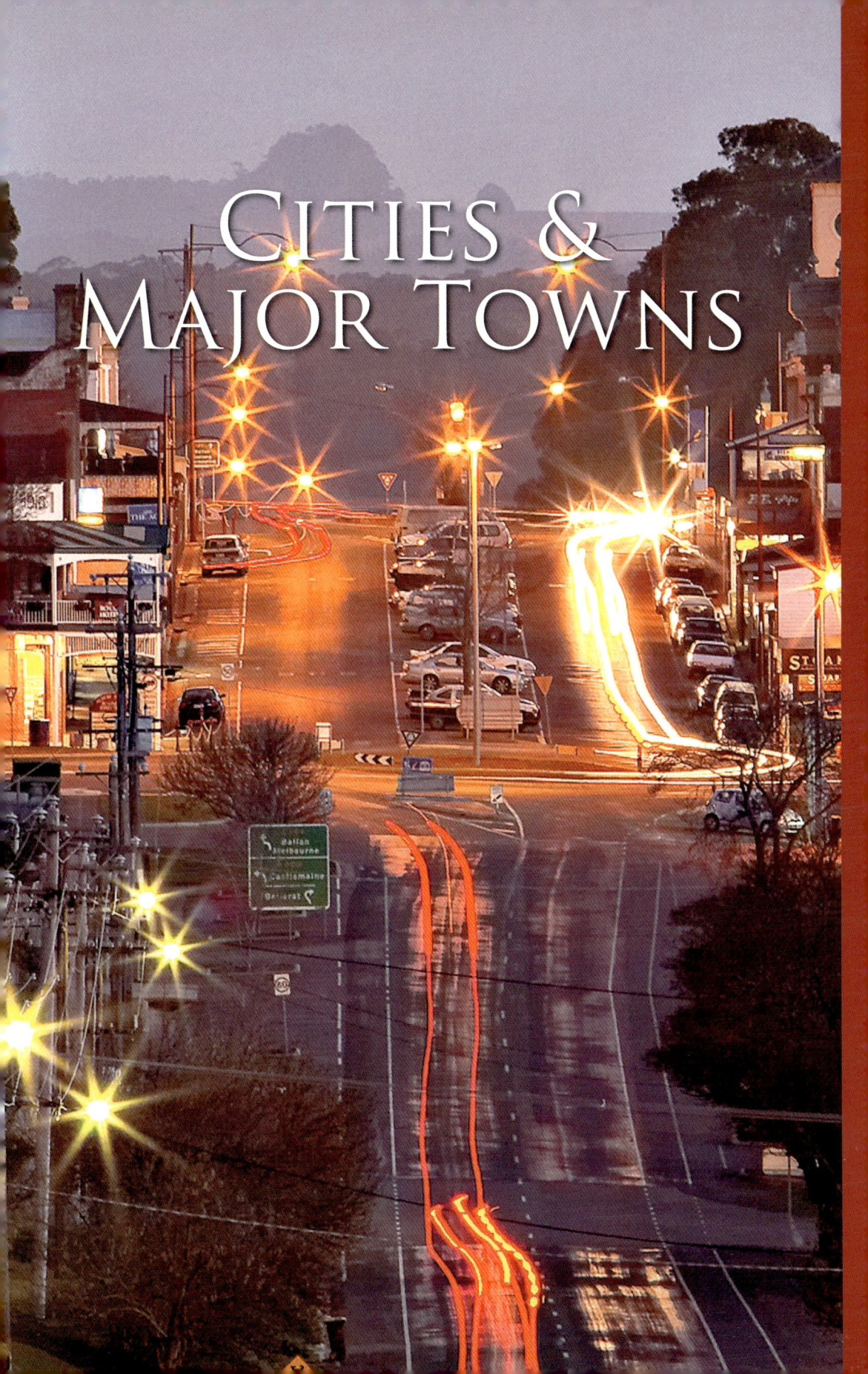

Ararat was the only goldfield in Victoria to actually be discovered by Chinese diggers and it proved to be one of the richest alluvial finds in Australia's history.

ARARAT

LOCATION: 200 KM WEST OF MELBOURNE ON WESTERN HWY
POPULATION 1857: 50.000
POPULATION 2006: 7000

Ararat is an attractive former gold mining town that, like Stawell further to the west, is now a major commercial centre for the surrounding agricultural region. The rich soil produces wheat, fine wool and wine.

More than most places on the Goldfields, Ararat is associated with the Chinese immigrants to Australia, both their triumphs and their woes. Their story is graphically told in the Gum San Chinese Heritage Centre on the western outskirts of the city.

China was not a happy place to live in the 1850s, what with the introduction of opium, European goods and political rebellion. Many young Chinese emigrated to earn money to repay family debts and pay for ongoing family welfare. The Victorian Goldfields became a magnet, particularly for those from Guandong Province and the Pearl River Delta area.

Tightly packed into ships Australia bound, many died on the voyage. A number of the survivors were offloaded at Robe in South Australia because they couldn't pay the Victorian gold and residence tax. From there they walked overland, aiming for Ballarat

and Bendigo. Ararat was on the way and one party of about 700 stumbled on success in a nearby creek.

In fact, Ararat was the only goldfield in Victoria to actually be discovered by Chinese diggers and it proved to be one of the richest alluvial finds in Australia's history. Five tonnes of gold were officially escorted from the field within six months of the discovery. However, the golden period was short and sheep graziers reasserted their influence on the economy within a few years.

In addition to the city's gold heritage, today's visitors to Ararat will be

Ararat and Mt Langi Ghiran from Pioneer Lookout, RE
Looking west from Pioneer Lookout, RE

quickly made aware of a dark side. The infamous J-Ward, associated with Pentridge Prison in Melbourne, became Victoria's most secretive institution, closed to society for more than 100 years. It is now a museum depicting the chilling story of the criminally insane.

A walk on the brighter side includes Ararat's botanical gardens, its art galleries, the surrounding vineyards, regional park and, from One Tree Hill Lookout, magnificent views of the Grampians to the west and Mt Langi Ghiran to the east.

BRIEF HISTORY

Thomas Mitchell was the first European to traverse the Ararat region during his epic journey of discovery through western Victoria in 1836. White squatters soon followed and took up runs across the fertile grazing land, displacing the Aboriginal owners.

In 1839 Horatio Spencer Willis, with his family and stockmen, drove 500 head of cattle and 5000 sheep into the region from the Murrumbidgee region to the north. At one point they camped on a hill which Willis named Mt Ararat, reportedly declaring: 'Like the Ark (in the *Bible*), we have rested here.'

Then the group of 700 hopeful Chinese diggers arrived in the area on their way to Bendigo. Stories are told that in 1857 this group was fossicking in an area called Pinky Point, about 10 km from Ararat, where gold had been discovered several years earlier. There they encountered European diggers who chased them off the field.

As the Chinese retreated over Mt Ararat, they stopped for a midday rest and coincidently discovered some rough gold in a stream. It proved to

be a fabulously rich find. Most of the party were from Canton (Gungzhou), so they called their discovery the Canton Lead.

However, after the hostility of the Europeans at Pinky Point they decided to keep the find a secret. That lasted a week before European diggers found out and the Ararat rush began. As the crowds of prospectors grew, so did hostilities. In June 1857 a group of about 20 European thugs attacked and bashed the Chinese, stole their belongings and set fire to their tents.

Justice was eventually done and the thugs were arrested, tried, convicted and sent to hard labour on the roads. Those Chinese with proof of claim (a residence ticket costing £1 per year) were allowed to return to their claims.

Mt Langi Ghiran from Ararat
back lane, RE
Town Hall, RE

Many of the Chinese turned to their original professions as doctors, druggists, tea dealers, merchants and interpreters.

Those without had to forfeit. Even though not all the Chinese had the proof, the number of Chinese with genuine tickets made up nearly 50% of the diggers at Ararat, an unusually high percentage compared to other goldfields in Australia.

Within five months of the original discovery Ararat residents began to build more permanent structures around the tent city. Over the next two years the settlement grew haphazardly around the gold diggings into an established town. The first land sales in the area began in 1858.

By 1861 the combination of more expensive and more strictly policed residence taxes (higher for the Chinese than Europeans) and the decline in gold finds caused many to leave the field. The population fell to about 1500 and gold was no longer the mainstay of the economy.

Realising the Chinese presence was needed to help maintain the population and keep businesses open, Ararat's business community petitioned the Governor to abandon the residence tax. It was finally repealed in 1865 and Chinese began to return, pushing the population back up over 2700 by 1881.

Chinese mining continued in Ararat until 1926, but by that time the town had become well established as a regional centre for other trades. Many of the Chinese themselves turned to their original professions as doctors, druggists, tea dealers, merchants and interpreters. Others took unskilled jobs as hawkers, carters and labourers or developed skills such as gardening, blacksmithing, cooking, groceries and butchery.

The last locally-known original Chinese miner died aged 93, but is still remembered in Ararat for pushing his vegetable cart around town to sell to his numerous customers.

Dominica House, RE

VISITOR INFORMATION

Ararat & Grampians
Visitor Information Centre
at the railway station complex
91 Vincent St, Ararat 3377
www.ararat.vic.gov.au
tel: 5355 0281
freecall: 1800 657 158

Bus and train services run to/from Melbourne.

FOOD & ACCOMMODATION

Ararat has a range of accommodation, including five motels rated between three and four star, and three hotels with comfortable budget rooms. There are also several B&Bs, self-contained cottages accommodating up to six guests and two caravan parks on the city's outskirts. Talk to the VIC for more information.

Ararat has several hotels which serve good bistro food. There are restaurants attached to the main motels and one excellent wine bar/café is located in the main street. There are also a number of lunchtime cafés and fastfood outlets.

SIGHTS

Ararat has preserved its early heritage by establishing museums with memorabilia from the pre-European days as well as the Chinese-influenced gold period and the dark years encountered by inmates of its prison system. The city has also fostered the arts and cultivated gardens. Forests have been preserved on the city's doorstep.

Brochures for a self-guided tour of Ararat, taking between one and two hours, are available at Visitor Information Centre.

J-WARD

One of the city's main attractions is the forbidding J-Ward. Guided tours are available on week days and every day during Victorian school holidays. The guides are very knowledgeable about the bluestone prison which began as a county gaol in 1861. They are also skilled at relating gruesome stories of the inmates, several of whom were hanged on the prison gallows. Floggings also took place prior to the prison being handed over to the Lunacy Department in 1887. Attached to the Ararat Mental Hospital it took on an even more sinister atmosphere by housing the criminally insane right up until its closure in 1991.

GUM SAN CHINESE HERITAGE CENTRE

The Gum San Heritage Centre is an authentic recreation of a traditional two-storey Chinese building, set in traditional Chinese gardens. The ornate roof with its curled up hips is typical of Chinese design. Adorned with dragons and other mythical creatures, it has become a unique and striking feature of the Ararat skyline. The tiles were generously donated by the Chinese Government and were fixed in place by four specialist tradesmen contracted from the city of Taishan.

Interactive and static displays bring the history of the Chinese miners to life, and explore the influence of Chinese culture on the economic, cultural and social development of Australia. Visitors can experiment with gold panning and calligraphy and view authentic artifacts, including Chinese dragons, costumes and a mining tunnel that was uncovered during construction.

J-Ward, RE
J-Ward, RE
Gum San Museum, RE
Gum San Museum, RE

LANGI MORGALA MUSEUM

The Langi Morgala Museum was built in 1873 as a wool and grain store. In 1968 the building was acquired by the local historical society and it now houses exhibitions that depict all aspects of Australia's history in which Ararat played a significant part. These include an extensive display of local Aboriginal artefacts, numerous costumes, photographs, equipment and other memorabilia from the pioneering squatters' days, the mining period, the railway and on into the present.

PYRENEES HOUSE

Pyrenees House with its imposing and ornate white tower was the second hospital built in Ararat. Constructed in 1886 on the site of the original 1860 hospital, it now adjoins the modern hospital. Pyrenees House has recently been restored to its former beauty and is used as a meeting and conference facility. It is listed by the National Trust.

TOWN & SHIRE HALL

The town hall and the shire hall stand opposite each other in the main city block which was originally the town's market square. The town hall, built in a Roman revival style, was opened in 1899. The shire hall was built earlier, in 1871. Both are magnificent buildings and listed by the National Trust. The town hall with its central tower and striking clock was renovated as a performing arts centre in 1980. The shire hall is now used as municipal offices.

POST OFFICE & COURT HOUSE

The old brick post office (1871) and nearby court house (1866) are also listed by the National Trust. The post office is now privately owned, but the court house is still in use.

DOMINICA HOUSE

Dominica House was the home of the Theo Grano and was built in 1899. It is named after the West Indian island of

Vindel House, KS
Shire Hall, RE
Pyrenees House, EG
Langi Morgala Museum, RE
Town Hall, KS

Dominica where the family originated. Theo, a barrister and solicitor, completed the house for his first wife and five children. It was one of the first two-storey residences in Ararat. The house was extended in 1925 to accommodate his second wife and nine children. The Grano family lived in the home until late in the 1990s.

ARARAT REGIONAL ART GALLERY

Ararat Regional Art Gallery holds one of the major contemporary fibre and textile art collections in Australia's regional galleries. It includes the Lady Barbara Grimwade Costume Collection which features fashionable dresses worn by the late Lady Grimwade, a noted Melbourne socialite between 1956 and 1989. Other collections include prints, glass, silver and ceramics.

AROUND ARARAT

ARARAT HILLS REGIONAL PARK

Surrounding the Pioneer Lookout is the 1000 ha Ararat Hills Regional Park where visitors can enjoy bushwalking, cycling, driving and picnicking in a natural forest setting. The park contains more than 200 species of native plants, including 34 species of orchids, and a wide variety of native animal and bird species.

Pioneer Lookout

Pioneer Lookout is reached via One Tree Hill Rd, a scenic drive just north of the city. It provides spectacular views of the Grampians to the west, while to the east the view looks out over the city to Mt Langi Ghiran and the Mt Cole Range beyond.

MT BUANGOR STATE PARK

Mt Buangor State Park has a varied eucalypt forest (including snow gums on the higher peaks) and takes in creek flats, a waterfall, and steep escarpments. The park is dominated by the 990 metre, granite Mt Buangor. There are picnic and camping facilities along the creek flats.

There is a 15 km network of walking tracks in the park, many of which extend into the adjoining **Mt Cole State Forest**. The Waterfalls Nature Walk, which begins at Ferntree picnic area, is a short walk suited to most people. The Cave Walking Track from Middle Creek campground is a steep and strenuous walk which takes you to a large rock overhang and extensive views to the south and west.

The park contains more than 130 species of birds.

LANGI GHIRAN STATE PARK

The Langi Ghiran State Park protects the granite peaks of Mt Langi Ghiran which reach a height of 949 metres. Langi Ghiran is an Aboriginal name for Yellow-tailed Black Cockatoo, sometimes seen in the forest below the park's rugged granite peaks.

Grampians Rd, near Ararat, KS

In the west, a gently sloping plain supports open woodland, rich in wildflowers. There are two reservoirs in the park, built from locally hewn granite blocks in 1880 to supply water to Ararat.

There is much evidence that Aboriginal people – the Ngutuwul Balug or 'mountain people' of the Djab Wurrung tribe – occupied this area. Four rock art sites, numerous shelters, scar trees and other artefacts have been recorded in the park.

In the south-east of the park, not far from the Western Highway, the Larne-Jeering Walking Track leads to an Aboriginal shelter displaying art unique to the area. Eighteen km east from Ararat the highway crosses the railway line; just after the crossing take a left turn onto Sandpit Rd. The short walk is about one km from the highway.

Eastern Grey Kangaroos inhabit the woodland and can often be seen from the roads. Echidnas, wallabies and a variety of birds are found throughout the park. Powerful Owls inhabit the northern and eastern slopes and Wedge-tailed Eagles are often seen soaring above the higher peaks.

LOCAL WINE & PRODUCE

Ararat is surrounded by vineyards of the Grampians-Pyrenees region. There are eight wineries within 40 km of the city, including Motara Wines, Mt Langi Ghiran and Cathcart Ridge. All provide cellar door tasting and sales, although Cathcart Ridge Estate is by appointment only.

Near Ararat, KS
Opposite: Vineyard, PW

AVOCA

LOCATION: 181 KM NORTH-WEST OF MELBOURNE;
70 KM NORTH-WEST OF BALLARAT ALONG THE SUNRAYSIA HWY
POPULATION 1854: 14.000
POPULATION 2006: 1400

Like many towns in central Victoria, Avoca began as a pastoral centre before it was dramatically changed by the discovery of gold. Today the emphasis has shifted back to agriculture and particularly wine – some would say liquid gold. The land is undulating with streams in the valleys shaded by River Red Gums. Rows of grape vines line up like regiments in the fertile soils of the surrounding hills with the blue ranges of the Pyrenees as a backdrop.

The emphasis has shifted back to agriculture and particularly wine – some would say liquid gold.

Avoca township itself has an airy, open feel, largely because the main street is extraordinarily wide – so wide, in fact, that there is room for parkland in the central median strip. Many of the shops and buildings along each side of the street have been restored to their former glory. One in particular, Lalor's Chemist shop, is the oldest chemist shop still operating in Victoria. The quaint sign 'Prescription Dispensary' is displayed outside. Other shops have local produce, crafts and pottery for sale.

Avoca is a pleasant rural centre and an ideal base from which to visit the surrounding vineyards, drive to the nearby goldfield hamlets, or just take walks through the countryside.

BRIEF HISTORY

In 1836 explorer Thomas Mitchell was the first European to pass through the Avoca area and he is said to have named the local river the Avoca after a river in County Wicklow, Ireland.

Based on his reports of good grazing land, squatters followed and took up runs in the surrounding countryside. In 1852 some diggers and their families travelling from Adelaide to Bendigo found a nugget near Homebush only three km from today's township. They settled in to look at the area more closely. Further discoveries brought more diggers and the population grew rapidly with more than 6000 working the nearby river beds by 1854.

Savings Bank, KS
Soldiers' Memorial, KS

A police camp of 50 troopers was established in 1853 and a lock-up was built the following year. The Bank of Victoria also came to town in 1854 to handle the new-found wealth and the same year saw the Avoca Hotel open its doors. This was followed by the Union Hotel in 1855, a Wesleyan Church in 1856, a school in 1857 and the court house in 1859.

However the gold reefs around Avoca proved to be intermittent and many diggers left without making any finds. The town's population fell back below 1000 by the early 1870s and it was clear the rush was over. Some lingered till the 1890s, but by then the farmers had begun to reassert themselves.

Some departing European diggers had remarked on the soils and climate of the area and thought it might have potential for vineyards, thus setting the scene for the region's more recent fame. Grapes were planted in the 1870s and with sheep, cattle and orchards, they began to dominate the surrounding hills.

Attempts to extract gold using dredging techniques continued on and off through the first half of the 20th century. The last of these operations closed down in 1957 and Avoca gave itself entirely to agricultural pursuits.

VISITOR INFORMATION

Avoca Visitor Information Centre
122 High St, Avoca, Victoria 3467
www.pyreneestourism.com.au
tel: 5465 3767

FOOD & ACCOMMODATION

Avoca offers accommodation at the town's two hotels, two three to 3½ star motels and self catering cottages. There are also several B&Bs and a caravan park. In the surrounding countryside there are a number of self catering cottages, and further north, 20 km out of town, the acclaimed Warrenmang Vineyard & Resort which offers a range of deluxe twin and four-share suites and cottages.

Good food is available at the town's hotels and a bistro as well as light meals at the bakehouse/café and a craft shop/café. Warrenmang Vineyard has a gourmet restaurant as does the nearby Taltarni Vineyard, Blue Pyrenees Vineyard and Redbank Vineyard. Most of the Avoca area vineyards also offer attractive picnic spots on their premises.

SIGHTS

Avoca has a number of historic buildings and a walking tour is the best way to appreciate the town's goldfield heritage. Several scenic drives from Avoca take in vineyards as well as some of the mining relics in the region.

LALOR'S CHEMIST & PRESCRIPTION DISPENSARY

The shop was built in 1854 and has been run as a chemist shop without a break. It still has old-style shelving that once held bottles and potions, but now holds modern packaged products. The shop is open on weekdays only.

Bank, KS
Savings Bank and Avoca Hotel, KS
Shop, KS
Lalor's Chemist, KS

timber was numbered so it could be put together… by numbers. Watford House is near the bridge on the Pyrenees Hwy to Ararat.

AVOCA HOTEL

The Avoca Hotel, built in 1870 on the site of the original 1854 hotel, is an attractive two-storey red brick building with a wide balcony on the top floor. It is one of the few buildings in town to have operated continuously since the goldrush days.

NATIONAL SCHOOL

The National School, built in 1857, is one of the few such schools in Victoria to survive. The attendence roll in 1872 exceeded 300 pupils. Over-crowding prompted building of a new school across the street in 1878. It became Avoca State School No.4. This building is now listed by the National Trust.

RAILWAY STATION

The Avoca Railway Station was built in 1876 when the line came in from Maryborough. By that time the goldrush was virtually over, so the line was used to transport wool, grain, livestock, fruit and firewood to markets in Melbourne. In 1890 the line was extended on to Ararat. Passenger services ceased in 1979.

VICTORIA HOTEL

The Victoria Hotel complex comprises a stone and brick hotel and a stone ballroom, dating from the 1850s (although the uninspired front section is dated 1930). The picturesque stone stables were built in 1872. Numerous Melbourne theatre groups performed in the ballroom which became known as the Victoria Theatre.

POLICE STATION, POWDER MAGAZINE & COURT HOUSE

The police station, lock-up and nearby powder magazine, built in 1860 at the request of the local council for safe storage of explosives, have been classified by the National Trust. The lock-up, built in 1867 is made of bluestone blocks and galvanised iron roof and replaced the original log gaol built when the police arrived in Avoca in 1854. It is next to the police residence, a red brick construction with a wide verandah built in 1859.

The Avoca Court House was also built during this period and was in use for 120 years before being officially closed in 1979. It is now the headquarters of the local historical society.

From the centre of town head south on the Sunraysia Hwy until you see the small court house on your right; turn right into Davy St and you'll see the residence and lock-up on your left.

WATFORD HOUSE

Watford House is a rare example of a prefabricated house imported from Europe. It was built for the proprietor of the Avoca Hotel; each piece of

Avoca Hotel, KS
Police Station, RE
Watford House, RE
Stables, Victoria Hotel, JM

CEMETERY

A little out of town on the road to St Arnaud is the Avoca Cemetery which has headstones dating back to 1857 laid out in denominational areas. In one section a Chinese burning tower and several slate headstones inscribed with Chinese characters, provide one of the few remaining signs that Avoca once had a large Chinese population. One interesting grave is that belonging to Eliza Crowhurst. Her mother carved the slate headstone and wheeled it from Percydale in a barrow.

AROUND AVOCA

Travelling by road to and from Avoca the visitor passes through a countryside of rolling fields and hills covered with vines. The Pyrenees Range is an ever-present backdrop.

PYRENEES STATE FOREST

Within the state forest, the **Governors Rock Lookout**, approximately eight km west of Avoca, gives spectacular views out across the river flats towards Maryborough and north to the Pyrenees Ranges. Box-Ironbark forests are dominant, and there is a rich flora and fauna. Gold was discovered in 1854 and slate (which was used for the footpaths of central Melbourne) was also quarried.

There is also the **Pyrenees Walking Track**, an 18 km hike beginning at the waterfalls picnic area and passing through a range of different landscapes and vegetation types. There are a number of bush camping sites along the route.

From Avoca, take the Pyrenees Hwy towards Ararat, turn right into Faraday St when you cross the bridge, then turn left at the football oval into Vinoca Rd.

PERCYDALE

The evocative ghost town of Percydale is reached by driving north along the Sunraysia Highway from Avoca for about five km and then turning off to the west on Percydale Rd. Percydale was originally called Fiddler's Creek after the first diggers in the area were discovered by the sound of the music they were playing by their campfire.

The town was once a busy centre with a Cobb & Co depot, hotels, butchers, banks, police, post office, a dressmakers shop, grocery stores, churches and an elaborate Chinese theatre, reputed to be one of the best in Victoria. Today the State Dairy is the only building still standing. It was built by Welsh miners using slate cemented with mud. The door surrounds and windows are held in place with wooden bolts.

Nearby is Dalys Cottage, an 1865 miner's cottage built of redgum slabs hewn with an axe. It has two prominent outside chimneys of brick and stone and there are still remnants of the original bark roof over the pantry.

LOCAL WINE & PRODUCE

Avoca is a hub for the Pyrenees wine region and a number of vineyards are within 20 km of the town.

Daly's Cottage, near Percydale, RE
Pyrenees near Avoca, JM
Deep Lead Mine near Avoca, DB

BALLARAT

LOCATION: 110 KM WEST OF MELBOURNE (WESTERN FWY); 81 KM NORTH-
WEST OF GEELONG (MIDLAND HWY); 91 KM EAST OF ARARAT (WESTERN HWY)
POPULATION 1871: 47,000
POPULATION 2006: 88,000

Thanks in part to the extraordinary living, outdoor museum Sovereign Hill,
Ballarat embodies the drama of gold. More than any other city, the story of the
Goldfields is knitted into its fabric.

Other alluvial goldfields were discovered earlier and some were richer – but
Ballarat was the first place in Victoria where prospectors got seriously rich
overnight. Miners on other goldfields railed against tyranny – but only miners
from Ballarat fought and died. Other goldfields had richer, deeper mines – but
the huge struggle to wrest gold from the buried rivers and quartz veins beneath
Ballarat was unparalleled.

*The city is a
vibrant energetic
place, not just a
golden survivor
living on past
glories.*

Other gold towns and cities have a
magnificent legacy of fine architecture
– but nowhere else does the anarchy of
the first goldrushes clash so stubbornly
and obviously with the order and
grandeur that was made possible by
great wealth. A brief glance at a street
map shows the fractured net of streets
in East Ballarat that was created from
the chaos of the diggings, contrasted
against the neat grid of West Ballarat.

Although the history of gold underpins
the story of Ballarat, the city is a vibrant

energetic place, not just a golden
survivor living on past glories. Ballarat
is a complex city with powerful and
competing social, economic and political
forces. The first memorial to the dead
of Eureka was erected to honour the
miners in 1856, but in 1879 another
memorial was built close by to honour
the soldiers. To this day you will find
locals who passionately support one side
or the other. Perhaps it is not surprising
the Federal seat of Ballarat has changed
hands many times.

Opposite: Ballarat Railway Station, BaT
North Lydiard St, KS

Ballarat is a city where open space is never far away, but it also has closely packed suburban streets, cafés, theatres and art galleries.

Flight From Pompeii, Botanic Gardens, BaT

The two elements so symbolically explicit in the town plan – conservative prosperity and vibrant struggle – continue to exist in an ongoing tension, a tension that has had important creative outcomes for the city, but also the nation.

Ballarat is a city of gold, but it is also a regional capital for one of Victoria's richest agricultural regions. It is a city that understands the rhythms of the countryside, but it also has a long history in manufacturing industry. It is a city where open space is never far away, but it also has closely packed suburban streets, cafés, theatres and art galleries.

Unlike most parts of Australia, Ballarat has four clearly defined seasons (or seven, according to the Wathaurong Aboriginal people). Those arriving from Melbourne or Geelong will be conscious of the climb to Ballarat which lies nearly 500 m above sea level and this, in turn, means you can often subtract a couple of degrees, and almost double the amount of rainfall.

Approaching from Melbourne your first impression of Ballarat will be of scale and civic pride, as you pass close by the site of the Eureka Stockade (the main Visitor Information Centre is at the Eureka Stockade Centre), tangle briefly with the one-way system as you cross the buried Yarrowee River and then emerge into the grand heart of the city: Sturt St.

Approaching from Geelong, it is best to turn right off the Midland Hwy at Buninyong (onto Warrenheip St, the C204), which brings you into Ballarat along the old Geelong Rd, and finally onto Main Rd, which runs through East Ballarat. Main Rd was originally the main street of Ballarat, lined with pubs and shops and running parallel to the extraordinarily rich diggings that followed the Canadian Lead. You also pass revitalised mining (Ballarat Goldfields), Sovereign Hill (where there is visitor information) and Golden Point – the site of the first discovery of gold.

Approaching from the west brings you into the urbane heart of the city through an arch commemorating Ballarat's service men and women, then past some magnificent old homes and into the commercial heart of Sturt St. There's visitor information at the town hall on Sturt St and at the art gallery on Lydiard St.

The best initial orientation point is Black Hill, an important Aboriginal site which was itself turned inside out as its quartz veins were ransacked. From the top you can get a sense of the complex topography of the town, including the basalt flows from surrounding volcanoes like Mt Buninyong and Mt Warrenheip. Volcanic flows pushed around the original rivers, sometimes burying them completely and damming the Great Swamp (now Lake Wendouree) which sits above the valley of the Yarrowee River.

It was the rivers that, over eons, washed gold out of the quartz veins in the hills. Golden Point (to the south of Black Hill, and just to the north of Sovereign Hill), where the first discovery of gold was made, was a gravel bank in the bend of a river before the volcanoes reshaped the landscape. The miners followed the course of these ancient rivers – known as leads – under the basalt cap to the deep leads, and then to the quartz itself.

From Black Hill – and particularly from the shores of Lake Wendouree – you can also get a sense of how beautiful this spot must have been before the miners arrived. The local Wathaurong people called it Ballaarat. The suffix 'arat' means 'place' and is common to Ararat; 'balla' is sometimes said to mean 'resting', sometimes 'elbow', so perhaps we can say it means 'resting on your elbow'! There can be no argument that with its permanent water and its position between the grassy plains to the west and the forest to the east, it would have made a perfect campsite…. But the debate over the spelling – whether there should be three or four 'a's – was not decided in favour of three until 1994.

Today Ballarat is a busy city with a large student population and an interesting selection of cafés and restaurants. The centre is still dominated by the magnificent chief thoroughfare, Sturt St, and a rich architectural heritage of churches, public buildings, hotels and banks. It is also home to Sovereign Hill, the finest living museum and historical park in Australia, which faithfully recreates the Goldfields and is as close as it is possible to come to time travel.

BRIEF HISTORY

Ballarat was close to the northern boundary of the Wathaurong's territory, which extended east to the Bellarine Peninsula, and west towards the Hopkins River. In 1837 the first squatters arrived in the region. They were the Learmonth brothers who settled near Mt Buninyong, which in turn became the location for the first township. A year later two more brothers, Archibald and William Yuille, took up almost all the land now occupied by Ballarat, including the land around the swamp, which became known as Yuille's Swamp before later being transformed into Lake Wendouree.

Town Hall, Sturt St, KS

They sold the gold for £3800, enough to set them up for life.

THE DISCOVERY OF GOLD

The first official discovery of gold was at Clunes. The second was just outside Buninyong, and this later discovery provoked the first significant goldrush. The goldfield at Ballarat was discovered in August 1851 by a young man, James Regan, who was returning to Buninyong after checking the possibilities at Clunes.

Regan began prospecting on a hill above the Yarrowee River. An ancient river had left behind a quartz-gravel terrace, richly salted with gold, approximately 800 m long by 200 m wide (north of Sovereign Hill and south of York St). Regan had stumbled on one of the richest goldfields in the history of the world, and it wasn't long before a serious rush began.

By 20 September there were 800 men (and perhaps a dozen women) at work at several sites. In two days the Cavanagh brothers from Clunes, (both of whom had Californian experience) took 952 ounces (27 kg) of gold from a depth of less than two metres They sold the gold for £3800, enough to set them up for life.

This kind of story was irresistible, and Geelong and Melbourne (but particularly Geelong) were emptied of men. A government camp took shape on a hill overlooking the diggings – and the collection of the hated licenses began. The intention was to deter people from leaving their normal employment, nevertheless by 20 October there were 6000 people on the field – out of a total population in Victoria of 77,000.

At the end of October a fabulous new field was discovered at Mt Alexander (Chewton and Castlemaine) 50 miles (80 km) to the north. The gold at this new field was even easier to win, and the Ballarat goldfield was almost deserted. In December 1851 the government surveyor, WS Urquart, rode in to lay out the new western half of a non-existent city. He avoided the bushy land to the east, where scaled-down mining activity was still continuing, and laid out a grid on the grassy plateau to the west. He then moved on to survey Castlemaine and other Goldfields towns.

Urquart was responsible for surveying both Ballarat and Malmsbury – it is no coincidence that the only street in the state to match Sturt St is the expansive Mollison St.

STRONG FOUNDATIONS

Ballarat gradually recovered with the discovery of the Eureka Lead (to the north-east of Golden Point, north of the Yarrowee) and Canadian Gully (to the south, along the east side of Main Rd). The number of miners grew steadily, but there were no sudden influxes, unlike the situation further north where spectacular discoveries created one rush after another. By the end of 1852 the community had begun to develop a sense of permanency.

The rewards were significant, but the Ballarat miners faced serious challenges that deterred many. They followed leads that increasingly took them deep below ground (20 m or more) through cement-like clays, drifts of waterlogged sand, and into areas where water and gases were a potentially fatal threat. The miners began to develop techniques to deal with all these challenges, as well as co-operatives (often partly financed by shopkeepers) to fund the process. A mine might take months to dig, but the returns justified the effort. On the Canadian Lead, for example, parties of eight averaged £2000 each.

By 1854 the population had reached 25,000 (more than the population of Melbourne three years earlier), permanent buildings were mushrooming, market gardens were thriving, tradesmen were pursuing their trades, and Cobb & Co was running daily coaches to Melbourne and Geelong.

Many businesses were concentrated on the river flat (today's Bridge St) beneath the government camp (which was bounded by Camp, Mair, Lydiard and Sturt Sts) and there were also some along Lydiard St.

EUREKA REBELLION

Despite continued success on the diggings, license hunts, obvious corruption and violent arrogance on the part of the authorities was building the pressure that would result in bloodshed. Eventually, under the leadership of Peter Lalor, the miners swore allegiance to the Southern Cross (or Eureka flag) and built a stockade near the Melbourne Rd. Early in the morning of 3 December 1854, government soldiers unexpectedly stormed the stockade killing 22 miners and suffering six fatalities themselves. See 'Eureka: the spirit of social and political democracy' in the introductory section for the full story about the rebellion.

Although the authorities won the battle at the Eureka Stockade, the license policy was entirely discredited and the subsequent Royal Commission replaced the license with miner's rights, costing £1 per year, which allowed their holders the right to vote and ultimately opened the way to manhood suffrage.

TOWN & THE DIGGINGS

The distinction between 'town' (the ordered grid of west Ballarat), and 'the diggings' (the chaos of east Ballarat) became entrenched. Entertainment venues (pubs, sly grog shops, brothels and theatres), restaurants and shops were close to their customers on the diggings; the banks, solicitors and other businesses whose interests were identified with the government camp were in town.

Ballarat continued to give up its gold, but genuine mining skills were required. In 1855 and 1856 a mine on the south-west corner of Sturt and Lydiard Sts went through 75 m of rock and clay to reach gold-bearing gravel – ushering in a new era. By the late 1850s quartz was also on the miners' menu. The new era required ever greater financial, organisational and engineering skills. Miners began to work for wages and the engineers and tradespeople of Ballarat began to create an industrial centre.

By the 1860s the bulk of the machinery required for quartz mining (steam engines, pumps, stampers and so on) was being produced locally. In 1862 the Victoria Foundry even

By 1854 the population had reached 25,000 (more than the population of Melbourne three years earlier).

Chinese Restaurant, ST Gill, Courtesy the
   Sovereign Hill Museums Association
Sovereign Hill, SV
Redcoats, Sovereign Hill, SV

In 1871 English novelist, Anthony Trollope, wrote: 'Ballarat is certainly a most remarkable town.'

produced Ballarat's first locomotive for the railway to Geelong. Building and clothing industries were growing, and agricultural equipment was being manufactured to cultivate the surrounding countryside.

The bustling activity on Main Rd extended from Bridge Rd at least two km towards Buninyong. Main Rd was famous for its mud, its confusion – and its multi-cultural vitality. There were 19 restaurants (including a number of Chinese), 13 hotels, many uncounted sly grog shops and brothels (especially around York St) and numerous shops, including David Jones' Drapery. Beyond the narrow strip of shops the ground was riddled with holes. The shops themselves often had impressive façades, but were nothing more than canvas behind. They were easy victims to fire and flood and they were often literally undermined, collapsing into tunnels. As the adjacent diggings (and the number of diggers) declined, however, Main Rd fell on hard times. By the 1870s the glory days of Main Rd were over.

GOLDEN DECADE

At the same time, Ballarat West thrived. Civic buildings and department stores lined Sturt St, while Lydiard St developed as the business stronghold.

Stockbrokers worked from the south-east corner of Sturt and Lydiard Sts. From the late 1850s, expensive brick and stone buildings multiplied. The Theatre Royal (1859), the railway station (1862), the first stage of the post office (1863) and the town hall (1871) were among them.

The 1860s were literally and metaphorically a golden decade. In 1869 there were more than 300 mining companies, giving employment to thousands of miners, not to mention 70 stockbrokers. The population of Ballarat was more than 60,000.

In 1871 English novelist, Anthony Trollope, wrote:

'Ballarat is certainly a most remarkable town. It struck me with more surprise than any other city in Australia… a town so well-built, so well-ordered, endowed with present advantages so great in the way of schools, hospitals, libraries, hotels, public gardens.'

By the 1880s the days of alluvial mining were over and the small mines were being overtaken by large operations. For example, the 'Star of the East' at Sebastopol reached a depth of 700 m and employed 340 men. As happened throughout Victoria, however, increasing costs and low gold prices led to a steady decline in

Mining Exchange, North Lydiard St, BaT
Sturt St, KS

mining. Fortunately, when mining finally ceased in 1918 Ballarat had sufficiently strong manufacturing and service industries to survive.

Today the city is within commuting distance of Melbourne but, thanks to its proud history and resilient culture, it is very much an independent city in its own image. Manufacturing is still important and service industries (especially education, health and tourism) are expanding. Well known companies operating in Ballarat include Mars Confectionery, McCain's Foods, Bendix Mintex, Timken and Selkirk Bricks. And last, but not least, gold is once again being mined not far from Golden Point by Ballarat Goldfields.

James Oddie

If any one man embodies the history of Ballarat it is James Oddie. Born in Lancashire in 1824 he emigrated to Australia in 1849, quickly found his feet and was soon a partner in a small foundry in Market Square, Geelong. When gold was discovered at Buninyong he collected two weeks rations and, leaving his foundry in his partner's hands and his wife in charge of their cottage, he set off with three other Geelong businessmen. When he arrived at Buninyong he sought out Thomas Hiscock (the discoverer of gold at Clunes and an experienced miner) and had a panning lesson before beginning his own attempts.

The work was hard and unrewarding at Buninyong, so his party decided to try the new goldfield that had been discovered to the north. He invited another Geelong man, Thomas Bath, and his wife Johanna (who was probably the first European woman on the diggings) to come along. On 1

September 1851, Oddie and the Baths reached Golden Point, Ballarat. There were seven tents pitched on the slope above the creek.

Oddie left Ballarat to chase gold at the new diggings at Mt Alexander, but returned to Ballarat in 1853 to set up a store on the diggings. He had one store at Smythesdale, then one in Eureka St, not far from the stockade, then another at the corner of Dana and Armstrong St. Around 1855 he began to concentrate on real estate and banking and in 1856 he was elected a member of the first Ballarat Council. He went on to become one of the most influential men, and most generous benefactors, in the city.

Oddie was involved in the establishment of many of the town's national schools, the Ballarat School of Mines, Ballarat Hospital, the benevolent asylum, the mechanics institute, the female refuge, the observatory – and, especially, the Ballarat Fine Art Gallery. In addition to donating many works of art, he organized the loan of the Eureka Flag to the gallery by Trooper King's family.

He also commissioned a statue of his old friend Peter Lalor, which stands in Sturt St. Apparently many of the surviving diggers stayed away from its unveiling; perhaps they wanted to remember Lalor at the height of his radical glory on the diggings, not as the conservative parliamentarian in robes and a wig that he became (and that the statue represents).

In 1867 Oddie began the Old Identities Association, later known as the Old Colonists' Club, and by 1909 he was the last survivor of the first days of Golden Point. On 1 September 1909, he defiantly celebrated the 58th anniversary of his arrival in Ballarat with a one-man banquet at Craig's

Gold is once again being mined not far from Golden Point by Ballarat Goldfields.

Original Digger's Wheelbarrow, Gold Museum, GM
Original Digger's Shovel, Gold Museum, GM

Ballarat has a growing reputation for fine food and wine.

Hotel (which had been founded by his friend Thomas Bath). After he had eaten, a number of people, including the city's leading citizens, came to pay their respects to the old man who had lived through such amazing times. Oddie died on 3 March 1911.

VISITOR INFORMATION

The excellent main Ballarat Visitor Information Centre is at the Eureka Centre (which is itself a sophisticated museum and interpretation centre on the site of the famous stockade just off Victoria St, the main entry into Ballarat from Melbourne.

Visitor information kiosks are also available at Sovereign Hill (open daily), Ballarat Fine Art Gallery (open daily), Ballarat Town Hall (Monday to Friday, 9 am to 5 pm excluding public holidays) and Buninyong (open Wednesday to Sunday).

In addition to supplying maps, brochures and souvenirs the main information centre offers a wide variety of services: accommodation referral, public internet, stroller and mountain bike hire, miner's rights, Eureka Pass (two-day attractions pass), Art 'n' fact Pass, and self-drive and walking tours.

Ballarat Visitor Information Centre
at The Eureka Centre
Corner Eureka and Rodier Sts
Ballarat 3350
tel: 5320 5741
freecall: 1800 44 66 33
email: information@ballarat.vic.gov.au
www.visitballarat.com.au

Fast train services run to/from Melbourne and buses link to most important Goldfields centres.

FOOD & ACCOMMODATION

Ballarat has a superb range of accommodation to suit every taste and budget. There are luxury boutique hotels in heritage buildings, old hotels (including the magnificent Craig's Hotel), serviced apartments, B&Bs (again in heritage buildings), self-catering cottages, many motels, caravan parks and a backpackers' lodge.

Ballarat has a growing reputation for fine food and wine. There's a wide variety of cuisines available – from Asian and Italian to modern Australian. There are some excellent pubs – ranging from restaurant quality to good-value – and there's a great range of coffee shops and cafés.

SIGHTS

Virtually nothing remains of the thousands of mines that were part of the first alluvial diggings, and little remains of the hundreds of mines that followed the deep leads and quartz underground. Instead a fine city – mostly dating from the days of deep leads and quartz – has taken their place.

The mining days are, however, dramatically captured by Sovereign Hill. Built on original goldfields and incorporating some original buildings, the goldrush days have been painstakingly reconstructed with extraordinary accuracy and attention to detail.

The living history at Sovereign Hill can be enjoyed on a number of levels: children will love panning for gold or riding in a Cobb & Co coach; everyone will enjoy the experience of being taken back in time; and those with a deeper interest in Goldfields history will absorb a vast amount of information that will make it possible to understand the real-world goldfields. It is the perfect place to begin an exploration of the Goldfields; experiencing such an authentic recreation makes it possible to interpret and (imaginatively) recreate ruins and relics when you come across them outside in the surrounding countryside.

The rivalry between Ballarat East and the City of Ballarat continued until amalgamation in 1921. Even though Ballarat East gradually became more orderly, it is still very different to the grandeur of Ballarat West. Sturt St largely reflects development that took place over 20 years from the early 1860s, after disastrous floods and fires had devastated East Ballarat's low-lying Main Rd.

In West Ballarat, Lydiard St, which crosses Sturt St at the crest of the hill that once overlooked the diggings, has one of the most intact 19th century streetscapes in Australia. Most of the buildings have played an important role in Ballarat's cultural life. The old government camp originally occupied the land bounded by Sturt, Lydiard and Mair Sts, so this is where you now find the post office and railway station. Lydiard St has a number of other significant buildings, including the art gallery, stock exchange and Old Colonist's Club.

There is much to see walking around the centre of Ballarat – especially in the quadrant bounded by the railway

Lydiard St has one of the most intact 19th century streetscapes in Australia.

Lydiard St from Sturt St, BaT

to the north, Lyons St to the west, Dana St to the south and Main Rd to the east. There are magnificent public buildings, hotels, churches, houses and cottages. Look for the craftsmanship in the detailing of the buildings and, if possible, duck down some of the side streets and lanes to get a less glamorous – but evocative – rear view.

To explore Ballarat:

- visit the Eureka Centre and Visitor Information Centre on the site of the Eureka Stockade

- walk down the centre of Sturt St to take in the sculpture and the architecture (and to note cafés that catch your eye for future reference)

- orient yourself from the lookout at Black Hill

- ride a tram around the western shore of Lake Wendouree, visit the botanic gardens, especially the sculpture pavilion, and have a coffee at the café/restaurant overlooking the lake

- explore Lydiard St – and duck into Craig's Hotel

- visit the oldest and largest regional art gallery in Australia – don't miss the Eureka flag

- allow the best part of a day to explore the superb recreation at Sovereign Hill – and save some energy for the fascinating Gold Museum across the road

Mine Tour, Sovereign Hill, BaT
Sovereign Hill, BaT

SOVEREIGN HILL

Sovereign Hill is an extraordinary recreation of life on the goldfields and is the ideal starting point for an exploration of the region. It is a vast outdoor museum that places the relics and ruins you will discover scattered around the region into a meaningful framework.

Sovereign Hill occupies a ridge above the original Main Rd on ground that was extensively mined during the 19th century, first by prospectors with shallow diggings, and later by larger underground quartz-mining companies. There are more than 60 buildings and tents that together recreate the ambience, lives and technology of the times.

Sovereign Hill interprets aspects of life on the goldfields surrounding Main Rd between 1851 and 1861 – Main Rd's heyday. There are diggings, shops and public buildings, weatherboard cottages and brick bungalows, an underground mine, factories, horse-drawn vehicles, steam operations, and heritage gardens…. They create a concrete reality that helps make sense of the ruins, relics and streetscapes around the region.

Sovereign Hill's Main St is busy with horse-drawn vehicles and people in period costumes. Craftsmen are at work and there are a variety of shops and businesses where you can purchase goods. At the Hope Bakery, pastries, cakes and breads are freshly baked each day in the wood-fired brick oven. You can pan for real gold, play bowls, be photographed in Victorian costume, ride in horse-drawn coaches, travel underground on a tour of the Red Hill Mine, or simply observe the working and living conditions that existed last century.

The Red Hill Gully Diggings is a reconstruction of a typical alluvial goldfield in the years before the Eureka rebellion when the majority of the population lived under canvas or bark, working small claims. The diggers were ingenious and inventive: they used whatever materials were to hand to make their dwellings and simple machines to extract and process the gold-rich earth. They used windsails like those on the migrant ships to get fresh air into their shafts. They baled water and timbered, dug and cradled, puddled and panned, day in and day out, to make a living – and, for the lucky ones, sometimes much more.

But luck was too often a fickle companion. It was hard and frustrating work, made harder by the brutal and arbitrary methods of the police enforcing deeply resented laws – especially the hated license system. At night, Sovereign Hill hosts a spectacular 80-minute sound and light show, *Blood on the Southern Cross*, which recreates the Eureka rebellion – the story of people pushed to extremes by insensitive authorities. Experiencing Sovereign Hill at night accentuates the experience of time travel. Soft lights inside the cottages and businesses, the street lamps, the sounds of people revelling in the hotels, the wash of a soft moon on a cloudless night, completely erase any hint of the 20th century.

Sovereign Hill has a stable of more than 50 horses to power its fleet of carriages, buggies and carts, each of which is made on site using original machinery. Each day, $70,000 worth of molten pure gold is poured at the Sovereign Quartz Mine Smelting Works. Visitors can also see tin implements being manufactured, a blacksmith operating at the forge, and the pottery turning out useful domestic wares.

Shopkeepers in Clarke Bros Grocery, in the Criterion Store, the Glasgow Saddlery and the Rees and Benjamin jewellery store all display the variety of practical and exotic wares that turned up in the young township from all parts of the Empire. Refreshments are available at the Charlie Napier Hotel or the United States Hotel, and there is always something happening at the magnificent Victoria Theatre.

Street theatre presents vignettes of incidents and issues in 19th century Goldfields life. The Redcoats – soldiers of the 40th Regiment of Foot garrisoned in Ballarat for gold escort duties and involved at Eureka – parade daily and fire a salute to the Union Jack.

In the mining museum, visitors can take a guided tour of an underground mine that intersects original late-19th century mine workings and artefacts. Sovereign Hill has one of the largest operating heritage steam plants in the world. It provides steam to power engines driving a quartz-crushing battery and a Cornish beam pump dewatering the underground mine.

Sovereign Hill has accommodation, and plenty of cafés, restaurants and picnic areas. Even the exterior of Sovereign Hill's on-site accommodation is carefully created to be consistent with the times – modelled on the court house and other buildings in Ballarat's Government Camp at the time of the Eureka rebellion.

Sovereign Hill is an extraordinary recreation of life on the goldfields and is the ideal starting point for an exploration of the region.

Main St, Sovereign Hill, SH
Heritage Pump, Sovereign Hill, SH
Poppet Head, Sovereign Hill, SH

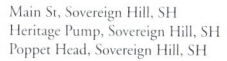

GOLD MUSEUM

The Gold Museum, opposite Sovereign Hill is often overlooked – despite having some superb collections that have been brought together to illustrate the development of Ballarat. Original paintings, drawings and photographs are interspersed with fascinating objects, so you are shown the broader picture and are then able to focus on everyday tools and objects that people actually used.

There is also an evocative, multi-media presentation, *The Land of the Wathaurong,* which presents the legends and lives of the Wathaurong people, whose lands include Ballarat. It's a creative and absorbing 10-minute presentation that successfully brings to life – and celebrates – the region's indigenous heritage.

The main exhibition includes an extraordinary panorama taken by the photographer William Bardwell from the top of the town hall tower in 1872. There are also paintings and illustrations by Gill, Cogné, Ham, Tulloch, Huyghue, Deutsche, Strutt, von Guerard and others. These, along with the museum's collection of documents, manuscripts and books,

Gold Museum, GM
Gold Museum, GM

have been vital tools and references in the recreation of Sovereign Hill's buildings and streetscapes.

The gold pavilion section of the museum has one of Australia's finest collections of gold coins, from ancient to modern times. There are also other precious golden relics and a fascinating collection of nuggets that have been gathered from around central Victoria (curiously, different areas and creeks produce nuggets with their own distinctly unique appearance).

KIRRIT BARREET (ABORIGINAL CULTURAL CENTRE, MUSEUM & GALLERY)

The Kirrit Barreet complex, at the bottom of the hill below Sovereign Hill and the Gold Museum, is dedicated to Bunjil and the traditional Aboriginal owners of the land. It is named after 'a place of creation' now known as Black Hill. The museum has comprehensive interactive displays and information about Aboriginal Australia and the local Wathaurong people. There's also a landscaped garden illustrating the seven climate seasons that the Aboriginals of the region recognised.

Lastly, the gallery and shop sell outstanding examples of contemporary Aboriginal art from around Australia, as well as didgeridoos, books and smaller souvenirs.

EUREKA CENTRE

The Eureka Centre is an iconic modern building, incorporating a huge Southern Cross flag, constructed near the site of the Eureka Stockade where one of the most significant events in the evolution of Australian democracy took place. The centre includes multi-media recreations of the events surrounding the rebellion. It is a short detour from Victoria St, the main entry into Ballarat from the Western Fwy, and it is also home to the main Ballarat Visitor Information Centre.

The centre is in the Eureka Stockade Gardens. Although the gardens were created just 16 years after the battle, there are no surviving physical signs. The Eureka Stockade Monument (1884) in the gardens continues to be a focal point for Eureka anniversaries and protests.

STURT ST

Sturt St (originally Raglan Rd) was a stock route, three chains (60 m) wide, when the first survey was made in 1851. There was nothing but a bullock driver's camp on the hill above the Yarrowee.

It is still 60 m wide, with an avenue of trees and sculpture running down its centre and three lanes of traffic on each side. It starts at narrow Bridge St (now a mall) in the east, runs through the business centre, past the church precinct on Dawson St, through a suburb of fine Victorian 'villas', and ends at the imposing Arch of Victory in the west.

Ballarat Mechanics Institute

The mechanics institute at 117-119 Sturt St, opposite Camp St, is one of Ballarat's oldest institutions and was established in 1859. It continues to thrive today; the subscription library has operated continuously for 140 years.

The mechanics institute movement was encapsulated in an editorial in the *The Ballarat Times*: 'It is the moral duty on the part of the better educated classes to confer, so far as in their power, upon those who have not been

Eureka Centre, BaT
Mechanics Institute, DB
Mechanics Institute, DB

so fortunate, these blessings by which they themselves have profited.'

In addition to providing classes in basic subjects (like mathematics, bookkeeping and grammar), there were well-attended lectures on subjects like 'Dr Livingstone', the 'Pleasures and Advantages of Scientific Studies', 'Life and Death Struggles of Italy', and 'Mnemonics'.

The building, with its beautiful façade (note the statue of Britannia with her spear), was constructed in two stages in 1860 and 1869. The institute has a nationally significant collection of 19th century print material, including many newspapers from the 1850s.

National Mutual Insurance Offices

Standing on the south-western corner of the intersection of Sturt and Lydiard Sts, these extraordinary former insurance offices with their octagonal turret were built in 1905 to a design by John Clarke and his son. The style is described as Venetian Gothic. John Clarke was responsible for many important buildings around Australia, including Government House and the Treasury Building in Melbourne, the tower of Brisbane Central Station, and the railway station at Maryborough.

Railings from Lal Lal, St Patrick's
Cathedral, KS
National Mutual Insurance Offices, BaT

Post Office

Across the road on the north-eastern corner of Sturt and Lydiart Sts, the post office, now part of Ballarat University, dominates the intersection. Built in stages from 1864, it has a grand Italianate style that was often applied to post offices of the era.

Ballarat Town Hall

The first foundation stone for the town hall is dated 1860, but the original building has been incorporated in a latter building dating from 1870.

In 1868 an architectural competition was held. The first prize was won by a local architect, Caselli; the second prize by a Melbourne architect, Percy Oakden.

The council appointed Oakden, but then instructed him to build an exterior that followed the plans of a third architect, Lorenz, and to use Caselli's plans for the interior! Could this possibly have been a decision by a local government committee? Surprisingly – thanks to Oakden – the end result is coherent and impressive; the interior – thanks to Caselli – is particularly fine.

Oakden later formed a partnership with Leonard Terry and was also responsible for the Creswick Town Hall.

St Patrick's Cathedral

Catholic services were conducted in Ballarat from 1851 onwards, but the foundation stone for the first church was not laid until 1858, and the building not completed until 1863. The miners presented the bishop with gold nuggets which were formed into a chalice and paten, displayed today in the Victoria & Albert Museum, London. The Gothic railings around the perimeter are made of iron dug and smelted at Lal Lal near Buninyong. Consecrated a cathedral in 1891, St Patrick's is Australia's oldest cathedral.

St Andrew's Kirk

Across Sturt St, also on the intersection with Dawson St, St Andrew's Kirk, now part of the Uniting church, is a fine church that has been built in stages, and it includes a magnificent spire (1864). Note the delicate railings and Romanesque chevron detailing at the entrance portals.

Mary's Mount Chapel

After a stretch of impressive houses, including some superb Victorian villas, you reach Mary's Mount, a Loreto convent school. Originally a lavish house, it was bought by the nuns in 1875.

The chapel, one of Australia's most magnificent convent churches, was built from 1898 to a design by WB Tappin, who also designed the Cathedral of the Sacred Heart in Bendigo. It was substantially funded through the estate of a young Austrian countess, Elizabeth Wolff-Metternich. The countess suffered poor health and, as a break from a world tour, she lived and taught at the abbey for several years. She died on her way back to Europe in 1899 leaving a large sum of money that was only released to the convent after a special dispensation from Kaiser Wilhelm.

Arch of Victory

The Arch of Victory is the starting point for a 20 km Avenue of Honour which commemorates all those from Ballarat who fought in WWI. A bronze plate at each tree bears a name and number of the unit of an enlisted serviceman or woman. The Ballarat tribute was the first of 128 Avenues of Honour that were planted throughout central Victoria.

The idea originated with the 500 employees of the Lucas factory – a women's clothing manufacturer that began in 1888 in a cottage in Ballarat East. The 'Lucas Girls' raised the funds for the arch and the trees; no doubt many had lost loved ones in the war. When the arch was opened in 1920 by the Prince of Wales, they presented the prince with silk pyjamas that each girl had helped sew.

The Ballarat Avenue of Honour is the longest in Australia and follows the Ballarat Burrumbeet Rd (formerly the Western Hwy). A total of 3332 trees of varying species were planted.

Mary's Mount Chapel was substantially funded through the estate of a young Austrian countess, Elizabeth Wolff-Metternich.

Victory Arch, BaT
Sculpture, Botanic Gardens, KS
Lake Wendouree, KS

The government camp was established above Camp St because it gave the troopers a clear view of the diggings below.

CAMP ST

The government camp was established above Camp St because it gave the troopers a clear view of the diggings below, although the diggings were surprisingly free of trouble. The troopers began their march to the Eureka Stockade from here on the morning of 3 December 1854; maps of the route they took are available from the Visitor Information Centre.

There are a number of interesting buildings including the Old Ballarat Police Station (1884-1886) with unusual decorative cast iron, the trades hall (1887) with its baroque façade) and the former State Savings Bank (1872) with its elegant Renaissance design.

LYDIARD ST – SOUTH

Lydiard St, north and south of the Stuart St intersection, is lined with historic buildings. Lydiard St south ends at the former gaol's front gate, which is now part of Ballarat University. The Anglican cathedral (constructed between 1854 and 1868) is one of Ballarat's earliest stone buildings and the Wesleyan church (1883) on the corner of Dana St, was designed by Percy Oakden.

Union Bank of Australia

Ballarat does not have a single architect who shaped the city in the same way that Vahland did in Bendigo. But Leonard Terry, who favoured a style known as Renaissance Palazzo, certainly had a significant impact. He ultimately designed more than 50 banks, most in Victoria, including six in Ballarat.

Five survive in Lydiard St: the Bank of New South Wales (1862), the Colonial (1860), the three-storey National (1862) and the Australasia (1864) stand together on the north-western corner of Sturt St. They all have completely different façades, but harmonize well. This group originally continued on the southern corner of Sturt St with the London Chartered Bank (1866), which no longer exists and, a little further west, the Union Bank (1863) which is sometimes regarded as Terry's finest design.

Terry also designed many Anglican churches and, as the official diocesan architect, also vetted designs by other architects. He was responsible for St Paul's Anglican Church (1861) in Humffray St, and St John's (1864).

Towards the end of his career he formed a partnership with Percy Oakden.

Craig's Royal Hotel

Thomas Bath who, with James Oddie, was one of the first arrivals at Golden Point built the first hotel in Ballarat in 1853 and this hotel was the site of the Royal Commission into the Eureka rebellion. In 1857 Bath sold the place to Walter Craig and Craig set about transforming it into a huge, grand hotel.

The hotel we see today was built in several stages from 1862, with colonnaded upper floors and a variety of towers, the most elaborate of which is

Craig's Hotel, BaT

on the corner of the site and is wreathed in flowers and fruit. As Ballarat's finest hotel, many famous people have stayed, including Prince Alfred, the Duke of Edinburgh in 1867.

Walter Craig was a keen race horse owner, and from 1867 to 1868 he employed the adventurous horseman and poet Adam Lindsay Gordon to run the livery stables. Sadly, Gordon's stay was marred by tragedy. The stables burned down, he was injured when he fell from his horse and his 11-month-old daughter died.

Gordon left Ballarat for Brighton, in Melbourne, but beset by money troubles, he committed suicide in 1870. The cottage Gordon and his family lived in now stands in the Ballarat Botanic Gardens and is used as a gallery/shop by the Craft Council of Ballarat.

Another sad story is told about Craig himself. Before the 1870 Melbourne Cup, Craig dreamed that his horse, *Nimblefoot*, won the race, but the jockey was wearing a black crepe armband. Craig died before the cup was run, *Nimblefoot* won and the jockey did indeed wear a black crepe arm band.

Her Majesty's Theatre

Opened in 1875, and restored to its original condition in 1990, this is the oldest continuously operating theatre in Australia. It has also been home to the Royal South Street Society and its famous Eisteddfod music competition.

Gaol

The gaol was built in 1857 and closed in 1965. One of its most famous inmates was Captain Moonlight, Andrew Scott. Scott led an incredible double life as a lay preacher by day

and a bushranger by night. He was arrested in 1872 for a bank robbery at Mt Egerton, near Ballarat, and put in the gaol. Scott soon escaped and continued his career of theft, which led to murder and his recapture and death by hanging in 1880. Today the remaining gaol buildings are used by the University of Ballarat.

LYDIARD ST – NORTH

Almost every building in the northern section of Lydiard St has historic interest, and it forms one of the most complete 19th century streetscapes in the world.

Banks

The four banks on the corner of Sturt St were all designed by Leonard Terry; see the section on the Union Bank (Lydiard St – North) for more information.

Stock Exchange

Initially the miners worked in partnerships or cooperatives, but quartz mining required large-scale capital, so share trading commenced. Initially there were very limited numbers of shares and stocks were traded informally opposite the post office (now the Commonwealth Bank) on a site that was simply known as 'The Corner'.

The adventurous horseman and poet Adam Lindsay Gordon ran Craig's livery stables.

Lydiard St North, BaT
Her Majesty's Theatre, BaT
Stock Exchange, KS

Stockbrokers soon got into the game and the Ballarat Stock Exchange was established. Wild speculation ensued, with fortunes won and lost. The exchange changed location several times, but in 1888 a vast new home was built to a design by CD Figgis. The façade is an impressive example of 'boom style' Classicism with a series of arches and windows, and balustrades with urns. This is one of the few surviving mining exchanges in Australia.

Old Colonists' Club

The Old Colonists' Club (1888) is on the site of the old police barracks and has a superb cast-iron balcony. The club was formed by old miners to assist any of their members who fell on hard times.

Ballarat Fine Art Gallery

The art gallery, completed in 1887, was the first provincial art gallery in Victoria. There have been many alterations and extensions since to house an important, and ever-growing, collection.

One of the most important connections the gallery has is with the

famous Lindsay family of Creswick. Encouraged by their grandfather, all the children (Lionel, Norman, Percy, Daryl and Ruby) frequented the gallery and its influence can be seen in their work. Apparently Solomon's dramatic painting of Ajax carrying off a nude Cassandra particularly impressed the young Norman Lindsay (who went on to paint a few classical nudes himself). The sitting room from the Lindsay's Creswick home has been recreated in the gallery and there are a number of works representing all the members of the family.

The gallery also has one of the finest collections of ST Gill's watercolour sketches of the Goldfields, and an important painting of the Goldfields by Eugene von Guerard. Later eras are also well represented and include work by Charles Conder, Walter Withers, Tom Roberts, Frederick McCubbin and Arthur Streeton, as well as contemporary artists like Rick Amor. Perhaps the most moving exhibit is the original Eureka flag, which survives, somewhat tattered, but still magnificent.

Fine Art Gallery, BaT

Railway Station & Hotels

At the north end of Lydiard St, the railway station was built in 1862 as the terminus for the Geelong-Ballarat line. The journey was scheduled to take six hours, but there were many teething problems and it often took much longer. The Melbourne line via Bacchus Marsh was not completed until 1889, and this was commemorated by the magnificent entrance: Roman colonnades topped by an Ionic portico and French-inspired tower!

Nearby are two extremely unusual hotels: the former Reid's Coffee Palace (1886), with its hand-painted ceiling in the central stairwell, and the flamboyant Provincial Hotel (1910) in its Edwardian Art Nouveau style.

MAIR ST

Mair St includes a number of fine buildings, including a number of significant warehouses dating from the 1860s. Look for Dunn's Warehouse (1861), JJ Goller & Co (1861), Lister & Angel (1868). Pratt's Warehouse (1869) is a rare two-storey masonry warehouse in its original form, with pulleys, hooks and hoists.

DAWSON ST

Dawson St is home to three of Ballarat's finest churches, and one of the finest religious precincts in Victoria. See the Sturt St section for more information about St Patrick's Cathedral and the St Andrew's Kirk. South of Sturt St and opposite the cathedral, the imposing Roman temple was formerly the Baptist church (1866). It's a rare example of the Classical Revival style.

LAKE WENDOUREE & BOTANIC GARDENS

The first squatter, William Yuille, settled close to the swamp that initially took his name before being transformed into Lake Wendouree by prison-labourers in the 1860s. The surrounding streets became fashionable residential areas, and there are some magnificent homes representing every architectural style that has been popular in Australia. For many years the lake hosted paddle steamers, and it was also home to the 1956 Olympic Games rowing and canoeing events.

In 1858, land on the western side, which had been used for grazing police horses, became the botanic gardens. Baron von Mueller sent cuttings of plants from Melbourne in 1860, and some of the trees date back to this distribution.

Today the gardens are famous for an annual begonia display, and for an extraordinary collection of 19th century statuary. Local mining entrepreneurs paid for the collection, including the famous *Flight from Pompeii* by Benzoni, which is housed in a special pavilion. There is also an avenue that includes a bust of every Australian prime minister, and a cottage, now home to the Ballarat Craft Council (there are crafts for sale), that was Adam Lindsay Gordon's house.

Lake Wendouree, KS
Baptist church, Dawson St, KS
Adam Lindsay Gordon Cottage,
BotanicGardens, BaT

murder at Bentley's Hotel triggered the rebellion.

The first recorded burial, of Charles May in 1853, lies behind the Eureka diggers. The Chinese section is in the north-eastern corner.

BLACK HILL

Black Hill was a significant site for the Aboriginal people before becoming important to the diggers because of its rich veins of quartz. The hill has therefore been extensively mined, but it still provides an interesting view over Ballarat, which helps make sense of the city's complex topography.

BALLARAT EAST

Ballarat East was a separate municipality until 1920 and many important public buildings located in Ballarat West are duplicated here.

Montrose Cottage

John Alexander, a Scottish stonemason, was an early arrival at Ballarat and he erected a tent on the Eureka lead. In 1856 he built a simple cottage of local basalt at 23 Eureka St. It's one of the oldest surviving cottages in Ballarat, and is now a B&B.

Ballarat East Fire Station & Mechanics Institute

At the corner of East and Barkly Sts, the fire station (1858 and 1916) has a magnificent octagonal tower; buttresses and crenulations give it the appearance of a fortress. It's one of the few remaining examples of 19th century fire stations and possibly the oldest in Victoria. Across the road is the rich polychrome brickwork of the mechanics institute (1861).

THE OLD CEMETERY

The old cemetery is in north Ballarat on Creswick Rd (A300), not far from Lake Wendouree. Follow the main driveway from the magnificent wrought-iron gates to reach the grave of the diggers who died defending Eureka Stockade; the monument was erected in 1856. Ironically, the graves of the soldiers who also fell are close by, along with a monument that was erected to their memory in 1879.

Appropriately, between the graves of the diggers and the soldiers lies a broken bluestone pillar erected to the memory of James Scobie whose

Ballarat East, Fire Station, BaT
Eureka Monuments, Old Cemetery, DB

BEAUFORT

LOCATION: 158 KM NORTH-WEST OF MELBOURNE ON THE WESTERN HWY
POPULATION 1855: 100,000
POPULATION 2006: 1000

Beaufort is a historic gold town that hosted a brief, but frenetic rush in the 1850s. Today the town is a modest service centre for the surrounding area engaged mostly in primary production – sheep, cattle, cereals, grape-growing and timber. It has several heritage buildings, including an unusually fine rotunda – so it is not surprising that Beaufort boasts one of the finest brass bands in Australia.

Beaufort is also the alleged home of an Australian icon – vegemite. It is said that this unique vegetable extract, which has spread its way around the mouths of millions of children over the years, was invented at Beaufort in 1923 by Dr Cyril Percy Callister.

The town can count radical activist, labour leader and poet, Bernard O'Dowd as a famous son. He was born in the town in 1866. But he may have been a bit too 'hot' for other community members because he failed in attempts to establish a local school. Apparently he was a sceptic when it came to religious affairs and his free-thinking attitude did not go down well with locals.

BRIEF HISTORY

The Kirkland brothers and a lone squatter named Hamilton were the first Europeans to take up land. Hamilton established the *Trawalla Run* in 1838, but sold to Adolphus Goldsmith in 1841. Goldsmith went on to develop a rich grazing enterprise, his prominence being recognised recently in the naming of the nearby Lake Goldsmith State Game Reserve.

Gold was discovered at Beaufort in 1852, followed by the Fiery Creek discovery close by in 1855. Suddenly the area had a huge population of diggers who produced a recorded 450,000 ounces (12.8 tonnes) of alluvial gold in the two years from 1855 to 1856.

In fact four settlements sprang up around the Fiery Creek diggings. Beaufort (originally called Yam Holes because native yams were plentiful) is the one that survived. The town was surveyed in 1857 and allotments were sold in 1858.

The alluvial gold didn't last past the mid-1860s and the population plummeted back to around 1200 in 1861. Deeper reef mining continued until 1914, although gold ceased to be the town's mainstay well before that time.

Gold's place in the economy was taken by a revival of grazing. Logging began in the surrounding forests and agriculture developed when some former miners took up selections in the district. A flour mill was built in 1865 and the railway from Ballarat arrived in 1874.

Beaufort is the alleged home of an Australian icon – vegemite.

Beaufort Fire Station, JM
Cemetery Hill Rd, JM

VISITOR INFORMATION

Beaufort Visitor Information Centre
72 Neill St, Beaufort 3373
www.pyreneestourism.com.au
tel: 5349 2604.

A daily bus service connects the town to Ballarat and Ararat. Interstate coaches will also pick up and set down passengers in Beaufort. The railway line has reopened with links to Melbourne, Ararat and Stawell.

FOOD & ACCOMMODATION

Beaufort has limited accommodation (two motels and two country hotels). There are also several self-contained cottages, nearby farm accommodation and one caravan park. The hotels provide counter meals and there are cafés, a tea room for light meals, a bakery and fast-food outlets.

SIGHTS

Beaufort is a small town and a short walking tour will take less than an hour.

BAND ROTUNDA

Occupying centre space in town is the octagonal band rotunda, erected in 1903 to commemorate the reign of Queen Victoria. The structure is notable for its four-faced tower and clock presented by Rear Admiral Bridges of Trewalla. The building also has some fine metal frieze work and an unusual 'cellar' where the band was able to practice. A central staircase leads up to the balcony performance area. The rotunda was fully restored in 1993.

MECHANICS INSTITUTE

The Beaufort Mechanics Institute was built in 1863 with the front rooms added in 1887 to celebrate Queen Victoria's Jubilee. Today the institute is used by the town's band as well as art and craft groups.

POST OFFICE

The post office, built in 1872, replaced an older building on the other side of the railway line. Post office customers received their mail through a pigeon hole. The building, still in use today, now has normal counter service.

BANK OF VICTORIA

The Bank of Victoria was built in 1862, later to become the CBC Bank and then the National Bank before closing in 2001. The building is still in use as a solicitor's office.

COURT HOUSE

The court house was built in 1864 as a warden's court for settling gold mining disputes. It was last used as a court in 1982 and is now the home of the local historical society.

PRIMARY SCHOOL

Beaufort Primary School is a particularly fine old building with a central three-storey tower. The original sections were built in 1864.

Beaufort Motors, JM
Band Rotunda, EG

AROUND BEAUFORT

EURAMBEEN WOOLSHED

Just west of town is the historic Eurambeen Woolshed on Eurambeen Station. It was built of bluestone quarried on the station in 1845. The building is no longer a shearing shed, but is still used for handling and drafting sheep.

MT COLE STATE FOREST

Coming from or going to the northwest of Beaufort along the Main Mt Cole Rd is the Mt Cole State Forest which adjoins the Buangor State Park. Small timber mills were established in the forest during the 1840s and they prospered during the goldrush period. Logging continues in Mt Cole Forest, but is banned in Buangor Park.

There are numerous picnic and camping areas in the forest, along with an extensive system of walking tracks which range from short family strolls to overnight hikes.

The forest also has several good lookouts, including Lookout Hill and Ben Nevis. Many of the 250 km of roads in the forest are 4WD only. Care is needed as they are narrow and used by logging trucks.

LOCAL WINE & PRODUCE

There are a number of wineries in the Grampians-Pyrenees region within 40 km to the west and north of Beaufort, making the town a good base for a winery visit.

Pyrenees Waterfall, JM
Mt Cole, KS

BENDIGO

LOCATION: 150 KM NORTH-WEST OF MELBOURNE (CALDER FWY);
35 KM NORTH OF CASTLEMAINE; 96 KM NORTH-EAST OF BALLARAT
POPULATION 1852: 40,000
POPULATION 2006: 95,000

Bendigo brings together a unique combination of elements: it is one of the world's richest goldfields, the city centre is dominated by opulent 19th century architecture, and the entire city is surrounded by natural bush.

The city has a certainty and confidence reflected in the size and scale of its cathedrals, civic buildings and art gallery. Unlike some regional cities it is a major planet in its own solar system, not just a minor satellite circling a state capital. Given it is only two hours from Melbourne visitors may well be surprised (and pleased) to find themselves visiting a vibrant city with its own distinct, and distinctive, character.

The city you see was fundamentally created by a single generation of late-19th century visionaries.

The city lies north of the Great Dividing Range, so it has an inland climate where blue skies are the norm. The main highway runs along a shallow valley surrounded by low, dry, forested hills and the city is strung along its length. At most points you do not have to go far to escape the traffic

and then to suddenly find yourself surrounded by forest.

Approaching from Melbourne, the road from Kangaroo Flat follows the creek flats that were the first incredibly wealthy alluvial diggings. After the Visitor Information Centre, your first stop should be the Lookout Tower (an old poppet head) in Rosalind Park overlooking the city centre. This will give you a chance to orient yourself and get a real feel for the city that is justifiably called 'Vienna in the Bush'.

Overlooking site of Hustler's Royal Reserve Gold Mine and Pall Mall, PW Colonial Bank (Mully's Café), RE

The city you see was fundamentally created by a single generation of late-19th century visionaries. They were migrants drawn from Europe, America and Asia, who believed in progress and who built for the future. In a little over four decades they created a city from nothing, and behind a remarkable cultural, economic and architectural heritage.

Bendigo takes its name from the famous British boxer, William Abednego Thompson, whose biblical name Abednego was shortened to Bendigo. A local shepherd, known as Bendigo because he was handy with his fists, lived near to the point where gold was first discovered. In 1855 the authorities introduced the name Sandhurst, but the new name was never popular, possibly because of its association with the military (and the military college of the same name). The name was officially changed to Bendigo in 1891 after a poll of ratepayers.

Today the name conjures up images not only of fabulous wealth but also of the blood, sweat and tears of thousands of miners. Bendigo's lively city centre is dominated by its splendid historical inheritance: domes and spires and porticoes, trams still moving along the principal boulevard, a vast gothic cathedral, and a great collection of old pubs. It's a very interesting place.

BRIEF HISTORY

Bendigo was part of the Dja Dja Wurrung's territory, but squatters seized the good land around Bendigo Creek by 1840. The run became known as *Ravenswood*. The Georgian-style brick mansion built in 1857 by the owner of *Ravenswood*, Frederick Fenton, still stands 17 km south of Bendigo just off the Calder. It is now a luxury small hotel.

DISCOVERY OF GOLD

Although there is plenty of evidence that various shepherds were collecting gold illegally for some time prior, the official discovery of gold at Bendigo was not until October 1851, five months after the first 'official' discovery at Clunes. Gold was found near a shepherd's hut (Bendigo's Hut) at a spot where a sandstone bar dammed the creek (now Maple St, Golden Point). It is said you could see the gold shining in the dirt and several *Ravenswood* workers and their wives (John and Margaret

Site of Government Camp from the air, BS

The End of the Rainbow, Golden Square
   1857, George Rowe, Newson Bequest
   Fund 2004, Bendigo Art Gallery
Sandhurst in 1862, Thomas Wright,
   Gift of the Bendigo City Council 1908,
   Bendigo Art Gallery
Bendigo School of Mines, KS

Kennedy, Patrick Farrell and his wife) were the first to secretly start work at the waterhole – famously storing their gold in their stockings.

Inevitably the secret didn't last long and the discovery was announced in Melbourne on 13 December. Miners from Castlemaine rushed to the site. The hordes followed the gold down the creek to Epsom and White Hills and up the creek to Kangaroo Flat. By June 1852 it was estimated there were 40,000 diggers on the field – an extraordinary number bearing in mind in February 1851 pre-goldrush Melbourne had a total population of only 23,000.

ALLUVIAL GOLD

At first a pan, a pick and a shovel were the only tools needed, and there were extraordinary stories of fortunes made. In one case a single bucketful of dirt yielded 600 ounces (17 kg) of gold; roughly equivalent in value to three year's wages, or 300 acres (120 ha) of freehold land. But many were not so lucky. It is estimated that only a small proportion of the diggers made fortunes and at any given time half

were struggling to survive.

By 1853 alluvial gold was becoming hard to find and this increased dissatisfaction with the mining license system. Bendigo was better administered than Ballarat, however, so although there were some tense stand-offs between protesting miners and the authorities, serious violence was avoided. Nonetheless, the Red Ribbon protests of August 1853 – when Bendigo miners wore a red ribbon to protest against license fees – were a direct precursor to the bloodier events in Ballarat.

The Bendigo diggings were administered from the Commissioners camp which was established in May 1852 on land now occupied largely by Rosalind Park, bounded by Camp St (roughly the location of Pall Mall), Commissioners Gully (now Barnard St), View Point (now View St) and Park Rd. The town grew up in the protective shadow of the camp, particularly around the western entrance and along the banks of the creek below, spilling over onto Bendigo Flat.

By 1854 shops were no longer tents,

there was a local newspaper, a range of churches, schools and the famous Theatre Royal (later the Shamrock Hotel) was in operation. More importantly a surveyor was appointed to plan and survey the town. Camp St, which meandered alongside the creek, was replaced with the thoroughfare that was to become Pall Mall. When the first land sales occurred the next year a single acre (0.4 ha) on Pall Mall sold for £3000!

Lucrative puddling activities continued throughout the 1850s, but thanks to skilled Cornish miners, the first quartz reef mines were also proving to be extremely rich. On Victoria Hill, Christopher Ballerstedt, a German veteran of Waterloo, was mining a reef to unheard of depths (90 metres) and making a fortune.

Quartz Mining

By the 1860s the days of ordinary people getting rich had passed. This ushered in a very different kind of city, one where capital, not sweat, was king. Bendigo's next incarnation was exciting, and it produced vast quantities of gold, but it was based on men working deep underground, very often in appalling conditions that took a terrible toll on their health.

In 1873 the New Chum Railway mine reached the deepest gold mined anywhere in the world at 4,225 feet or nearly 1300 metres. The working conditions at that depth were horrendous. The temperature was over 38°C the humidity approached 100% and silica dust from drilling and blasting was thick in the air. Tuberculosis and lung damage – both of which were very often fatal – were rife. A study carried out in 1906 showed that Bendigo had the highest incidence of lung-related disease in the world.

The Booming 70s

Some of the entrepreneurs who controlled the vast new mines had made their first fortunes on the goldfields and companies were often at least partly owned by local shareholders – so some of the wealth did trickle down. By 1862 the town was sufficiently attractive that Lola Montez the famous courtesan and entertainer even bought a house with the intention of settling permanently.

There was an investment spree in the early 1860s and again in the 1870s. The share market based in the Beehive Building in Pall Mall was at the centre of the speculative frenzy. At its peak the share market was one of the largest exchanges outside London and thousands of people would spill out on the road as they traded their shares.

Money flowed freely and this was reflected in a building spree that still defines the city. The man who takes most credit is Willam Charles (Carl Wilhelm) Vahland. Vahland had been a practising architect in Bremen and Hamburg prior to an unsuccessful stint on the goldfields. He set up an architectural practice with his friend Robert Getzschmann (who died in 1875) and was responsible for many of the city's most famous landmarks.

Vahland

Vahland's career spanned 50 years and his influence extended to the next generation of architects who followed in his footsteps.

He designed and directed the building of major public architecture including the town hall (1885), the hospital, the school of mines, now the Bendigo Regional Institute of TAFE (1887), the mechanics institute, the masonic hall,

Lola Montez, the famous courtesan and entertainer, even bought a house.

Pall Mall from Post Office, PW

now the Capital Theatre (1873), the Cascades and the Alexandra Fountain.

He was responsible for many private homes, from mansions such as Fortuna to simple miners' cottages. During the 1870's Vahland designed a simple cottage for miners with a four-posted verandah. It could be mass-produced and simply erected, and examples can still be seen around Bendigo today. In 1857 he designed and built his own residence at 58 Barkly Terrace.

The churches he designed included the Eaglehawk Wesleyan Church, the Lutheran church, St Liborius Catholic Church at Eaglehawk, St Killian's Catholic Church, and the Anglican, Methodist and Congregational churches in Forest Street. Commercial premises, ranged from small shops to banks and large hotels, including the National Bank (now Mully's Café), the City Family Hotel (1872) and the Gold Mines Hotel at Victoria Hill.

In 1858 Vahland was one of the founding members of the Bendigo Land & Building Society, the institution that eventually grew into the Bendigo Bank. He was also a pioneer in the wine industry. His 300 ha Charterhouse Estate, at Elmore exported wine and vacuum-packed grapes to Germany, but like many others was decimated by phylloxera.

Vahland, who considered himself originally Hanoverian, not German, took British (Australian) citizenship in 1857. Despite this, and his enormous contribution to the city, in 1914 after the outbreak of WWI he was declared an enemy alien, and was forced to surrender his passport and assets. He died in 1915.

Vahland was one of the founding members of the Bendigo Land & Building Society, the institution that eventually grew into the Bendigo Bank.

BACKHAUS & WINE

When it became clear that many diggers were not going to make their fortunes finding gold, a significant number turned to making wine. They were encouraged by Rev Dr Henry Backhaus, a Catholic cleric who played a key role in the growth of the city. He was the first priest on the goldfields and played a major role in developing the hospital, water supply, railway … and the wine industry. Although he lived simply and frugally, during his life he amassed a considerable amount of property. His estate is still administered by Sandhurst Trustees Ltd.

Thanks to Backhaus' encouragement and the involvement of many of his German compatriots, the local wine industry flourished through the 1860s. Unfortunately, phyloxera was discovered in 1877 and the Victorian Government ordered all vines in a 20 mile (32 km) radius to be destroyed.

In 1959 a new vineyard, Balgownie, was established by Stuart Anderson about eight km from Bendigo. There are now dozens of wineries in the region which is particularly famous for its full-bodied reds. Some of the original vineyards have been replanted. At Mandurang, Chateau Dore, which was originally established in 1866 by Jean Theodore deRavin, a Frenchman born on Martinique, has been re-established by his grandson.

LANSELL

For a time, the deeper the quartz mines went, it seems the richer the pay-off. One of those to benefit was George Lansell who bought Ballerstedt mine at Victoria Hill – the first quartz reef to be worked – after Ballerstedt had presumed it was finished. Ultimately the mine, named the Big 180, reached a depth approaching 1000 metres.

Lansell arrived in Bendigo about 1853 initially setting up a butcher's shop before becoming a major mine investor. He made and lost several fortunes and ultimately became one of Bendigo's best-known and most successful entrepreneurs.

Lansell built his famous villa, *Fortuna*, close to the Big 180 – his favourite mine. Designed by Vahland, it was truly palatial. Expensive antiques – including a gold-plated piano – filled the building.

Following the death of his first wife in 1877, Lansell returned to England, remarried and settled in London. This coincided with a downturn in mining that some attributed to his absence, so a petition with 2628 signatures begged him to return. He made a triumphant return to Bendigo and the field began to prosper again, at least for a while. He died in 1906 leaving generous bequests to all his many workers and staff – as well as to all the surviving pioneers of the 1850s. A statue to his memory stands in Pall Mall between the RSL Hall and the post office.

DECLINE & RECOVERY

Quartz mining is expensive and with Victoria's major economic depression in the 1890s higher costs collided with a reduction in the availability of capital. As a result there was a continual decline in gold production until the 1930s when government assistance and higher gold prices underwrote a revival.

The Central Deborah mine was one of many operating in the 1930s period. In 15 years of operation it produced about 30,000 ounces (850.5 kg) of gold before a fixed gold price and high costs

once again saw the demise of mining. On 16 November 1954, the last ore truck was taken to the crushing battery and operations ceased. The nearby North Deborah Gold Mine closed a few weeks afterwards bringing over 100 years of continual gold production in Bendigo came to an end.

Over 500 gold mining companies operated on the Bendigo field from 1854 to 1954. Together their historical gold production was 22¼ million ounces (631 tonnes), making Bendigo the largest gold producer in Victoria, and second only to Kalgoorlie in Western Australia as Australia's greatest gold producer. Today, Bendigo remains among the top 10 gold producing fields in the world and a new mining company, Bendigo Mining Ltd, plans to add a further 12 million ounces (340 tonnes) to the total.

VISITOR INFORMATION

The Bendigo Visitor Information Centre & Interpretative Centre should be the starting point for any exploration of Bendigo. The centre is in the magnificent old post office building in Pall Mall, just past the Alexandra Fountain, and disseminates information on walking tours, guided tours of the town hall, as well as brochures covering arts and entertainment, food and accommodation.

Bendigo Visitor Information Centre
51-67 Pall Mall, Bendigo 3550
www.bendigotourism.com
tel: 5444 4445
freecall: 1800 813 153
email: tourism@bendigo.vic.gov.au.

Train services run to/from Melbourne and buses link to most important Goldfields centres.

Bendigo remains among the top 10 gold producing fields in the world.

Tram Pole Detail, DB
Law Courts, Detail, RE

Bendigo has a reputation for fine food.

FOOD & ACCOMMODATION

Bendigo has a superb range of accommodation to suit every taste and budget. There are luxury boutique hotels in heritage buildings, old hotels (including the magnificent Shamrock Hotel) serviced apartments, B&Bs (again in heritage buildings), self catering cottages, many motels, caravan parks and a backpackers' lodge.

Bendigo has a reputation for fine food, with a number of acclaimed restaurants that feature the outstanding local produce and wines. Again there's a big variety from Asian and Italian to modern Australian. There are some excellent pubs – ranging from restaurant quality to good-value – and there's a great range of coffee shops and cafés.

SIGHTS

Bendigo's city centre has strong echoes of European cities. Outside the centre, however, the rest of the city is resolutely Australian with boom-style mansions on the central hills, rows of brick houses with shady verandahs, and sprinklings of timber miners' cottages. There are few completely fine streets – all is chaotic, the legacy of mining boom-and-bust. The fine buildings at the heart of the city, however, have

a self-conscious opulence that reflects the fabulous wealth that came from beneath their very foundations.

Entering from Melbourne the main road passes through modern suburbs that were in fact early mining sites. A few older buildings survive to tell the story: little redbrick churches and schools, former stables and coach-builders' shops. Soon the most prominent building in Bendigo, Sacred Heart Cathedral, comes into view – first the 87-metre spire, then the impressive façade. On the right is the poppet-head of Central Deborah Mine and a tramline enters the street from the adjacent tram terminus.

At the turn into High St, the view is toward the Alexandra Fountain and the towers and domes of Pall Mall, the heart of Bendigo. The City Family Hotel has a long frontage to the street, facing former insurance buildings and banks on the left. A glimpse up the steep rise of View St towards the site of the government camp reveals an impressive row of classical buildings.

The central view of the city used to be dominated by mining operations, especially by Hustler's Royal Reserve Gold Mine which was just beyond the fountain. The poppet-head and crushing plant were not replaced until 1921 when the RSL hall was built, with its frontage a domed belvedere intended for use as a bandstand.

There is much to see and experience, walking around the streets and viewing close-up the splendid buildings of a past era – look for the intricate craftsmanship in the detailing of buildings. The buildings and their magnificent interiors demonstrate the high seriousness of those who designed and built them, as well as their eye for beauty.

Bazzani's, KS
Interior Colonial Bank (Mully's Café), BeT

To explore Bendigo:

- orient yourself from the Lookout in Rosalind Park

- take a tram ride down Pall Mall

- get in the elevator at Central Deborah and go down a mineshaft into the dark unknown

- Rest in the Chinese Gardens, then wander through the museum next door and look Sun Loong - the largest Imperial Dragon in the world – in the eye

- be dazzled by the golden ornament inside the old town hall

- explore the treasures of the art gallery (and have a coffee in the adjoining café)

- take a walking tour – and detour along McKenzie St so you can look in the backyards of the banks lining View St to see the chimneys of their gold smelters

VIENNA-IN-THE-BUSH

Sophistication came early to Bendigo, arriving with the German architect William Charles Vahland in 1857. Vahland's buildings generally speak a classical architectural language with a baroque or mannerist accent.

Of course, many other architects made contributions to the city, and include names such as Charles Webb (Melbourne's Windsor Hotel) and Joseph Reed (Royal Exhibition Building, Melbourne Town Hall) – Melbourne's best. Reed's partner WB Tappin designed Sacred Heart Cathedral in the 1890s. WG Watson was chief architect in the Office of the Public Works and designed Bendigo's final glorious post office and court house during the 1880s.

There were also local architects. RA Love designed the redbrick Cathedral of St Paul in Myers St – an eccentric Gothic construction. John Beebe built a number of brick and stucco buildings late in the century in a distinctive 'arts & crafts' idiom, such as the YMCA (High St). His brother William designed the old fire station (View St).

What Bendigo still has, which few other towns can boast, are some splendid interiors to match the grand exteriors:

- old town hall, between Hargreaves and Lyttleton Sts

- Supreme Court House, Pall Mall

- Shamrock Hotel, Pall Mall

- Mully's Café Gallery (former National Bank), Pall Mall

- Bendigo Art Gallery, View St

- Wine Bank on View (former Union Bank), View St

- Capital Theatre (former masonic lodge), View St

- Chinese joss house (temple), Emu Point – end of tramline

- Golden Dragon Museum, Gardens & Kuan Yin Temple, Pall Mall & Bridge St

Bendigo still has, splendid interiors to match the grand exteriors.

Bazzani's, KS
Interior Colonial Bank (Mully's Café), BeT

The mine's historic mining equipment is still in working order and buildings such as the blacksmith's shop and miners' change rooms convey the mine's essential character during its working life from 1939 to 1954. The functioning machinery, including a Cornish boiler, winding engine and two compressors, forms a unique industrial collection. The Central Deborah also has a stunning collection of portable artefacts including rare mining stretchers which are still situated in the mine's first aid room.

The mine reopened as a tourist mine over 30 years ago and has won a reputation as a place where people can experience what life was like as a miner in the period from 1930s through to the 1950s.

CENTRAL DEBORAH GOLD MINE

BY ALAN BODDY

The Central Deborah Goldmine provides a glimpse into the amazing history and human endeavour that lies below Bendigo. The mine is now one of the Goldfields' most popular attractions and is a living chapter in the story of gold mining in Australia. Visitors step back in time as they experience the frantic early days of the discovery of gold through to the mining revival of last century.

The main event is a guided tour of the old mine workings. Visitors are introduced to Bendigo's underground world by knowledgeable tour guides. They are 'kitted up' with a hardhat and lamp and then taken down to Level 2 in a modern, spacious, state-of-the-art lift. (There are 17 levels in the original mine which is 422 m deep). Once below the surface visitors see the original shaft, the crib room where the miners ate, various mining drills, the bogger, the use of explosives (safely of course!) – and real gold in the quartz reefs. Those eager to experience more may undertake a three-hour adventure tour to Level 3 of the mine.

SACRED HEART CATHEDRAL

Sacred Heart Cathedral is one of the largest cathedrals in Australia, dominating the entrance to the city centre when travelling from Melbourne. Designed by Victorian-born architect William Brittain Tappin for the Roman Catholic Diocese of Sandhurst, it took over 80 years to complete. The form and details of the cathedral are identifiably Early English Gothic. The entrance and nave, opened in 1901, were in use for more than half a century before increased income from real estate during the 1960s and 1970s allowed the completion of the cathedral. Transepts, presbytery and chapels were built according to original plans over 12 years from 1966, doubling its size. The summit cross of the huge (87 m) central spire was put in place in 1977, making it the third highest in Australia.

Sacred Heart Cathedral, David Preece
Interior Sacred Heart Cathedral, PW
Central Deborah Mine, KS
Central Deborah Mine, PW

PALL MALL

Some of Bendigo's most important commercial businesses have a Pall Mall frontage, but it took nearly 30 years for Pall Mall to assume its present form. Like many goldrush towns, the centre of Bendigo was a chaotic heap of mud in the earliest years. Richard William Larritt was given the task of surveying the township, straightening wagon-tracks, and moving huts and tents aside so roadways could be clearly defined. From 1859 Larritt occupied the Lands & Survey Office in View St, now Dudley House.

Larritt's vision was for the centre of town to be dominated by an avenue with civic buildings backing onto Bendigo Creek. But pressure from council and public for housing lots in the area delayed his plans. He played a waiting game, vindicated when a severe downpour and flood duly proved the unsuitability for housing. The creek was given a wide masonry channel and the banks were built up. Larritt's vision was finally realised in 1882 when the government began planning the present post office and court buildings.

Pall Mall Trams

Bendigo's current tram system is part of a larger system dating from 1890 – steam at first, then electrified in 1903 – that went from Golden Square to Emu Point, and from Eaglehawk to Quarry Hill, crossing at Charing Cross. Today it travels from Central Deborah Terminus to Lake Weeroona via Pall Mall, a pleasant return journey of less than an hour past some of the city's finest sights. Run as a tourist venture by the City Council since 1972, when the system was threatened with closure, the Bendigo tramway is now the home for early tramcars from Melbourne, Adelaide and Sydney. It has some of the oldest and rarest vehicles surviving in the world.

Trams, as they are known in Australia, were originally horse-drawn carriages set on rails. The advent of electricity around 1900 brought about self-propelled vehicles, invented and known in America as streetcars or trolley-cars. A cheap form of railway running along the streets, they soon spread to every city of a certain size. In Victoria, tramway systems operated in Geelong, Ballarat and Bendigo from the earliest years of the 20th century. Melbourne itself had horse-tramways from the 1870s, followed by cable trams (as in San Francisco), until electrically powered vehicles took over all the routes. Apart from Melbourne, Bendigo's are the last trams in Australia to be seen running along the streets.

City Family Hotel

The handsome City Family Hotel was designed by Vahland & Getzschmann for Jean Baptiste Loridan, a local miller and investor. Unfortunately, in the late 1960s an ornate double-storey verandah was removed, along with the balustrade and urns of the distinctive tower. A superb triple-flight corkscrew staircase survives as witness to the original quality of this landmark building. It is actually just before the fountain (south-west) on High St, which becomes Pall Mall.

The Bendigo tramway is now the home for early tramcars from Melbourne, Adelaide and Sydney.

Pall Mall Trams and Beehive Building, KS
Family Hotel, JK

The Bendigo Stock Exchange was for a time the busiest in the Australian colonies.

Princess Alexandra Fountain

The Princess Alexandra Fountain forms a striking element at Charing Cross, where High St intersects with View and Mitchell Sts and Pall Mall begins. The fountain was designed by Vahland and built of polished Harcourt granite. The great bowl is a single piece and the total weighs 20 tonnes. Local contractors provided stonework, lighting and plumbing, while the figures were modelled in marble and bronze by E Temper of Melbourne. It was declared open in 1881 by Princes Albert and George, sons of the Princess and Prince of Wales (later Edward VII).

Beehive Building & Stock Exchange

A handsome three-storey building in Pall Mall facing the RSL hall, the Beehive Building's restrained façade conceals the triple-level arcade behind. Built as the Bendigo Stock Exchange after a fire in 1871, it was designed in Melbourne by Charles Webb, architect for the Windsor Hotel and the Royal Arcade.

The Beehive Stores operated at street level from 1854 to 1987, giving the building its name. The Bendigo Stock Exchange above was for a time the busiest in the Australian colonies, where millions – even billions – of dollars worth (in today's currency) of business in goldmining stocks changed hands. At the peak of speculative fervour, excited crowds would spill out onto the street.

Myers

The Myer Emporium on Pall Mall is an uninspiring 1960s façade, rebuilt after a fire and an extension into neighbouring stores, but this is a very important part of the history of Myer. In the late 1890s Simcha Baevsky, who later changed his name to Sydney Myer, arrived in Bendigo from Russia. He hawked merchandise around the region in heavy suitcases until he could open his first drapery store in View St. Other stores soon followed in Hargreaves St, Pall Mall, then, of course, Melbourne and many other cities and towns.

Colonial Bank (Mully's Café)

Erected in 1887 for the Colonial Bank, this small yet palatial building was designed by Vahland on the model of one of the stone tombs of Petra, in Jordan. There are ground level Ionic columns, first storey Corinthian columns, and an attic storey with female Terms supporting exaggerated cornices with huge urns and a broken pediment. Altogether, it is one of the most sensational façades in Australia.

The interior does not disappoint – a high (9.5 metres) coffered ceiling with light-access panels at the centre, and elaborate pilasters on the walls. There is a stairway to the upper-level balcony overlooking Pall Mall. Today it is Mully's Café, a busy eatery with some pictures for sale.

Post Office & Law Courts

Bendigo's centrepiece is the pair of civic buildings standing prominently on Pall Mall's broad avenue. Built of brick, with cement-stucco facing and ornamentation, in a French Renaissance style, double-storied with high Mansard roofs and truncated domes at the centre, they appear identical. But the

Princess Alexandra Fountain, RE
Stock Exchange, Beehive Building, PW
Beehive Building, PW

once – at the request of Dame Nellie Melba who was trying to sleep in the Shamrock Hotel opposite.

Entry to the law courts is through the arched loggia where notices are displayed. A side entrance leads to a small vestibule, from where a glimpse is gained of the first of two superb double flights of stairs. Tall arcades rise on three sides of the stairway. All is finely detailed with Classical Corinthian ornament, creating one of the grandest public spaces in Australia. The supreme court is of a similar scale, with matching detail.

post office (opened in 1887) is slightly larger and has a clock tower, while the court house (1896) has an extra storey at the centre above its vestibule and the supreme court.

Like the town hall, which they emulate, each building stands monumentally entire, to be viewed from all four sides. Raised on bluestone podiums, they each have a deep basement level. The post office basement was for safe handling of precious cargoes, while the one at the court house was for the confinement of prisoners. Solid basalt and cast-iron fences surround the buildings, enabling light and ventilation to reach the basements. These fences are punctuated by bluestone corner-piers with elaborate light stands supported by chimeras.

The post office is now home to the Visitor Information Centre and an exhibition, *Making a Nation* which explores Bendigo's role in Australia's Federation. The town clock chimes quarters and hours, to a curious sequence unique to Bendigo. The chimes have been turned off only

Shamrock Hotel

The Shamrock Hotel is deservedly famous as one of the grandest hotels in Victoria. Standing opposite the post office tower, it complements the style of the civic buildings with its French Second Empire Mansard roofs. There is an international flavour in its High-Victorian assertiveness, although it was designed in Bendigo in 1897 by Philip Kennedy, a pupil of Vahland.

Standing four storeys high plus attics, the Shamrock has an ornate verandah for the two lower levels and the bravado style of a European or American resort hotel of the late 19th century. The entrance is through Italianate arches on curly cast-iron columns, where the magnificently tiled floor leads the eye

Post Office, BeT
Interior Law Courts, PW
Shamrock Hotel, KS
Interior Shamrock Hotel, PW

to an ornate staircase with wonderful stained glass. No expense was spared.

The Shamrock Hotel has played an important role in Bendigo's social and cultural life and has hosted a range of important visitors – including Prince George (son of King Edward VII), Dame Nellie Melba and Donald Bradman.

The first building on the site was a restaurant and music hall started in 1852. When it sold two years later to two Irishmen, the name changed from Theatre Royal to The Shamrock. From 1859 it was the office for the Cobb & Co stage coach and telegraph company.

School Of Mines

The tower of the School of Mines is the third prominent landmark of the city centre, after the post office and town hall towers, about 250 m east of the Shamrock on Pall Mall. The impressive group of cement-stucco buildings reads as two distinct parts, the main entrance in a two-storey façade of 1879 with a central arcaded loggia, and the three-storey block with tower of a decade later. The earlier section is soberly classical, while the later is an example of Vahland's German style, where refined classical elements are enriched by elaborate friezes.

In 1887 an unusual Octagonal Library was added behind the entrance front. It has a centralised form with the gallery following the classical orders – base level Tuscan Doric, first floor Corinthian – with enriched cornices and a shallow dome. This is an elegantly beautiful interior retaining its original gaslight-fitting. It is one of Vahland's most surprising interiors, now used by hospitality and catering students.

St Kilian's Church

St Kilian's Church looks rather plain and austere from the outside, but it is particularly beautiful inside. The first Catholic mass was heard here in 1852 by the Rev Dr Henry Backhaus, whose canny investments ultimately enabled the construction of the Sacred Heart Cathedral. This area was originally the largest red light area in the district. A stone church was built in 1857, but the ground was unstable and it was pulled down and replaced with the lighter wooden structure that survives today. The interior woodwork is magnificent. The church is about 400 m east of the information centre on McCrae St, as the continuation of Pall Mall is known.

Old Town Hall

There was already a municipal chambers from 1859 on site when, in 1885, Vahland began his greatest work: the main hall and all four facades of the town hall. Completely enclosing the earlier structure, the building has a two-storey classical scheme. There are paired Tuscan columns on the lower storey, Corinthian above, and open pediments alternately angled or curved. Everywhere, there are large arched windows.

Each corner rises into a truncated

Old Town Hall, KS
Interior Old Town Hall, BeT
School of Mines, KS
Interior, St Kilian's, PW

pyramid with attic windows, except the north-west corner where there is a substantial tower 36.5 metres high. Muscled Atlas figures crouch to take the weight of the clock.

Nothing can prepare the visitor for the impact of the hall within the building. There is a vast amount of gold, and original murals of mythical female figures and cherubs and other decorations by German artist, Otto Waschatz (who also decorated the royal palace in Copenhagen, Denmark). There are gilt plaster cherubs above each doorway, a coffered ceiling with gilt pendants, and Gothic arches above each window. It is overwhelmingly beautiful, and expresses in an instant the fabulous wealth of the gold boom. Tours are a must; make inquiries at the Visitor Information Centre. The town hall is a short walk from the information centre, at the end of Bull St, which is opposite the law courts.

GOLDEN DRAGON MUSEUM

There were more than 4000 thousand Chinese miners in Bendigo at the height of the goldrush. The role and popularity of the Golden Dragon Museum signals the city's recognition of the importance of the Chinese community. The museum and associated Chinese Gardens and Kuan Yin Temple have been created since 1991 on the site of old Chinatown in

Bridge St, with a handsome entrance gate from Pall Mall.

The ceremonial dragons have long been a part of street processions in the city, especially the Easter Festival (a hospital fundraising event dating from 1871). These dragons are spectacularly displayed in the circular museum – Sun Loong taking several revolutions of the building to contain his great length (over 100 metres).

Bendigo and the Easter Festival have been a long-standing focus for Chinese families throughout Victoria, bringing large numbers who participate in the street processions with the two Imperial Dragons and attendant figures. The splendour and intricacy of these ceremonial figures and masks can be seen up close in the museum.

The museum is within easy walking distance of the information centre; heading east, veer left off Pall Mall down Bridge St, past the historic conservatory; the museum is on the left.

BUSH'S PRODUCE STORE

Bush's Store, at 94 Williamson St, was built in stages between 1857 and 1890. It is a remarkably intact 19th century store, notable for the integrity of its interior and remaining machinery. The information centre is on the corner of Williamson St and Pall Mall; cross Pall Mall and it's a short walk to the store through the busy commercial centre.

VIEW ST

View St once formed the western boundary of the government camp and rises steeply to the north of Alexandra Fountain. Almost every building has its own points of interest giving the street a rare historical coherence. Lined with

The interior of the town hall is overwhelmingly beautiful.

Golden Dragon Museum, KS

Williams, Penleigh Boyd and Grace Cossington Smith gives strength to the outstanding Australian collection.

The 19th century European collection includes painting and sculpture. The British works give a sense of the sentimental culture of the late 19th century.

Walter Withers, Rupert Bunny, Grace Cossington Smith, Rah Fizelle and Margaret Preston are among the many major 20th century Australian painters and sculptors represented in the collection. The gallery has recently acquired the series of sculptures made by Patricia Piccinini for the Australian Pavilion in the 2003 Venice Biennale.

former banks and insurance houses, it was for 100 years the financial heart of the city. The National Bank is the only survivor in its original premises. The former post office is now Sandhurst Trustees; then in order up the hill are the Temperance Hall, Penfolds Gallery (a former hotel), and the Trades Hall.

The Bendigo Art Gallery stands newly revealed, with its entrance and extension in the shade of the magnificent Capital Theatre. Opposite are government offices, occupying the former site of Bendigo's Princess Theatre.

Bendigo Art Gallery

Bendigo Art Gallery, founded in 1887, is one of Australia's oldest galleries. The collection includes important art from the Bendigo goldfields along with 19th century European paintings, sculptures and decorative arts.

These works are housed in a beautiful series of restored 19th century spaces with polished wood floors, ornate plaster arches and cornices, and diffused natural sky-lighting through rooftop lantern towers. There's also a new contemporary wing, including an excellent restaurant overlooking the park. The contemporary building allows flexible spaces for travelling exhibitions.

Work by ST Gill, Thomas Ham, Louis Buvelot, Eugene von Guerard, Walter Withers, Rupert Bunny, Sir William Dobell, Tom Roberts, Arthur Streeton, Hans Heysen, Arthur Boyd, Fred

View St, KS
Temperance Hall, RE
Art Gallery, John Gollings courtesy
   Bendigo Art Gallery
Union Bank (Wine Bank on View), KS
Interior Union Bank (Wine Bank on
   View), KS

Union Bank (Wine Bank on View)

Built in 1875, the former Union Bank on View St is one of the most striking Classical fronts in Bendigo. It is now a wine and beer bar with a range of local and imported wines. The small vestibule leads to a magnificent chamber roofed internally as a square dome. There is little ornament except for a hugely glorious rose at the centre of the ceiling and a fireplace at one side. The effect is in the proportions.

At the rear is a two-storey accommodation block for single members of bank staff, possibly as a deterrent against robbery. Beyond is the original smelter house with its chimney. Banks on the goldfield used to smelt gold into ingots as a secure means of preparing it for transportation to Melbourne. The miner's weight of gold dust or nuggets was first recorded in an inventory. This property retains all the parts of a working bank on the goldfield, and is a rare survival.

Capital Theatre

Built in 1875 as the masonic hall to the design of Vahland & Getzschmann, the Capital Theatre was saved from threatened demolition in 1990 and gradually refashioned into Bendigo's Performing Arts Centre.

There is a Baroque ripeness in its classical manner. The Corinthian portico – the most imposing in Victoria – is set halfway into the façade of the building. It is all in stuccoed brick, resembling stone. From the top of the rise, the Capital Theatre dominates View St. Inside, the elegant foyer leads to a stair hall, left of which is the banquet room and the very beautiful (though not large) Bendigo Bank Theatre, formerly the masonic room.

Upstairs the theatre was originally a

ballroom and occupies the full frontage of the building. Vahland was asked in 1890 to add a proscenium and make a stage using the building next door. The auditorium has an 18th century German-Bavarian character.

Dudley House

At the top of the hill, Dudley House was opened in 1859 and is the oldest surviving government building in Bendigo. It was used by Larritt, the city's first surveyor.

BENDIGO BENEVOLENT ASYLUM (ANNE CAUDLE CENTRE)

Once you reach the top of the View St hill, you can continue a walk around the boundary of the old government camp. The northern half is taken up by sporting facilities, but if you continue down Barnard St, you'll see some fine Victorian housing and eventually come to the majestic Anne Caudle Centre, which dates from 1859, but owes most of its design to Vahland. It was the original beneficiary of the fundraising activities centred on the Easter Festival and was originally intended to care for the poor and for women in childbirth. It is now amalgamated with the Bendigo Health Care Group.

Interior Capital Theatre, PW
Capital Theatre, KS
Dudley House, KS
Bendigo Benevolent Society (Anne Caudle Centre), PW

GAOL RD

Gaol Rd, behind the Rosalind Hill Lookout is home to Camp Hill Primary School (now Bendigo Senior Secondary College), the police barracks and Bendigo gaol The impressive school was built in 1877, and its tower was used as a fire lookout; signals were relayed to the old fire station in View St. The police barracks were completed in 1860, the gaol in 1861. The barracks are one of the city's oldest buildings.

AROUND BENDIGO

BENDIGO GAS WORKS

The gas works are on Weeroona Avenue, near the lake and the tram terminus. Established in 1859 they ran until 1973 when natural gas arrived in town, ending coal-gas operations. The plant is maintained in working order, waiting for the day when the site is developed to allow visitors to explore coal gasification and the provision of power and light in the early industrial age. It is now the only surviving 19th century coal gas plant.

JOSS HOUSE

The joss house at Emu Point (at the end of tramline past Lake Weeroona) is a survivor from the 1860s. It is in three sections - the temple, a caretaker's residence and an ancestral hall. The temple is for Chit Kung Tang, a god of war and one who also brings peace and prosperity. Red and gold, the colours of happiness, are the principal colours of the temple's interior. The complex has been restored by the National Trust, advised by a Chinese historian through the auspices of the Bendigo Chinese Association.

BENDIGO POTTERY

Bendigo Pottery is one of the most interesting survivors of the many manufacturing industries that sprang up to service the goldfields. It was established in 1857 by George Duncan Guthrie and now has one of the most significant collections of historic kilns in the world. An interpretive museum has been built around the old kilns and there are fascinating displays of equipment and products. Over the years the pottery has made everything from teapots to tiles, from bottles to massive industrial pipes – everything that can be made from clay. Today the pottery continues to produce all sorts of domestic pottery in a variety of styles. It is just off the Midland Hwy (signposted to Echuca) at Epsom, 6.5 km north of the famous Alexandra Fountain in the centre of the city.

Camp Hill Primary School (Bendigo Senior Secondary College), PW
Bendigo Gas Works, KS
Interior Bendigo Pottery, BeT
Interior, Joss House, PW
Bendigo Pottery, RE

VICTORIA HILL HISTORIC RESERVE

Victoria Hill is the symbolic heart of Bendigo's quartz mining, and was one of the richest and busiest areas on the goldfield – thanks to Chistopher Ballerstedt.

As a teenager Ballerstedt was drafted into the Prussian army and fought at the Battle of Waterloo in 1815 when Napoleon was finally defeated. Like many others he arrived at the goldfields with his possessions in a wheelbarrow. In 1854 Ballerstedt bought a claim on Victoria Hill. After working the quartz reef from an open-cut mine he sank a shaft 90 m and struck gold. In the 1860s he was known as the father of quartz-reefing and was one of the richest men in the state.

Believing the mine had bottomed, he sold out to Lansell who proceeded to prove him wrong. A number of exceptionally rich mines operated on the hill, and in 1910 the Victoria Quartz became the deepest mine in the world, reaching a depth of 1400 metres.

Today it is possible to wander around a peacefully deserted Victoria Hill, through the original open-cut mine to a poppet head you can climb, past the remains of a crushing battery and many other signs of intense mining activity. There is interpretive signage. The reserve is on Marong Rd (the Calder Highway signposted to Mildura) a 10 minute drive from the city centre. The beautiful Gold Mines Hotel is just across the road from the reserve.

BENDIGO'S CEMETERIES

There are four significant cemeteries around Bendigo: Bendigo (Carpenter St), Eaglehawk (Grenfell St, adjacent to Neangar Memorial Park), Kangaroo

Flat (Helm St), and White Hills. All four are picturesque and fascinating, giving a sometimes horrifying insight into life and death on the goldfields. All four are divided by religious denomination, and the White Hills Cemetery is notable for its Chinese Burning Tower and burial ground. The Bendigo Cemetery has an unusual Gothic Mortuary Chapel and a major monument in memory of the explorers Burke and Wills.

EAGLEHAWK

Turning into the Loddon Valley Highway a few minutes north-west of Bendigo, you might think Eaglehawk was merely an outer suburb of a sprawling regional city. Indeed, since the 1990's, this elegant goldmining town has been a part of larger Bendigo. Despite amalgamation, however, the once independent borough still thrives on its strong working-class local identity and an individual goldmining past.

In 1852, whilst searching for stray horses, Joseph Crook found a nugget that began a new goldrush. The town's name is owed to an eaglehawk flying near his claim.

By 1853, the alluvial gold was exhausted, but quartz mining continued. As was the case elsewhere in the region, working conditions for the miners were appalling and many died of respiratory disease at an early age. A local doctor, Stewart Cohen, organised a movement against the Bendigo Mine Owners Association, seeking to improve ventilation and other working conditions.

Novelist Rolf Bolderwood had a farm to the north of town and owned the local butcher's shop. His novel, *The Sphinx of Eaglehawk* (1895) draws on first-hand knowledge. The town is

Victoria Hill, PW
Gold Mines Hotel, Victoria Hill, RE
Ballerstedt's Open Cut, Victoria Hill, RE
Chinese Burning Tower, Bendigo
   Cemetery, DB
Burke and Wills Monument, Bendigo
   Cemetery, DB

Adjacent to the court house is the old log lock-up which was built around 1855 for prisoners and those awaiting trial. It is constructed of rough-hewn logs stacked and crossed at the corners. The museum can arrange admission.

Canterbury Park Gardens

also mentioned in one of A.B. 'Banjo' Paterson's better-known poems, *Mulga Bill from Eaglehawk* (1902) who 'caught the cycling craze'.

Eaglehawk, centred mainly around High St, Spring Gully Rd and Peg Leg Rd., is a conglomerate of lovingly preserved timber and stone cottages and grand public buildings, like the ornate town hall fronted by historic cannons.

Brassey Square

The opulent town hall, flanked by the former post office and mechanics' institute in Brassey Square, is the civic heart of old Eaglehawk. It was built in 1901 on the site of an earlier town hall. An eaglehawk is motif is used in one of the building's stained-glass windows. It is one of the most notable and dramatically sited of all Victorian provincial town halls and is almost identical to the Castlemaine Town Hall.

Eaglehawk Court House Museum & Log Lock-Up

Behind the town hall is the redbrick court house, which contains a display of local history, particularly photographs.

Transformed from a denuded ruin of alluvial gold workings adjacent to Napier St, today this recreational complex includes botanic gardens with a lake, a leisure centre with sporting facilities, picnic grounds and a wetland haven for waterfowl. At the far end of the park, Lake Neangar is named after the local Aboriginal tribe. The area has some good examples of well-kept miners' cottages.

Municipal Offices

The municipal offices are in a superb building built in the 1890s as a private residence.

St Liborius

Liborius was the patron saint of Paderborn in Germany, the birthplace of Dr Henry Backhaus who laid the foundation stone. The church was designed by Vahland.

Harveytown

In Clarke St you will find four sandstone cottages built around 1875 by Cornish stonemasons as part of a significant Cornish mining community.

Eaglehawk Town Hall, KS
Eaglehawk Log Lock-up, KS
St Liborius, KS
Canterbury Park Gardens, KS

Greater Bendigo National Park

The Greater Bendigo National Park and the Bendigo Regional Park virtually surround the city. Stately Red Ironbarks with deeply furrowed trunks stand tall on the ridge tops. Further down the hills are Yellow Gum, Grey Box and Red Box. The brilliant winter-flowering Whirrakee Wattle is unique to the Bendigo area, and there are spectacular spring wildflowers. In the northern sections of the park, especially the former Kamarooka and Whipstick state parks, low mallee trees, including the rare Kamarooka Mallee dominate.

The Aboriginal association with the parks dates back thousands of years and there are also many relics relating to gold mining and eucalyptus oil production. There are alluvial diggings, old mining dams and water races, shafts and mullock heaps. The eucalyptus oil industry dates to the 1860s and old stills can be seen in the area; most of the picnic sites have been built on old eucalyptus distillery and homestead sites.

Mining generated a huge demand for timber so the forest was extensively cleared. Fortunately many trees survived or regenerated so today's forest is the result of 150 years of coppice regrowth. There is reputed to be one of the greatest concentrations of songbirds in Australia – including the Grey Shrike-thrush, the Crested Bellbird and Gilbert's Whistler. Over 170 bird species have been recorded. The animals most likely to be seen are the Eastern Grey Kangaroo, Black Wallaby, Echidna and Common Dunnart.

Shadbold Picnic Area

There are a number of walks from the Shadbolt Picnic Area, north of the city, to the viewing area at Flagstaff Hill (1 km) and to Old Tom Mine (5.3 km return). The Old Tom Mine walk takes in relics from alluvial mining and eucalyptus oil production.

One Tree Hill

One Tree Hill, to the south of the city centre, has panoramic views over Bendigo. The remains of the One Tree Hill Pioneer Mine, which was sunk to a depth of 90 metres, are 600 metres northeast of the lookout tower. Edwards Rd takes you to the lookout and a two km walking track.

Diamond Hill Historic Area

Diamond Hill Historic Area also has many relics from mining days: mullock heaps, water races, foundations, shafts, and remnant gardens. Diamond Hill was once home to 8000 people, many of German descent. They were initially attracted to a gold-bearing quartz outcrop, now long gone. Excavations under the hill created a cavern known as Aladdin's Cave. Diamond Hill is five km south of the city centre near to the new Bendigo Mining operations. The major access roads are Diamond Hill Rd, Burns St and Kangaroo Gully Rd.

Bendigo Bushland Trail

The Bendigo Bushland Trail has 65 km of walking-cycling trails which link many of the city's parks and reserves, as well as taking in the surrounding bushland. A map is available at the visitor information centre.

Bark, PV (AP)
Aboriginal Grinding Stone, PV (AP)
Bark, PV (AP)

CASTLEMAINE

LOCATION: 119 KM FROM MELBOURNE (CALDER FWY, THEN PYRENEES HWY)
POPULATION 1852: 30,000
POPULATION 2006: 7000

Castlemaine is the golden city at the centre of the Mt Alexander goldfield. It was built at the strategic intersection of several valleys where tonnes of gold were found, literally just below the surface. Today Castlemaine and the surrounding valleys and hills give an extraordinary insight into the hectic era of alluvial goldrushes 150 years ago.

Telegraph Office, Mostyn St, KS
Oriental Bank, Mostyn St, KS
Cottages, Campbells Creek, KS

From 1852 this area was overrun by miners turning over the top two metres of soil of the richest shallow alluvial goldfield ever discovered. Initially known as the Forest Creek Diggings, then more generally as the Mt Alexander Goldfield, the population of this region soon exceeded that of Melbourne.

Many of the buildings that characterise Castlemaine and the neighbouring villages were built during the next 20 years, for after that time the population dwindled as the surface gold gave out – from a peak of 30,000 to about 7,500.

Unlike Ballarat and Bendigo, Castlemaine did not have a major follow-up boom in deep-lead or quartz mining. As a result it is much smaller, and it does not have such an ostentatious architectural legacy. Nearby Maldon does have a long history of quartz mining, but Castlemaine has remained the commercial and administrative heart of the district. It is a substantial regional city with a strong sense of its own history and importance.

Castlemaine is rich in historical buildings, but the scale and style of the city is more modest and egalitarian than its two big relatives. There are some fine civic buildings but only a couple are more than two storeys high, and the streetscapes tend to sprawl. There are some fine villas tucked away, but no boom-style mansions.

The population – approximately 7000 – is around that magic number where everyone knows everyone, but it is also large enough to encompass a diverse range of people and an energetic cultural life. Alluvial gold has been described as the democratic mineral, and some of that democratic temper has survived. With its large population of hot-rod enthusiasts, artists, writers and musicians, Castlemaine is more an (ancient) Athens-in-the-bush than a Vienna-in-the-bush (as Bendigo is sometimes described).

The city and the nearby townships are closely integrated with surrounding forests that are rich in mining relics – from the first days of small-scale panning to the last days of widespread sluicing. To safeguard this heritage, much of the bush to the east of Castlemaine is protected by the unique Castlemaine Diggings National Heritage Park.

BRIEF HISTORY

Major Mitchell was the first European to appreciate Castlemaine's wide, grassy valleys with their crystal-clear streams and red gums – but he missed the gold that lay centimetres under his feet. The Major and his party 'overturned a cart through bad driving' and camped in a ravine on 29 September 1836 very close to where gold was first discovered. He named the ravine Expedition Pass – today the site for a tranquil reservoir.

Squatters soon moved in to take up the land and, in on July 1851, a shepherd found gold at Barkers Creek. A small group of workers from Dr Barker's Mt Alexander run managed to keep the discovery secret, but it was finally made public in the Melbourne *Argus* on 8 September. Within eight weeks it is estimated there were 8000 men at work along Barkers, Forest and Campbells creeks. Ballarat (which was publicised a month earlier in August) was almost deserted, and hundreds more arrived from Melbourne every day.

The first government camp was at Deadman's Gully, Golden Point, half way between Chewton and Faraday. The diggings spread rapidly down the creek and at the end of October the camp was relocated near the site later occupied by the Mt Alexander Hotel. Chewton was the first township on the Mt Alexander Goldfield, but the commissioner's camp continued to move: to Moonlight and Pennyweight Flats and lastly to Castlemaine.

The Castlemaine Camp was built near the junction of Barkers and Forest creeks, to the south west of today's commercial centre. Only a couple of original buildings remain; the old court house is on Goldsmith Crescent.

Castlemaine is more an (ancient) Athens-in-the-bush than a Vienna-in-the-bush (as Bendigo is sometimes described).

American Dream, Castlemaine, KS
Forest Creek, ST Gill, Courtesy the
    Sovereign Hill Museums Association
Cottage Detail, KS

Bond Store (Antique Centre), KS
Thompson's Foundry (Thompson's Kelly & Lewis), DB
Cottage, KS
Lyttleton St, KS

By March 1852, there were an extraordinary 25,000 people on the goldfield. They turned the place into something that looked like the surface of the moon. It was a far cry from today's wide, tree-lined streets, fine historic buildings and well-kept gardens. During the first 10 years, the field yielded an estimated total of 3.5 million ounces (110 tonnes) of gold.

Mining was at first an individual exercise, each miner working his small claim – or pooling efforts with his mates to dig and wash more effectively. The riches were extraordinary, and were soon famous around the world: claims eight feet (2.4 m) square yielded 3,600 ounces (112 kg), which at over £3 per ounce was more than enough to set a man up for life.

It's hard to imagine the intensity of life on the diggings. The degree of deprivation, the sense of adventure, the hatred of the police and soldiers trying to enforce the iniquitous gold licenses, the profound despair when a hole gave up nothing but dirt and the wild excitement when gold was discovered….

The pattern changed when large companies entered the field during the 1860s, sluicing areas of hillside and gully and filtering the residue mechanically. As this happened many traces of workings from the first decade disappeared, and many early buildings were demolished and recycled.

From 1874, water was reticulated from Malmsbury to help gold-sluicing operations, which continued until the mid-20th century. The frantic search for gold changed both the landscape and the surface soil. Almost all the stony and clay soil – where gold was likely to be found – was washed from the surrounding ranges.

Today light box and ironbark forests grow on depleted hillsides, with undergrowth composed of hardy native varieties and persistent exotic weeds and herbs. The trees are mostly scrubby regrowth from the roots of larger trees that were cut down to meet the needs of the miners, or the local foundries. The scars and regeneration are still visible today.

MANUFACTURING INDUSTRY

From the late 1850s Castlemaine also developed its own manufacturing industry. The oldest surviving brand is Castlemaine Rock, a distinctive hard candy, made continuously by the Barnes family since 1853. Perhaps the most famous brand, however, now lives on as the beer Queenslanders consider their own: Castlemaine XXXX. In 1859 two Irish brothers, the Fitzgeralds, founded the Castlemaine Brewery. They expanded to Brisbane in 1877 (taking their recipes with them) and, although their business is now part of Lion Nathan, their famous brand survives.

More significant as a long-term contributor to the economy of the town is Thompson's Foundry, which dates back to 1875. There have been a number of important foundries in the town, including Vivian's and Horwood's Albion, but Thompson's survives to the present day, now known as Thompsons Kelly & Lewis. The foundry has produced everything from locomotives to the main towers for Melbourne's West Gate Bridge.

The largest employer today is also over 100 years old: KR Castlemaine (a huge bacon and smallgoods manufacturer) was founded in 1905.

VISITOR INFORMATION

The Castlemaine Visitor Information Centre is in the city's most striking building – the old Market Building looks as if it has been scooped up from the shores of the Aegean Sea. There are often exhibitions, and visitor services include free planning advice and accommodation bookings. There are maps and brochures that are particularly useful if you plan to explore the city or the national park.

Castlemaine Visitor Information Centre
Market Building
44 Mostyn St Castlemaine, 3450
tel: 5470 6200
freecall: 1800 171 888
email: visitors@mountalexander.vic.gov.au

Train services run to/from Melbourne and buses link to most important goldfields centres.

FOOD & ACCOMMODATION

Castlemaine has a good variety of accommodation, including motels, caravan parks and an excellent selection of B&Bs, ranging from homely to stylish and luxurious. There are a number of traditional country-style pubs with counter meals and, especially during the day, some excellent cafés serving meals and good coffee.

SIGHTS

Travelling from the Calder Fwy, the road descends quickly through a winding series of gullies. At peaceful Chewton, miners' cottages, chapels and shops hug the road, with creek flats behind. Approaching from Daylesford, the experience is similar as you wind your way through straggling Campbells Creek.

Castlemaine's skyline is defined by Gothic gables and spires. They appear on local churches, but the Gothic style extends to the earliest public schools and many local houses – here and there a steep gable or frilly barge-board peering across a garden fence.

To orientate yourself, do a quick loop around the city:

1. start at the Market Building and Visitor Information Centre

2. head to the Burke & Wills Monument at the eastern end of Mostyn St, with its great views (Burke lived and worked in Castlemaine)

3. drive to the western end of Mostyn St (note Christ Church on Agitation Hill)

4. turn left, then first right into Forest St, past Goldsmith Crescent opposite Camp Reserve (the site of the Government Camp)

5. turn right into Bowden St, past the gaol (which has good views back over the railway station and the town)

6. turn right into Parker St (past the Botanic Gardens and Thompson's Foundry)

7. and last, turn right into Hargraves St which will bring you back into the centre of town for a cup of coffee

Mostyn St, KS
Barker St, KS
Burke and Wills Monument, DB

The Castlemaine Market Building, which opened in 1862, is one of Australia's finest heritage buildings.

(note the old Savings Bank and court house between Templeton and Lyttleton streets)

To explore Castlemaine:

- wander around the centre on foot (don't miss the Restorers' Barn in Mostyn St)

- visit *Buda*, a historic home and garden

- take in the Australian collection at the Castlemaine Art Gallery & Historical Museum

- visit *Tute's Cottage*

- explore the Castlemaine Diggings Heritage National Park, especially Pennyweight Flat Children's Cemetery, Eureka Reef, and Garfield Water Wheel

TOWN CENTRE

Castlemaine Market Building

The Castlemaine Market Building, which opened in 1862, is one of Australia's finest heritage buildings.

During the goldrush days, Castlemaine's market square was crowded with people, tents, carts, horses, fruit, vegetables and every conceivable product and tool. There were also several large trees where people could pin messages.

Castlemaine quickly became a commercial and administrative centre and two market halls were built in 1858. The third market hall, designed by W Downe, is the one that survives today.

The design of the north front is unique in Australia. The entry to the building is a grand Doric portico attended by domed corner-pavilions. Appropriately, at its peak there is a statue of the Roman harvest goddess, Ceres, holding a cornucopia (or horn of plenty). The arched main entry has ornate cast iron gates that were cast locally.

The interior takes the form of a Roman Basilica with a clerestory roof; the market stalls were located in the side halls. Many local materials were used in the construction: sandstone in the base, and the typical red bricks of Castlemaine.

Victory Park

Victory Park, on the corner of Barker and Mostyn Sts, has a remarkable

Castlemaine Market Building, Geoff Hocking
Fountain, Victory Park, KS

bluestone drinking fountain. Under the canopy there is a drinking fountain for people, nearest the corner there's a trough for horses and at the sides there are lower troughs for dogs! The design was by Thomas Fisher Levick, an art teacher at Kyneton and Castlemaine. The fountain was built as a memorial to Sir James Patterson, who went from Mayor of Chewton to Victorian State Premier (1893 to 1894).

Barker St

Several handsome bank buildings face the park. Perhaps the best is the former **Oriental Bank Chambers** (now solicitors' offices) in Barker St – a redbrick two-storey structure with exquisite proportions and Georgian detail.

Further to the south, also on the west side, a large three-storey brick warehouse behind a petrol station is one of the few surviving industrial buildings from the 1850s. It has superb double-arched windows. Built as flour-mill in 1857, it has been the site for a range of industrial activities, becoming Cornish & Bruce's Railway Foundry in 1859. A quartz-crushing plant was added at the rear for miners to bring in their raw materials, and the Fitzgerald's Castlemaine Distillery also used the building for a time.

Telegraph Office

The electric telegraph reached Castlemaine in 1857. The telegraph office is solidly constructed of local sandstone, and is just up Barker St from the post office.

Theatre Royal

The Theatre Royal in Hargraves St is obviously the local cinema – its Art Deco front gives it away. But it's more than that – it's the oldest continuously-operating theatre on the Australian

mainland. Its doors opened in 1855 when it was surrounded by tents and muddy diggings. The famous temptress Lola Montez performed her dances here in 1856, and many other notables 'walked the boards', including Dame Nellie Melba.

Photographs from the early days show the theatre has been renovated at least three times. The late 19th century interior had a horseshoe balcony, deep arched proscenium and a pretty, flowery ceiling. Its present appearance is the result of a complete makeover in 1937.

Lyttleton St

The finest group of buildings in Castlemaine form the administrative heart in Lyttleton St, between Barker and Hargraves Sts. This comprises a row stretching from the 1877 classical court house westwards to the Post Office. The 1889 school of mines façade displays subtle classical motifs and confidence of proportion – the hallmark of its designer, Bendigo architect WC Vahland.

However, Castlemaine's great centrepiece is the **post office** of 1875; designed by James J Clark, it is a classical building that ranks with the best in the state. Clark was also responsible for Government House, Melbourne, the Maryborough Railway Station and the Melbourne City Baths.

Next to it is the old **drill hall**, solidly timber-built and handsome. The **town hall** (1898) has been improved by recent restoration, but in such distinguished company it seems less convincing than its neighbours.

Across the street are two buildings familiar to viewers of the TV programme, Blue Heelers: the 1960s blonde-brick police station and the former **Beck's Imperial Hotel**

Banks, Barker St, KS
Telegraph Office, Barker St, KS
Post Office, KS
Town Hall, KS
Imperial Hotel, KS

with its extravagant balconies and mansard roof. Built in 1861, the hotel stands out as a very early use of the French Second Empire Style, a style associated with the opulence of Paris and particularly popular for hotels and public buildings in America. The Imperial was a predecessor to the larger Second Empire buildings in Bendigo, including the Shamrock Hotel and the town hall.

Restorer' Barn

The Restorers' Barn at the eastern end of Mostyn St is an extraordinary treasure trove of junk, collectibles and antiques – and is much bigger than it appears from the street.

AROUND THE TOWN CENTRE

Castlemaine Art Gallery & Historical Museum

The Castlemaine Art Gallery, at the western end of Lyttleton St, has focussed on Australian art since it was founded in 1913. It has a handsome Art Deco façade designed by Percy Meldrum with sculptured figures in the panel above the entrance, and it houses an impressive permanent collection, particularly of the Australian Impressionists. There are representative works by artists such as Louis Buvelot, Fred McCubbin, Tom

Roberts and Arthur Streeton. Artists of the early 20th century include Walter Withers, Rupert Bunny and the Lindsay brothers. More recent artists include Russell Drysdale, Fred Williams, John Brack, Clifton Pugh, Albert Tucker, Lloyd Rees, Margaret Preston, and Roger Kemp.

In 1990 the gallery began collecting photographs of Australian artists by Australian photographers and holds works by photographers such as Max Dupain, Olive Cotton and May Moore.

The **Castlemaine Historical Museum** is housed in the lower level of the gallery building. It offers a significant collection of historical artworks, photographs, costumes, decorative art objects and artefacts relating to the history of the Mt Alexander district. Much of the Gallery's fine collections of porcelain and gold- and silverware are also on show in the museum area.

Lyttleton St churches

The former Congregational church (now Presbyterian) built in 1861 has astonishing presence – from its triple gables framed by immense pinnacles to the windows filling most available wall space.

Across the street is the original Presbyterian church (now Uniting), one of the finest Arts & Crafts compositions in Australia, designed in 1894 by CD Figgis of Ballarat. The architect described it as an adaptation of Florentine or Paduan Gothic. The main entrance is a spectacularly large arch containing a pair of doors with chequered tiles above.

Savings Bank

In Hargraves St near the corner with Templeton St, the old Savings Bank of 1855 has an interesting Mannerist (classical with attitude!) façade, with an extraordinary gorgon's head as keystone of the arched doorway. This was designed by local architects Burgoyne & Poeppel. After 1921 the building was used as a police station for 50 years; it is now the local office for Parks Victoria.

St Mary's Church

Across the road, from the old savings bank, is the Gothic St Mary's Roman Catholic Church. The site gives it prominence enough, but being painted white gives it even more. St Mary's was also designed by Burgoyne & Poeppel, yet it is different from their work across town at Christ Church. More Germanic, it has a beautiful interior with good stained glass.

Poeppel, as his name suggests, came from Germany – and although his impact was not as great as his more famous compatriot, Vahland, in Bendigo, he definitely made a significant contribution to Castlemaine.

Burke & Wills Monument

The Burke & Wills Monument looks down Mostyn St from the east. It is a huge obelisk made of Harcourt granite in 1862 to honour the ill-fated explorers, especially Robert O'Hara Burke, who was stationed here as Police Superintendent for four years.

Buda

The best known house in Castlemaine is Buda. The original bungalow was built in 1861 for a retired Indian Army colonel, but when Ernest Leviny, a celebrated gold and silversmith, bought it two years later, he named the house for his hometown, Buda (the sister town to Pest, in Hungary). Leviny's daughters left their home and its contents to the townspeople in 1981 and it provides a fascinating insight into the cultured and artistic lifestyle of the family.

The historic gardens are a cool retreat on a hot day; many of the plants plants brought into the garden in the 1860's remain. There are elaborate Victorian-style flower beds, formal gardens reflecting early 20th century garden design, a garden pavilion and an ornate aviary. Huge hedges and shady walks invite visitors to linger.

The house and gardens are now administered by a trust, and are open to the public.

Christ Church & Agitation Hill

The Anglican Christ Church in Mostyn St was founded in 1852. It is the most traditional example of the

St Mary's Church, KS
Buda Entrance, KS
Buda Garden in Winter, BaT
Buda Garden in Summer, KS
Christ Church, Agitation Hill, GS
Interior Christ Church, GS

Gothic style in town. Although it looks very English, the church was designed by the locals, Burgoyne & Poeppel. It is usually open and is worth a visit for the stained glass, old woodwork, and especially for the flags and standards of the regiments that served in Castlemaine in colonial days.

The church stands prominently on Agitation Hill, the site where diggers grouped together to air their grievances – in full sight of the Government Camp just across Barkers Creek. Miners gathered where they had the visual advantage over the camp, to express their complaints, chiefly about the mining licenses and corruption.

Many of the agitators were dismayed when the Church of England was allotted the site after the town was surveyed in 1853. (The church certainly made its allegiances clear by displaying the regimental flags.) The gatherings at Agitation Hill, together with those on other goldfields such as the Monster Meeting at Chewton, the Red Ribbon protests in Bendigo and, ultimately, the Eureka Stockade, are celebrated as the beginnings of social democracy in Australia.

Camp Reserve

Camp Reserve was once the administrative centre for the Central Goldfields and 300 men were based here in a tent town of their own. Unusually, the town, and particularly institutions like the post office, were built some distance away.

The **old court house** in Goldsmiths Crescent was built in 1852 and the first criminal court was presided over by Judge Redmond Barry. It is possibly the oldest surviving public building of the Victorian goldrushes. When the local lock-up was full, prisoners were chained to several of the huge nearby trees.

Just around the corner, **Broadoaks**, at 31 Gingell St, was for a short time occupied by Robert O'Hara Burke, the explorer, who was the local Police Superintendent. It was built in 1852.

Castlemaine Gaol

The impressive gaol was built between 1857 and 1864, a remarkable period of building activity that saw eight prisons constructed in Victoria. It was used as a prison until 1990; nine executions were carried out between 1865 and 1876.

Tute's Cottage

Tute's Cottage is a five-room dwelling, constructed of local stone, with an attached weatherboard kitchen, bathroom and laundry. Like many Goldfields homes it has evolved through various stages, beginning with a two-roomed stone cottage. From 1855, the Miner's Right entitled diggers to a ¼ acre (0.1 ha) of land for a house and garden. Tute's Cottage has been occupied under this right since construction. The Miner's Right of the cottage's last occupant, Gladys Power, the daughter of a miner, expired in 1996.

Gaol, KS
Tute's Cottage, GS
Old Court House, GS

Castlemaine Botanic Gardens

The gardens are one of Victoria's oldest regional botanic gardens. Built from 1860 on a moonscape left behind by alluvial diggings, they are now a green and beautiful oasis. Monumental pillars, a carriage drive and large lake allow the visitor to experience a 19th century designed landscape. One of the oldest known cultivated trees in the state – an English Oak planted in 1867 – still thrives here.

The first curator was Philip Doran, a protégé of Baron Ferdinand von Mueller, the botanist and explorer who was director of the Melbourne Botanic Gardens. Apparently the baron frequently came to stay with Ernest Leviny at Buda so it is likely he had some direct involvement with the gardens.

AROUND CASTLEMAINE

BARKERS CREEK

Barkers Creek is a small roadside settlement on the Midland Hwy just north of Castlemaine. It's near the western end of Speciment Gully Rd, one of the most interesting short drives in the Goldfields. See the separate section in Small Towns & Villages.

CHEWTON

Chewton is virtually a suburb of Castlemaine, but it has always been fiercely independent. It lies at the heart of the Castlemaine Diggings National Heritage Park. See the separate section in Small Towns & Villages.

VICTORIAN GOLDFIELDS RAILWAY

The Victorian Goldfields Railway runs vintage trains and rolling stock between Castlemaine and Maldon via Muckleford. Trains are hauled by steam locomotives, except on days of total fire ban. They run from the normal town railway stations. It's best to check the timetables, but generally they run on weekends and Wednesdays. For information phone 5475 2966, or visit www.vgr.com.au.

CASTLEMAINE DIGGINGS NATIONAL HERITAGE PARK

Chewton is an integral part of the Castlemaine Diggings National Heritage Park; see the Chewton section in the Small Towns & Villages chapter for details on this unique and outstanding park. The park brings together natural heritage (Box-Ironbark forests) and cultural heritage (Aboriginal sites and the ruins and relics left behind by 19th century gold rushes). It includes the Garfield Water Wheel, Pennyweight Flat Cemetery, the Eureka Reef Walk and Spring Gully Junction Mine. The Castlemaine Visitor Information Centre can provide detailed information, including maps. Also see the sections for Fryerstown and Vaughan.

Castlemaine Botanic Gardens, KS
Victorian Goldfields Railway, GC
Moonlight Flat Diggings, KS

CRESWICK

LOCATION: 18 KM NORTH OF BALLARAT ON THE MIDLAND HWY
AND 129 KM NORTH-WEST OF MELBOURNE.
POPULATION 1855: 30,000
POPULATION 2006: 2300

Creswick is an unexpected gem. The town combines mining heritage with a rich artistic and cultural history, and a pioneering tradition in land conservation and forestry. It also set the pace of national workplace reform and was the birthplace of an extraordinary number of well-known identities in Australia's recent past.

Cobb & Co rattled through the throng and the town boasted 37 pubs along both sides of the promenade.

To the north lie rich volcanic plains still dotted with impressive mullock heaps and other relics of the gold mining days. Volcanoes once poured molten lava over the ancient landscape, burying the rivers where the alluvial gold had concentrated. Miners called these underground rivers deep leads.

Creswick itself has a wide, picturesque main street featuring a number of carefully maintained heritage buildings. Given the town's sedate pace today, it needs a little imagination to conjure up the roaring goldrush days of the 1850s when the horse-drawn coaches of Cobb & Co rattled through the throng and the town boasted 37 pubs along both sides of the promenade.

Just three pubs remain. The Farmers' Arms has a central position while, intriguingly, the American Hotel stands at the south end of town and the British Hotel occupies the northern end. Was there an international rivalry? The town's history is silent on this question.

Not so a later era at the dawn of the 1900s when the extraordinarily talented Lindsay family of artists, writers, musicians and orators grew up in the town. Norman Lindsay's humorous novel *Saturdee* depicts many of his boyhood friends and escapades. *Redheap*, another of his novels, was banned for many years because of its explicit sexuality (for those days) and the locals were unimpressed by the unflattering portrait of the town.

Other famous sons include World War II Prime Minister John Curtin born in Creswick in 1885 and three times Victorian Premier Sir Alexander Peacock, born in 1861. Additionally, the town is the birthplace of Australia's landcare movement due to the work of John La Gerche who successfully carried out rehabilitation and reforestation of the mine-ravaged countryside during the late 19th and early 20th centuries. isitors may also be surprised to learn that the Creswick miners' union, led by William Guthrie Spence in the 1880s, was crucial in establishing better workers' pay and conditions throughout Australia.

Brief History

European settlement in the Creswick region began in 1839 when Captain John Hepburn took up land at Smeaton and extended his run south to encompass what became the Creswick townsite.

Hepburn was followed in 1842 by the Creswick brothers (John, Charles and Henry) who took up an area along a nearby creek that, with the town, now bears their name. Agricultural pursuits took a back seat 10 years later with the discovery of gold in 1852. The first pub on the new field was licensed in 1853 and the Creswick town site was surveyed in 1854. Within a year the population had reached 30,000.

The alluvial gold was worked for about 20 years before the diggers turned their attention to deeper deposits, beginning with the discovery of the Broomfield deep lead in 1872.

An enterprising bank manager named Lewers set the tone for the mining to come by allowing prospectors free access to sink shafts on his land provided they did no lasting surface damage. If they found gold he charged a 10% royalty on its value. Other freehold farmers followed suit and by 1878 about two dozen companies were working around Creswick.

The most successful mines were sunk on the Berry Deep Lead to the north of town. Over a period of several years more than 1.7 million ounces (48.2 tonnes) of gold was won from the river gravel buried beneath the surface basalt.

The Ristori Incident

However tunnelling under and then up into these buried river beds was dangerous and uncomfortable work. Pay and conditions were important to the miners and when, in 1878, the Ristori Freehold Company attempted to reduce the daily rate of pay, the miners staged mass protests.

At that time the mines were at their most prosperous and management was keen to maximise output, so the directors of Ristori quickly agreed to restore the pay rate. Nevertheless, this dispute brought to the fore two local union organisers – William Guthrie Spence and John Sampson. Spence in particular went on to wider acclaim in the union movement.

The Ristori incident raised the profile of the Creswick Miners' Union and led to a wider unity of Australian miners. The Creswick stand also helped moves towards regulation and safety in the mining industry, including limits on working hours, standards for ventilation and use of ladder-ways in the mines.

Spence himself became Secretary of the Amalgamated Miners' Association of Australia in 1882 and 12 years later General Secretary of the Australian Workers' Union.

Bandstand, KS
Quartz Tailings near Creswick, KS

White Shop, KS
Brick Shop, KS
Gold Bank, PW
Salvation Army Hall, PW

NEW AUSTRALASIA DISASTER

Unfortunately the improvements did not prevent a tragedy in Creswick late in 1882. On 12 December that year two miners working in a drive connected to the New Australasia No.2 shaft struck water which immediately flooded into the mine.

In the ensuing panic five men managed to beat the rushing water to safety. Twenty seven others took refuge in a rise, hoping that the water level would not reach them. But the air soon became stale and the miners began falling, unconscious, into the water.

On the surface, men feverishly deployed all available engine power to pump out the water at over 500 gallons (2273 litres) per minute. Even so it took more than two days before rescuers could reach the trapped miners. By that time 22 of the 27 had died.

The New Australasia tragedy has gone down as the country's worst ever gold mining disaster.

FORESTRY

A looming tragedy of a different sort on the surface around Creswick in the 1880s had a happier ending. Early alluvial mining damaged and often destroyed vegetation and laid the ground bare down to bedrock. This made large areas of the diggings into a stark moonscape of eroded hills and gullies, while chemicals used in the gold extraction process seeped into the waterways to do damage further afield.

In the 1880s some natural revegetation did occur, but indiscriminate tree felling continued as the need for mine timbers and housing materials continued. At this point the Victorian Government sent Jersey-born immigrant John La Gerche to Creswick as Crown Land Bailiff with a mandate to police timber-cutting by issuing licenses and imposing fines on illegal loggers.

La Gerche went much further. He established a nursery to supply seedlings for a revegetation program in the Creswick area. The plantations flourished and by the turn of the century the idea for a wider application to other mine sites in the country gained support in government circles.

Creswick-born parliamentarian Sir Alexander Peacock, then Victorian Minister for Mines and Forests, supported the establishment of the Victorian School of Forestry. The government bought and renovated the former Creswick hospital to serve as a campus and the first students enrolled in 1910.

Today the school is part of the University of Melbourne and the adjoining nursery and land care centre propagates more than 650,000 native and exotic plant species each year for planting in gardens and rural areas.

The mines of Creswick passed their peak in the late 1800s and by the start of the 20th century the region's focus returned to farming.

VISITOR INFORMATION

Creswick Visitor Information Centre
Cnr Cambridge & Raglan Sts
Creswick 3363
www.hepburnshire.com.au
tel: 5345 1114.

The town has a bus service daily from Ballarat and an airport for light planes.

FOOD & ACCOMMODATION

Creswick has a mix of old-style and modern accommodation which makes the most of the town's goldfield heritage. Dining in town offers a choice of light meals in several cafés and tea rooms. Pub bistro-style meals are available in the three hotels and an á la carte menu restaurant at the Forest Resort.

SIGHTS

The best way to enjoy Creswick is at a leisurely pace. Perhaps begin with a walking tour around the town and then a short drive to the lakes and forest attractions just out of town. Maps and information are available at the Visitor Information Centre.

The oldest building in Creswick was built in 1856. It originally served as the stock exchange, later became Spargo's Hotel and more recently served as the **Cosy Corner Café.** The old stables are still at the rear.

Tait's Store was built in 1861 by a shipbuilder and blacksmith from the Clyde in Scotland. Its façade has been carefully maintained as an example of ornate architecture of the period.

The **band rotunda** (1897) was built to mark Queen Victoria's Diamond Jubilee. The town band had been formed to provide marching music for the Creswick Volunteers, a local brigade formed as a defence against a feared Russian invasion.

The **Lindsays' house** (*Lisnacrieve*) was demolished in the 1960s. Today a plaque marks the site. Fortunately one room has been saved and reconstructed in the Ballarat Art Gallery.

POLICE STATION & COURT HOUSE

The police lock-up, built in 1860, police station (1861 and remodelled in 1901) and court house are just opposite the information centre.

TOWN HALL & GOLD BATTERY

The strikingly unusual town hall was built in 1876 is now the historical museum. Its ornamental façade and clock tower have been restored. The interior of the council chambers is retained, including ornate carved stone pillars, original furniture and a winding bluestone staircase. Displays include photographs and items from the gold mining past, along with historical artworks, including those from the famous Lindsay family.

Behind the museum is the 1890 gold battery, still operational and run by volunteers of the museum. Tours are by appointment.

Rotunda and Town Hall, PW
Town Hall Detail, PW
Town Hall, GS

BANKS

Both the Gold Bank and the old Bank of New South Wales (1854) are now private residences. The Bank of NSW owners run an antique shop in the front of the building. Internal stairs with polished wooden banisters lead up to the private residence.

HOTELS

The American Hotel across the main street from the information centre was built by American publican Thomas Anthony in the early 1850s. Originally it was a long two-storey timber building set back from the main street. However the old weatherboard front was demolished in the 1920s and replaced with a red brick single storey with a wide verandah over the footpath.

At the other end of the main street is the British Hotel, a two-storey whitewashed brick building with dark paint picking out door and window surrounds more in the English pub style – except for the galvanised iron roof.

MELBOURNE UNIVERSITY – VICTORIAN SCHOOL OF FORESTRY

The original Goldfields Hospital (1862) and *Tremearne House* (1884) are now occupied by the Victorian School of Forestry (established in 1910) and the Creswick campus of Melbourne University. *Tremearne House* is a particularly grand two-storey red brick building with fine cast-iron lacework along the verandahs.

CRESWICK RAIL BRIDGE

Creswick rail bridge (1874) is a rare and dramatic example of iron rail bridge construction in Victoria. Its tapering hollow cast-iron pillars were sunk 18 m below the creek bed and filled with concrete. The deck is built of iron girders.

AROUND CRESWICK

Two of the Goldfields best secrets and most fascinating sights can be found around Creswick: the Berry Deep Leads and the La Gerche Walking Track.

The **Creswick Miners' Walk** takes in the old goldmining sites from Creswick to Ballarat and follows the route the miners took to join their colleagues at the Eureka Stockade uprising in 1897.

BERRY DEEP LEADS

The mines that exploited the deep leads north of Creswick have created an extraordinary landscape. Huge mullock heaps and the remains of abandoned engine houses litter the fertile volcanic plains. They give vivid evidence of the vast enterprise that produced nearly 50 tonnes of gold

Streetscape, GS
British Hotel, PW
Court House and Police Station, GS

– although at an enormous cost to the health and lives of the miners. To explore, take the Clunes Rd, and turn right to Allendale and Smeaton. The mines lie to the west.

Continue to Smeaton and turn left towards Clunes on the Daylesford-Clunes Rd. Two of the most striking ruins are the engine houses at Hepburn Estate, just to the north side of road and, a bit further on, the Berry No.1. They both have the dramatic simplicity of medieval castles. See the separate Smeaton section and 'The Buried Rivers of Gold Heritage Trail' for more information.

LAND CARE CENTRE (NURSERY) & LA GERCHE WALKING TRACK

A few minutes' drive out of town on the Daylesford Rd is the nursery and land care centre which adjoins the forestry school. About 650,000 plants (native and exotic) are grown here each year. Note the extraordinary wooden office that was built by forestry students, and the bizarre fountain they also created.

While at the land care centre you need only an hour to stroll around the loop of La Gerche Forest Track. This two km trail takes you through introduced trees that make up a forest which is well over 100 years old. The trail is marked with information plaques that tell the tale of John La Gerche's efforts to develop the plantation and rehabilitate the mine fields.

In the 1880s La Gerche began planting a mixture of native and exotic trees (mainly pines and deciduous species like oaks). The plantations he established behind the land care centre have never been logged and create a spooky, European-type woodland. One of the features is the first *Pinus Radiata* planted in Victoria, the mother tree to countless plantations. It's a huge and spectacular tree.

CRESWICK REGIONAL PARK & SPENCE HOUSE

The Creswick Regional Park lies south-east of the town on either side of the Melbourne Rd, and encompasses the land care centre and St Georges Lake. A plaque at Jackass Gully within the forest marks the site of the house of mine unionist W. G. Spence.

NEW AUSTRALASIA MINE MEMORIAL

Going north from town on the left hand side of Clunes Rd (about one km past the turn-off to Allendale and Smeaton) there is a cairn erected at the New Australasia Mine site in memory to the 22 miners killed in the tragedy of 1882.

The plantation behind the land care centre creates a spooky European woodland.

Land Care Centre, RE
La Gerche Forest, RE
La Gerche Forest, RE

Daylesford & Hepburn Springs

LOCATION: 111 KM NORTH-WEST OF MELBOURNE,
45 KM WEST OF BALLARAT, 39 KM SOUTH OF CASTLEMAINE
POPULATION 1865: 12,500
POPULATION 2006: 5500

Daylesford and Hepburn Springs grew up alongside each other and are now popular spa towns. They both have a significant gold-mining heritage, but this is less obvious than is the case for many other Goldfields towns. They have a distinct European flavour. This is partly because of the European street plantings, significant surviving heritage buildings and two small man made lakes.

Replace the Convent Gallery on Wombat Hill with a castle, and Daylesford could easily be a small European market town. The sense of history, the European trees and the close scale of the landscape all contribute to the illusion. On the other hand, if you stand anywhere on Wombat Hill (actually an extinct volcano) and look to the north, you could only be in Australia.

Daylesford and Hepburn Springs lie 600 m above sea level, in the rain-washed hills of the Great Divide, so the climate can be cool. They are beautiful towns, and over the last decade they have turned themselves into thriving rural retreats with a large number of weekenders and day-trippers. The visitors come to pamper themselves and to enjoy an extraordinary variety of outstanding restaurants and cafés, walks, views, galleries, bookshops, lakes, spas and mineral springs.

BRIEF HISTORY

The first European settlers in the area include the Clowes brothers (*Holcombe*) and Alexander Mollison (*Coliban*) to the east of today's Daylesford, Egan (*Corinella*) and Captain John Hepburn (*Smeaton Park*) to the west, Alexander Kennedy (*Bough Yards*), Dugald McLachlan (*Glengower*) and Lachlan McKinnon (*Tarrengower*) to the north.

Opposite: Autumn, PW
Daylesford Lake, RE

Tunnels criss-crossed under the town, and some burrowed into Wombat Hill.

Wombat Hill and Convent Gallery from
the Air, BS
Daylesford Lake, KS
Fountain, RE

GOLD

Around August 1851 Egan and a small group of companions found alluvial gold on Wombat Flat on ground now covered by the man-made Lake Daylesford.

In 1852 Egan made another discovery at Spring Creek (Hepburn Springs) which came to be called the Jim Crow Diggings, a name that was later used to describe the entire Daylesford-Hepburn goldfield.

By 1856, Daylesford had 45 hotels and boarding houses, including the two-storey brick building now known as the Central Springs Inn, which was originally John Goodwin Howe's produce store. In 1859 about 3400 diggers were on the local diggings, 800 of whom were Chinese.

The surrounding forests were home to dozens of mills and hundreds of men who provided wood to the nearby towns for building and to the surrounding mines, particularly in Ballarat (to run their steam engines and to support the tunnels and shafts). The Wombat Forest sawmills also supplied timber to feed the Melbourne building boom of the 1880s. Timber towns like Koweinguboora and Dean sprang up but by 1890 the timber resources were veavily exploited – prompting a government enquiry and a big drop in production.

Novelist Joseph Furphy, the author of one of Australia's earliest and greatest works, *Such is Life*, worked as a threshing-machine operator in the district in the 1860s. He was married at Daylesford in 1867 and took over the lease on a small local farm and vineyard belonging to his mother-in-law.

By the 1860s the heyday of surface alluvial gold was over and operations shifted to deep lead mining under the volcanic lava. Investors also began to develop deep shaft quartz reef mining. This ushered in a golden era, where the town prospered. Tunnels criss-crossed under the town, and some burrowed into Wombat Hill – one caused the steeple of St Peter's church to crack and to be pulled down. Significant underground mining continued until 1949.

MINERAL WATER

Today mineral water has taken the place of gold. Mineral waters are the result of a rare hydro-geological process that occurs in the hills around Daylesford; more than 80% of Australia's mineral water springs are in the region.

It is uncertain whether the Aboriginal people used the mineral waters, but the European arrivals from Italy, Switzerland and Germany were accustomed to 'taking the waters' in their own countries and immediately realized the significance of the local springs.

The Hepburn Bath House was developed in the 1890s and subsequent hotel and lodge accommodation in the area marked the beginning of spa tourism. Popularity waned during the Great Depression in the 1930s and it was not until the 1980s that the old spas were redeveloped.

Thousands of people now visit the area each year. In conjunction with the springs and spas there are many businesses and individuals offering complementary services, including every kind of massage known to humanity, as well as reiki, shiatsu, acupuncture, aromatherapy, reflexology, spiritual healing, tarot readings….

VISITOR INFORMATION

The excellent Daylesford Visitor Information Centre has information about all the surrounding mineral springs and spas, as well as maps for a range of excellent short walks.

Daylesford Regional
Visitor Information Centre
98 Vincent St Daylesford 3460
tel: 5321 6123
email: visitorinfo@hepburn.vic.gov.au

FOOD & ACCOMMODATION

Vincent St is one of the busiest and most sophisticated regional main streets in Australia, with cafés, bars, and restaurants busy day and night. There are also several excellent restaurants around town, including the internationally acclaimed Lake House.

Hotel and motel accommodation is available close to the centre, and there is an enormous variety of high quality B&Bs, and self catering villas and cottages.

SIGHTS

Vincent St, which continues on to Hepburn Springs, is where you will find the visitor information centre, the greatest concentration of significant buildings, and a range of excellent cafés and restaurants. The best orientation point overlooking the town is the observation tower in Wombat Hill Botanic Gardens.

Wombat Hill has a crop of church spires, and the towers of the Old Convent. The small, but elegant and well-maintained Classical Revival court house (1863) is next to the police station in Camp St. On all sides there are charming timber cottages with elegant verandahs and well-kept gardens, as well as occasional grand mansions such as *Mt Stuart House* and *The Manse* (both now B&Bs).

The Uniting Church (former Methodist, 1867) has the tallest spire, and a good polychrome brick nave and tower. Next to the present Uniting and Anglican churches are the original church buildings, relics of the earliest years of the goldrush. Nearer the hilltop is the former Presbyterian church, a neat red brick building with an excellent octagonal spire.

To explore Daylesford and Hepburn:

- visit the superb botanic gardens

- walk around Lake Daylesford, sample the mineral springs, and go for a row

- visit the Hepburn Mineral Springs Reserve (remember the water from the first few pulls of the pumps will be flat!)

- wander up and down Vincent St (look out for historic buildings, cafés, Avant Garden Bookshop, Pantechnicon Art Gallery)

- visit the outstanding Convent Gallery

- pamper yourself with good food, a spa, or a massage….

Vincent St is one of the busiest and most sophisticated regional main streets in Australia.

Springs Retreat, AP
Vincent St, PW
Cottages, KS

VINCENT ST

Vincent St has a number of striking buildings including grand old pubs (with and without their lace verandahs), solid commercial premises, a former cinema, and an impressive fountain, slightly off the main spine of the town.

Museum & Primary School

Just down the road from the visitor information centre on Vincent St is the Daylesford Historical Society Museum, which has an interesting pioneer section. It is in the old school of mines which was built in 1890 to develop deep-lead mining skills. Displays include photographs, clothing, Aboriginal artefacts and items related to local sawmilling, agriculture and goldmining. Next door to the museum is the impressive old primary school (1874).

Post Office

The post office was built in 1867 to an Italianate style that was popular throughout regional Victoria.

Town Hall

The town hall has a splendid façade – confidently baroque with giant Corinthian pilasters. It was designed by George Johnson in 1882. Below street level is the hall itself, a magnificent traditional town hall with balconies on three sides and a proscenium stage. Johnson specialised in civic buildings and gave Melbourne some of its grandest town halls – including Collingwood, Fitzroy and North Melbourne.

DAYLESFORD BOTANIC GARDENS

The town wraps itself around the northern and western slopes of

Wombat Hill, at the crest of which is a dramatic profile of mature conifers. These firs and redwoods from Europe and North America are one of the chief glories of the gardens, which date from 1861. Baron Ferdinand von Mueller, the inaugural director of Melbourne's Botanic Gardens, helped select the original plantings.

The gardens also feature a fern gully laid out by William Sangster in 1887. Sangster went on to design many private gardens around Melbourne, as well as the city's Exhibition Gardens.

Wombat Hill offers the most exciting views in town from the historic all-concrete observation tower (1937).

THE OLD CONVENT

The Old Convent (from 1860s) is an extraordinary, rambling building that has been superbly restored. It started life as the private residence of the local gold commissioner, became the presbytery for St Peter's Catholic Church then, in 1891, was acquired by the Sisters of the Presentation who developed it into a convent and girls' boarding school.

Today it once again hums with activity. There are weddings in the restored chapel, receptions in the extensive dining and reception rooms, a bar, a restaurant – and an extraordinary range of high quality artworks, jewellery and local food and wine. It is remarkably picturesque – and fascinating – with its beautiful grounds and its haphazard assembly of former schoolrooms, dormitories, chapel and onion-topped tower.

DAYLESFORD SPA COUNTRY RAILWAY

The old railway station is at the southern entrance to the town and

Post Office, RE
Convent Gallery, MH
Daylesford Railway Station, PW
Botanic Gardens, PW

an active preservation society runs the Daylesford Spa Country Railway. The railway has collected rail motors of the Victorian railways system from the early years of the 20th century. Trains run every Sunday on a preserved section of track through the Wombat State Forest as far as Bullarto (16 km).

DAYLESFORD MINERAL SPRINGS

The Central Mineral Springs are at the northern end of Lake Daylesford alongside the creek, and there are three more within a short distance, each with a different flavour.

TIPPERARY WALKING TRACK

The Tipperary Walking Track starts at the southern end of the Lake Daylesford, near the causeway, and continues 16 km all the way to the Hepburn Mineral Springs Reserve. You don't have to walk the entire track; there are a variety of loop walks of various lengths. The visitor information centre can provide details.

HEPBURN SPRINGS

A special part of the Daylesford experience is taking the waters at Hepburn Springs. The road from Daylesford descends rapidly along a narrow ridge lined with old miners' cottages, with steep valleys on either side. The village of Hepburn Springs has some interesting buildings that show the distinct influence of northern Italy and Switzerland. One of the earliest is **Parma House** (1864) built by Fabrizzio Crippa.

OLD MACARONI FACTORY

Pietro Lucini came from the Italian Alps close to the Swiss border. He arrived in Melbourne in 1854 as a political refugee and established Australia's first macaroni factory in Melbourne. He soon realized his best market was at the goldfields so he moved to Hepburn Springs found gold and established his bakery. Pietro also knew the value of mineral water and was among community leaders who convinced the government to set the springs aside as public land.

Guided tours of the Macaroni Factory give a rare insight into the story of the Lucini family – told through Giacomo Lucini's (Pietro's brother) unique frescoes, which he painted on the ceilings and walls. These colourful frescoes are Australia's oldest and reflect political statements as well as the brothers lives and scenes from their Italian homeland.

There is also a café that specializes in… pasta. Hours are a little complicated and bookings for both the tour and the café are recommended (tel 5348 4345).

HEPBURN SPRINGS SPA COMPLEX & MINERAL SPRINGS RESERVE

Captain Hepburn discovered the mineral springs in 1836. After the local goldrush had subsided a bath hut was built in 1895. The current Edwardian pavilion is the third structure to have been built over the main spring.

In the reserve, mineral water is supplied by hand pumps or free-flowing pipes. Each outlet has a different taste, depending on the minerals present – Some taste sweet, others sharp, some are very subtle in flavour, others strong.

Central Springs, JM
Villa Parma, KS
Old Macaroni Factory, DB
Hepburn Springs Spa Complex, PW

MARYBOROUGH

LOCATION: 70 KM NORTH OF BALLARAT ON THE PYRENEES HWY;
166 KM NORTH-EAST OF MELBOURNE.
POPULATION 1854: 50,000
POPULATION 2006: 8000

Maryborough will never be allowed to forget Mark Twain, the American novelist, satirist and traveller, who visited in 1885 and described the city as 'a station with a town attached'. He added facetiously, 'You can put the whole population of Maryborough into it, and give them a sofa apiece, and have room for more.'

There is no doubt that the railway station is the largest and grandest of Maryborough's many fine old buildings, but it was the discovery of gold in the surrounding district that first put the city on the map and attracted the railway to its door.

In a golden triangle formed by the towns of Dunolly, Wedderburn and Inglewood, an amazing 80% of the largest gold nuggets in the world have been found.

The town was named in 1854 after Maryborough in Ireland, the birthplace of gold commissioner James Daly. Even the name of Havilah which appears on the city's seal has connotations with gold. The Book of Genesis in The Bible refers to the land of Havilah ...'where there is gold; and the gold of that land is good'.

Maryborough was deep lead country and yielded over a million ounces of gold (28 tonnes) from the buried river sands within six years in the 1850s. Some quartz mining in the region added to the booty.

Not far to the north, in a golden triangle formed by the towns of Dunolly, Wedderburn and Inglewood, an amazing 80% of the largest gold nuggets in the world have been found just below the surface. Only 40 km north of Maryborough, at Moliagul, the *Welcome Stranger* nugget was discovered a mere one inch down (less than 3 cm) in 1869. The daddy of them all, this bonanza weighed 2380 ounces (74 kg)!

Today though, Maryborough is a busy rural centre surrounded by grazing land interspersed with eucalypt forests of Red Ironbark and Grey Box. The rich gold past can still be seen in the form of white mullock heaps stained with streaks of orange. The occasional glimpse of a windlass in amongst the trees indicates that prospectors are still trying their luck in the ground beneath.

The city encompasses several blocks with well-maintained period buildings such as the Flagstaff Hotel and the Bull and Mouth Hotel standing beside modern shops in the main street. Even more impressive is the grand architecture of government buildings like the town hall, court house and post office which line McLandress Square.

Maryborough succeeds in integrating a wealth of goldfield history with the needs of a modern rural centre.

BRIEF HISTORY

The Maryborough region was originally inhabited by the dja dja wurrong people who called the area Tuaggra.

The first Europeans to settle were the Simson brothers, who in 1840 established a sheep station, *Charlotte Plains*, which occupied a vast tract of land extending east all the way from present day Castlemaine to Avoca. The station homestead was built about 15 km north-west of Maryborough. Unfortunately it fell into disrepair and little trace remains.

The Simsons' peace was shattered in 1854 when gold was discovered at White Hill, just four km north of where Maryborough now stands. Within a few months 25,000 diggers had rushed into the area and the number reached 50,000 at the height of the fever.

So swift was the population build-up that Maryborough township was surveyed and named that same year. A police camp, Methodist church and hospital were amongst the first infrastructure. One of Victoria's earliest newspapers, *The Maryborough Advertiser*, was also established in 1854.

Land sales began in 1856 and Maryborough quickly became the administrative and commercial centre of the area. It became a borough in 1857.

The railway came to town in 1875, linking Maryborough with Castlemaine and Melbourne, a recognition of the importance of this thriving centre.

Gold mining continued well into the 20th century, the last mine closing down in 1918. One of the biggest deep lead mines was the North Duke at Timor a little to the north of town. Other alluvial diggings were known as Four Mile, Chinaman's Flat (a reference to the large contingent of Chinese miners) and the Bet Bet.

Reef mining took place at The Leviathon, the deepest and richest in the area at 330 m deep with a production of over 50,000 ounces (1.4 tonnes), and The Mariner's Reef which produced over 40,000 ounces (1.1 tonnes).

When the miners left, commerce reverted to agricultural pursuits, particularly sheep grazing. Maryborough also developed a strong manufacturing base which became a key to resurgence of the local economy in the 1950s. The Maryborough Knitting Mills opened in 1924 and established the town as a centre for the wool industry.

McPHERSON'S PRINTING GROUP

The book you are reading right now was created in Maryborough by the McPherson's Printing Group!

Maryborough is known as a centre for printing and publishing and produces more paperback books than any city in Australia. The beginnings of the McPherson's Printing Group can be traced back to a chance meeting between James Henry Hedges and Gordon Ernest Bell in 1946. Hedges owned a grocery store in Maryborough, and printed some of his own brand labels as a hobby and sideline, while Bell was a printer by trade. Together they launched Hedges & Bell in 1946.

Over the years, through a combination of mergers, sales and acquisitions, today's company emerged. McPherson's became 50% owners in the late 1940's, and acquired full ownership in 1956; the Dominion Press was acquired in 1981; Globe Press was acquired in 1991 and the Australian Print Group was acquired in 2001.

Coffee Palace, KS
Streetscape, RE

VISITOR INFORMATION

Maryborough Visitor Information Centre
cnr Alma & Nolan Sts
Maryborough, Vic 3465
www.visitmaryborough.com.au
tel: 5460 4511.

Maryborough's annual New Year's Day Highland Gathering is the longest running athletic festival in Australia. First held in 1857, it incorporates the Maryborough Gift sprint, along with various Highland games, music and dancing.

FOOD & ACCOMMODATION

Accommodation in Maryborough offers a range of options, including five modern motels rated three and and four star. There is also a hotel/motel and traditional old-style pub rooms in the other hotels in town. Three B&B venues provide excellent service in pleasant surroundings and there are several self-contained cottages for those who want more independence. In addition the town has a venue that caters for groups of up to 10.

Maryborough has two caravan parks and country farm accommodation can be found in the surrounding district.

There is good dining in town at the hotels as well as several restaurants and a wine bar at the station. Light meals are supplied by cafes and bakeries.

SIGHTS

Maryborough has a lot to offer the traveller interested in Goldfield heritage as well as those visitors keen to browse the galleries and antique shops. In some cases you can combine both interests at once. A walking tour is recommended for the city centre, while several driving tours take in places of interest in the surrounding district. Maps and guides can be obtained from the visitor information centre.

Maryborough's oldest surviving building is the Court of Mines, built in 1858 and now preserved virtually unchanged on the outside as part of a doctor's surgery

Railway Station, EG

RAILWAY STATION

Taking a lead from Mark Twain, most visitors head straight to the town's most famous building, the railway station. Built in 1890, it replaced an earlier station that had been on the site since 1874. Local legend suggests the design was originally meant for Spencer St Station in Melbourne, but like all good stories this is a myth. The architect was James Clark who also designed Castlemaine's magnificent post office, the Melbourne City Baths and much of Melbourne's Government House.

The intact ornate brick building with its turrets, gables and tall clock tower has a grand entrance framed by tall pillars.

Inside it has tessilated tiles made by the same firm who supplied the tiles for Parliament House in Melbourne. The ceilings are high and beautifully finished in timber. It is a prime example of the peak of lavish expenditure on railways infrastructure built following the Railway Construction Act of 1884 (sometimes called 'Octopus' Act) which authorised the building of 66 new rail lines in Victoria.

The verandah-covered platform is one of the longest in Victoria. In the 1950s there were up to 48 trains a day passing through and four main lines radiating from the town. Sadly, passenger trains ceased to operate in 1993.

Today the central station area houses an antique emporium, a café and a regional wine centre. The latter offers a quiet nook to enjoy a glass of wine or coffee sitting in an adapted first-class seat taken from one of the old passenger trains. More art and woodworking galleries are located in the former station offices further along the platform.

FLOUR & KNITTING MILLS

Not far away beside the rail line is the Maryborough Flour Mill established in the early 1880s as one of Maryborough's earliest industries.

On the way back towards the main city blocks is the old Maryborough Knitting Mill which began with 100 employees and grew to 600 employees by the early 1970s. It no longer operates.

HOTELS

The **Bull and Mouth Hotel** has occupied the same site since 1854, making it one of the first established businesses in Maryborough. The original pub was demolished in 1904 and rebuilt into what today is one of the finest buildings in the main street with its complex façade and ornate individual balconies for each of the upstairs rooms.

Two other hotels along the main street, the **Flagstaff Hotel** and the **Albion Hotel**, also add an early goldfields atmosphere to town.

CIVIC PRECINCT

One street back is McLandress Square with the post office (1878) flanked on either side by the court house (1893) and the town hall (1887) – all from the Victorian era and representing the heart of Maryborough's civic precinct. The size and quality of the buildings reflect the importance of Maryborough as one of the colony's premier cities built with the wealth from gold mining.

The post office was the original court house until the new building was built alongside. The original town hall was built in 1858 then superceded by the present building in Queen Victoria's

Post Office, KS
Court House, RE

Jubilee year. The fountain in front of the Town Hall is a memorial to the first soldier from Maryborough to die at war.

PHILLIPS GARDENS.

Phillips Gardens, named after Neville Henry Phillips, Maryborough's Town Clerk 1888-1935, contain a small lake that was the city's first water supply. It was purchased by the town council from the local miners for £70 in 1860 so that the firemen and horses could use the water as required. This was discontinued in the 1870s when the Goldfields Reservoir was constructed nearby. Planting of gardens around the original lake was completed in 1875.

FIRE STATION

One of the most striking buildings, the old fire station, is also one of the oldest, built in 1861. The existing high brick and wooden bell tower is actually the third on this site and was constructed in 1888. The fire brigade moved out in 1982 and the building is now an art gallery showcasing mostly local artists in changing exhibitions. However two prized possessions on show are Pro Hart's 1975 oil painting entitled *Building the Headframe* and the same artist's undated oil on masonite called *Grasshopper*.

WORSLEY COTTAGE & MUSEUM

Worsley Cottage & Museum was built of brick and stone by stonemason Arthur Worsley. He completed the two front rooms for his bride Agnes in 1894 using stone quarried from the Avoca Rd. A detached kitchen out the back was made of timber. Worsley added two more rooms in the early 1900s.

The cottage has been preserved as the headquarters of the local historical society and furnished with period furniture along with displays of maps, photographs and mining relics.

BRISTOL HILL & MARYBOROUGH CEMETERY

On Bristol Hill overlooking the town is a whitewashed bluestone tower built in the 1930s. Bluestone for the spiral staircase came from the old town gaol buildings. At the bottom of the hill lies the site of the original 1850s town cemetery. There were 300 graves, but now a memorial plaque is all that marks the spot.

The 'new' Maryborough Cemetery (used from 1859) is located on the south-eastern edge of town and is neatly divided with a section reserved for each Christian denomination. The McCulloch mausoleum is a noticeable feature – a large edifice which houses the remains of a family, who, with the exception of the father, all died young in the 1800s.

Also prominent is the tall fluted and wreath topped pillar of the Simson Memorial, dedicated to the founding graziers in the district.

A number of other gravestones give an insight into life on the goldfields, particularly the children who often met early deaths due to epidemics and the seemingly impossible task of achieving general hygiene.

Of interest, too, is the Chinese section, unfortunately with only a few remaining headstones carved with Chinese script. A memorial commemorating the dead takes the place of the missing headstones. Beside it is a funeral oven where offerings were burnt in honour of those who died.

Fire Station, EG
Worsley Cottage, EG
Chinese Burning Tower, EG

AROUND MARYBOROUGH

The road north of Maryborough passes through the Golden Triangle between Wedderburn, Dunolly and Inglewood where some of the world's largest gold nuggets have been found, including the *Welcome Stranger*. Some good finds are still being made, although nothing like those of the past.

ABORIGINAL WELLS

About 11 km south of Maryborough there is a good example of Aboriginal wells, dug prior to European settlement. This example is one of four sets recorded in the Maryborough area. They are dug into the sandstone to gain full advantage of the natural rainwater catchment formed by a rock ledge above.

The wells have narrow mouths to reduce evaporation and fouling by animals and wind blown debris. Locals in the area have never known them to be dry. The maximum depth is 130cm and they have a total capacity of about 160 litres.

MARYBOROUGH REGIONAL PARK

Wedged between the Paddy's Ranges State Park and the outskirts of Maryborough this park is noted for its impressive wildflowers and avifauna and has developed tracks for local informal recreation.

PADDY'S RANGES STATE PARK

The park has been used for grazing, gold mining, timber harvesting, eucalyptus oil and honey production – and there are relics from all these activities. There are impressive spring wildflower displays.

Walking tracks radiate from the picnic area where an information board gives further details on features, history and activities.

Most of the 1675 hectare park was burnt in January 1985 by a wildfire which created prolific regeneration, especially of wattles.

There are more than 140 native bird species in Paddy's Ranges including the rare Painted Honeyeater, which visits in the spring and summer. Swift Parrots visit the area in winter each year but they return to Tasmania in spring. The Crested Bellbird, Wedge-tailed Eagles, and Peregrine Falcons are also present.

Aboriginal Rock Wells, EG
Reservoir, RE
Paddy's Ranges State Park, RE

St Arnaud

LOCATION: 246 KM FROM MELBOURNE ON SUNRAYSIA HWY;
64 KM FROM MARYBOROUGH
POPULATION 1911: 4000
POPULATION 2006: 3650

St Arnaud has retained much of the Goldfields heritage in its distinctively charming and gracious streetscapes. Many of the main street buildings have retained theiroriginal façades. Wide cast-iron lace verandahs and balconies are a feature of many shop fronts and hotels, while the windows still have original timber or metal frames. Leadlight, glazing and tiled surfaces are also a feature of the stunning architecture.

St Arnaud, RE
Balcony, KS
Horse Trough, RE

This well preserved old-style character is today combined with St Arnaud's role as an important regional centre for the surrounding area. The mines have shut and the gold has mostly gone, although prospectors and fossickers still find gold in the nearby creeks.

Today the town thrives on agriculture – grain, fine merino wool, beef cattle. There is a growing contribution made by vineyards and olive groves of the nearby Pyrenees region. The town is a busy place, increasingly attracting commercial interests, and yet retaining its rural and old-world charm.

Curiously, St Arnaud is named after

Jacques le Roy de Saint Arnaud, a Marshall of France. St Arnaud commanded the Allied armies (French, British and Turkish) against the Russians in the Crimean War. Under his command the Allies won the crucial battle of Alma in 1854, but St Arnaud died of fever a few days later. All this happened to be about the same time that New Bendigo (the original name of St Arnaud) was experiencing its goldrush. The French Marshall was a hero and the town decided to honour his name.

Today a bronze statue of the Frenchman, sculpted by local panel-beater, talented artist and former mayor Maurice

McGrath, stands near the botanic gardens on the main street at the eastern entrance to town. McGrath has also sculpted a smaller bronze of explorer William John Wills (of Burke & Wills fame) which is in the former land office, now the town's Visitor Information Centre. It is said that Wills, a surveyor of Crown Lands before he went exploring, helped survey the streets of St Arnaud in the mid 1850s.

BRIEF HISTORY

The story goes that seven men camped on New Years Day 1855 in what they considered a likely gold bearing spot. They spread out into the surrounding hills, panning the creeks and sinking shallow shafts and they soon met with success. They called the place by the unimaginative name of New Bendigo. But names weren't really on their minds at the time. Rather, they wanted to scour the area and stake their claim before others found out about their strike.

In three months work they found the Gap and the Wilson Hill Reefs before the secret was out and the rush began.

The Victorian Lands Department surveyed a site for the proposed village along what is now St Arnaud Creek in 1856, but the plans were ignored. It seems the locals had already decided the place for their town. Maybe it was this knock-back that prompted Surveyor Wills to go exploring with Burke soon afterwards.

St Arnaud's gold was won over a longer period than some rushes in the region. The population didn't peak until 1911 when several mines were in operation at once.

The Lord Nelson Reef is right alongside town. Others, like New Bendigo and Garibaldi, occupy a tract of country about 10 km wide. There was also a silver mine near Bell Rock, about one km north of town.

The grade of the gold ore bodies varied greatly at different depths and the deepest mines followed the quartz reefs 800 m below ground. The Lord Nelson mine tunnelled right under the township and it is said that in one particular tunnel a miner could stand several hundred metres below the main bar of one of the town's pubs.

Rising water was a constant problem in the mine workings, defeating the best efforts to pump out the shafts. Some of the smaller mines consequently had short lives. However, the town's principle mine, the Lord Nelson, operated until 1918. The last mine in the area, the Welcome Nelson, closed in 1926.

By that time St Arnaud had turned to agriculture to earn its living. The townspeople brought in soil to rehabilitate the Lord Nelson mine waste land and turned the area into a sporting park, including trotting track, football and hockey ovals, along with various club facilities and picnic area.

Tiles, KS
St Arnaud, KS
Old Shire Hall, RE

VISITOR INFORMATION

St Arnaud Visitor Information Centre
4 Napier St, St Arnaud 3478
www.ngshire.vic.gov.au
tel: 5495 1268.

FOOD & ACCOMMODATION

Visitors to St Arnaud can find a comfortable bed in any one of the town's seven hotels. There are also four motels, three B&Bs and a caravan park. Two of the town's hotels – the Botanical and La Cochon Rose (the Pink Pig) – offer elegant dining. Several others have bistro meals. The Old Post Office B&B has a luncheon restaurant. There are several cafés and fast food outlets in town.

SIGHTS

A street walk around St Arnaud reveals many substantial historic buildings, the result of wealth and prosperity created by the discovery of gold as well as by the products of a rich agricultural industry.

A street walk around St Arnaud reveals many substantial historic buildings.

Old Victoria Hotel, KS
Court House, RE
Old Victoria Hotel, KS

CROWN LANDS OFFICE & COURT HOUSE

A good place to begin is at the Visitor Information Office which is housed in the former Crown Lands Office. Built in 1876 of red brick on a double foundation layer of granite blocks, it came at a time when agriculture was in the ascendency and land selections in the district were booming.

Behind the lands office is the court house, a good example of the Victorian Free Classical architectural style. The original building comprised a courtroom and three ancillary rooms. Major additions in 1882 included a verandah, new rooms for judges, barristers and witnesses as well as a jury box and retiring room. The court house is still in use.

SHIRE OFFICE & TOWN HALL

The shire office, built in 1902 of locally-made bricks, is a rare example of a Federation style local government building in rural Victoria.

The imposing town hall was built in 1869 of red brick and white stone with granite steps up to the entrance. Later additions around 1900 included a large concert hall and in 1932 a two-storey section all of which add to the present day grandeur.

FIRE STATION

The fire station, also built with locally made bricks is the oldest fire station in Victoria. An old hand-drawn fire cart equipped with leather buckets and some short ladders is housed inside.

HOTELS

The **Botanical Hotel** has a wide surrounding balcony decorated with exceptionally fine cast-iron lace. It was built in 1905 and replaced a rough slab and bark building that was the Jones Family Hotel & Store destroyed by fire. The **Victoria Hotel**, built in 1873, is another good example of lace decoration on balconies and verandah.

PIONEER PARK

The Pioneer Park near the site of the Lord Nelson mine was designed by Edna Walling beginning in 1939 and continuing after WWII. It features a wide variety of trees, a pergola, music bowl and a wide expanse of lawn and is dedicated in memory of local pioneers in St Arnaud.

A dam for the mine was located just above the park. This was converted into a swimming pool in 1939 and was filled with water pumped from the mine. The heavily mineralised water was unsuitable for gardens. The pool was closed in 1971 and is now filled with fish and used by the local angling club.

QUEEN MARY BOTANIC GARDENS

Queen Mary Botanic Gardens designed in 1884 featured exotic trees, an ornamental lake, pathways, garden beds and a picket fence perimeter with hedges inside and lockable gates. The gardens were restored in the 1990s.

LOVE'S COTTAGE

On the edge of town is Love's Cottage, built in 1868, the oldest remaining miner's dwelling in St Arnaud. Its 30 cm-thick walls were made with flat stone gathered from a mine site. Originally the cottage comprised just two rooms. The kitchen and adjacent room came later. All the floors are packed earth over which boards have been laid. In the 1940s the building's

exterior was rendered in concrete, although the back wall was left to show the original brickwork.

AROUND ST ARNAUD

St Arnaud services the region between Ballarat to the south, Horsham to the west, Swan Hill to the north and Bendigo to the east. The journey in from any direction offers wide panoramas across undulating fields with some high points and a forested area in the St Arnaud Ranges National Park (formerly the Kara Kara State Park) about 35 km south of the town.

PEBBLE CHURCH

A short drive south of St Arnaud on the Dunolly Rd brings you to the hamlet of Carapooee and **St Peter's Anglican Church.** Known as the **Pebble Church**, it was built between 1867 and 1869 using stones gathered by the miners from the nearby mine shafts. Due to sound workmanship the original building has stood the test of time. It is still used for weddings and baptisms and an annual service held each October.

WOOLSHED

At Tottington between St Arnaud and Navarre, a shearing shed built in the 1840s for Laurence Rostron who founded *Tottington Station* in 1844. Unsawn red gum slabs stood vertically were used for the walls and bark for the roof. Today to roof is of tin, but the rest of the structure is much as it was in Rostron's days. It is the oldest surviving woolshed in Victoria.

Fire Station, RE
Queen Mary Botanic Gardens, RE

The park has one of the largest intact areas of old-growth Box-Ironbark vegetation.

PROSPECTING

Gold prospecting is still popular around St Arnaud. Cherry Tree Creek, Carapooee Creek and Strathfellion Creek are recommended as good for gold panning. Metal detectors are also useful and they can be hired in town.

ST ARNAUD RANGES NATIONAL PARK

St Arnaud Range National Park has 13,900 ha of mainly steep, forested terrain and is an ideal place to experience what the forests were like before the goldrushes. It encompasses the former Kara Kara State Park and much of the St Arnaud Range State Forest.

The park has one of the largest intact areas of old-growth Box-Ironbark vegetation. When eucalypts are about 80 years old, hollows begin to form in their trunk and branches. The hollows are used by a variety of native birds, mammals and reptiles as nesting sites and for shelter. Hollow-dependent species include the Kookaburra and Crimson Rosella and nocturnal animals such as the Yellow-footed Antechinus and Sugar Glider.

Alluvial and shallow reef mining for gold commenced during the 1860s and a number of significant sites can be found around the park. Stock grazing continued in some areas until 1995 and sheepyards built from bush timbers also remain.

The **Centre Road Nature Drive** is approximately 50 km long, starts from the Wimmera Highway just outside St Arnaud and continues to Redbank. Two-wheel-drive vehicles are advised to exit at Stuart Mill.

The forest also contains the **Wax Garden**, named for the profusion of Fairy Wax flower growing in the area. The garden covers nearly four hectares and contains a wide variety of local floral species. It is best in October when the flowers are in bloom.

ST ARNAUD REGIONAL PARK

Rich with natural and cultural features this park, one km west of St Arnaud, offers great opportunities for bushwalking, picnics, and scenic drives in Box-Ironbark country, with magnificent views from Bell Rock and View Point. It provides habitat for many threatened species.

Bell Rock is a huge boulder shaped like a bell. In the 1860s miners called it Sebastapol Rock. Several mines were located in this area, including a large silver mine. The crushing plant was later converted first into a flour mill and more recently into **Ekersley's Eucalyptus Oil Distillery**.

LOCAL WINE & PRODUCE

St Arnaud is at the northern end of the Pyrenees wine region. Olives are also grown and eight major commercial groves are now established in the area, each with over 4000 trees. Essential oils such as eucalyptus, lavender and rosemary are found in the district.

Old Growth Forest, St Arnaud National Park, RE
Mechanics Institute, KS

STAWELL

LOCATION: 250 KM NORTH-WEST OF MELBOURNE ON THE WESTERN HWY
POPULATION 1857: 25,000
POPULATION 2006: 7000

Stawell is synonomous with the world famous foot race (the Stawell Gift), run every Easter, that attracts an international field of professional and amateur runners. However, the town does have a number of other strings to its bow.

It is Victoria's premier gold mining town. Founded on gold, it is still producing from a mine that burrows deep beneath the city streets 150 years after the initial strikes. At the same time the city has preserved some of its fine buildings and documented its mining heritage.

Today Stawell is also a major service centre for the surrounding district with a vibrant economy in its strong industrial, agricultural, tourism and retail sectors. What's more, there is an active art and cultural base. In addition, the city is the gateway to the Grampians National Park as well as being on the south-western edge of the Pyrenees wine growing district.

Miners' races had been run for recreation since the gold diggings began in the 1850s. But the more formalised Stawell Gift, as the race is called, began when a group of men formed the Stawell Athletic Club and held its first sports contest on Easter Monday 1878 in the town's botanical reserve (now the Grampians Gate Caravan Park).

The race was over 120 yards (110 m) of straight track and the winner collected a princely sum of 20 sovereigns. The race has been run every year since, with the exception of four years during WWII. In 1898 the event was moved to its current home at Central Park Stadium.

Prize money has increased to $40,000 for the winner, but hundreds of thousands of dollars more are won and lost in the betting ring. Because of the money involved and the fact that it is a handicap event, the Gift has attracted its share of controversy and at times, particularly in the 1930s,

Lookout, JM
View of Stawell, KS
Streetscape, RE

The Magdala Mine is the oldest mine still working in Australia.

Rear of Town Hall, KS
Building, JM
Town Hall, KS

figures from Melbourne's underworld have been known to take an interest in proceedings.

Some runners have even staged mediocre times in the months before the event to obtain a favourable handicap mark, while the rumours of a dark horse or two have cause betting plunges over the years. However the 2005 winner, Joshua Ross, has earned a special place in the race annals. He is the only man to have performed the double feat of winning the Gift off scratch and to win the race twice (2003 & 2005). Huge crowds now flock to Stawell for the Easter Carnival and the Stawell Gift deserves its fame.

BRIEF HISTORY

As in other parts of the Western District of Victoria, Mitchell's reports of 'Australia Felix' quickly encouraged squatters to follow in his wake. The first run to be taken up in the vicinity of Stawell was John Allen's 23,000 ha *Concongella Station* in 1841.

The pastoralists enjoyed a dozen years of calm before a shepherd named William McLachlan discovered gold at Pleasant Creek in May 1853. An estimated 20,000 miners descended on the diggings by 1857 and they set up a canvas and wood settlement known as Pleasant Creek.

The settlement was renamed Stawell in 1858 after Sir William Foster Stawell, Chief Justice of Victoria at the time. Alluvial gold began to run out in the 1860s and activity shifted a short distance to the north-east around a prominent geographical feature called Big Hill where quartz reefs were found to contain payable gold.

Administrative buildings were built in the new area and the original Pleasant

Creek townsite became known as Stawell West. By 1869 the two areas had been amalgamated into the Borough of Stawell.

Stawell experienced a boom period during the 1870s as mining activity diminished in other areas like Ballarat. Capital and experienced miners flowed into the town and the reef mines flourished. The arrival of the railway in 1876 further boosted social and economic life.

Reef mining continued until the 1920s when the main high-yielding Magdala mine ceased operations (temporarily as it has turned out). By that time, close to 58 tonnes of gold had been extracted.

Stawell survived the mining closures between the two world wars because it had become a service centre for the farming community and because it had established other industries such as a flour mill, brickworks, a tannery and woollen mills.

Big advances in mining techniques and higher gold prices sparked a revival of the Magdala Mine in 1981. It has the distinction of being the oldest mine still working in Australia – totalling 85 years of operation. Stawell's mine is also the richest gold producer in Victoria.

Today the mine produces 100,000 ounces (2.8 tonnes) of gold a year from a sulphide ore and new underground ore bodies are being exploited. The mine is more than one km deep and has hundreds of km of tunnels running under the city.

In addition, companies are drilling exploratory holes to look for other payable ore bodies in the area. Combining gold mining with the wealth from local commerce and agriculture, Stawell has become one of the Western District's most prosperous cities.

VISITOR INFORMATION

Stawell & Grampians
Visitor Information Centre
50-52 Western Hwy, Stawell 3380
www.ngshire.vic.gov.au
tel: 5358 2314.

Bus and train services run from Melbourne.

Stawell's premier event is the Stawell Gift Carnival held over three days from Easter Saturday to Easter Monday.

FOOD & ACCOMMODATION

Stawell is a large rural centre and it offers a wide range of accommodation including eight motels (three to four star) and five comfortable old-style hotels. The city also has four B&Bs, holiday apartments and self-contained cottages and cabins. There are two large caravan parks.

Dining options range from fully-licensed á la carte restaurants to wholesome pub bistros, luncheon cafés and bakeries. Many of the restaurants are associated with the city's motels.

SIGHTS

The sights of Stawell can be seen at a leisurely pace with a walk around the streets to look at the heritage buildings followed by a drive along the five km Golden Trail through Time which includes Big Hill, which overlooks the city and the majestic Grampians beyond.

Tracks in the Deep Lead Nature Conservation Reserve just out of town are also accessible by 2WD vehicles for most of the year.

CENTRAL PARK

Central Park, the walled, spacious oval that is the venue for the annual Easter Stawell Gift, is also used as a cricket, football and netball ground. The magnificent Memorial Gates, designed and built by a local foundry, were erected in 1903 to commemorate soldiers in the Boer War.

The grandstand, first built for the 1899 Gift meeting, was restored in 1990 using the original plans and is as close as possible to the early splendour of its covered, tiered seating.

Just along from the Memorial Gates is the Gift's Hall of Fame. It was opened in 1987 and features memorabilia of the Stawell Athletic Club's history, including newspaper items, photographs and paintings. There is also a series of continuously running videos of past events.

ST MATHEW'S UNITING CHURCH

St Mathew's Uniting Church is a Stawell landmark, built as a Presbyterian church in 1869. The imposing spire is 50 m high.

OLD FIRE STATION

The Old Fire Station had its tower built first to hang what was then the largest bell in the colony. It was cast in England. The fire station itself was opened in 1883.

QUARTZ GOLD MEMORIAL

The Quartz Gold Memorial lies at the foot of Big Hill and marks the location of the rich reefs discovered in 1856. A small arched brick construction nearby is one of the few remaining mine powder magazines left in Victoria today. It was built in 1875.

Gate, Victory Park, JM
Stadium, Victory Park, JM
Fire Station, JM
Powder Magazine, KS

DANE'S MEMORIAL SEAT

Dane's Memorial Seat has been erected at the summit of Big Hill in memory of Robert Dane, a miner who arrived in Stawell with his wife in 1856. Mrs Dane was the first white woman to set up home in the town and their daughter was the first white child born on the Goldfields.

MAGDALA MINE

On the eastern base of Big Hill is a viewing point for the operating Magdala Mine. Information boards and an audio commentary have been set up overlooking the crushing plant and the entrance to the mine's new decline.

RAILWAY STATION

The railway station, with stationmaster's

residence above, was built in 1876. The station closed in 1993 and has been converted into an art gallery with paintings, sculpture and photographs on display.

TOWN HALL HOTEL

The two-storey Town Hall Hotel, built in 1874 by local builder William Candy who became the establishment's first licensee, is one of Stawell's few original hotels to have survived.

JOHN D'ALTON MEMORIAL

In the centre of town the John D'Alton Memorial Fountain commemorates the efforts of John D'Alton to bring water to Stawell. Water was piped from Fyan's Creek in the Grampians, a distance of 35 km, using wooden flumes, a tunnel and piping. The project began in 1875 and finished in 1881. Much the same system (with more modern equipment) is in use today.

PLEASANT CREEK

The original settlement at Pleasant Creek is a few km west of today's main city centre, but it still contains several heritage buildings.

CourtHouse

The court house was built in 1860, the first of Stawell's public buildings. When the administration moved into the new centre, the building became the police barracks. It now houses a large collection of district records and memorabilia for the local historical society.

Police Residence

Nearby is the police residence, built in 1889. It is an elegant house and grounds and some of the original trees remain. The Police Department occupied this building for 100 years.

SCIENTIFIC & LITERARY INSTITUTE

The other heritage-listed building at Pleasant Creek is the Scientific & Literary Institute. Built in 1868 it had a library and meeting/reading rooms on the ground floor. An upstairs room was used as a lecture room and classroom while the local grammar school was being built. Constructed of hand-made red bricks it has single brick cavity walls and no internal supporting walls.

DIAMOND HOUSE

On the way back into central Stawell the Diamond House presents an intriguing sight. It was built as a private residence in 1868 by John Hearne. White and brown quartz stones were cut and painstakingly framed with timber to create the diamond pattern of its walls. Apparently there were no nails used in the building's construction. The original stonework remains untouched. Unusually for the period, it has a flat roof. The building has National Trust classification and is now a restaurant and motor inn.

AROUND STAWELL

Apart from the many vineyards located on all approaches to the town, the **Grampians National Park** lies 25 km south of Stawell. Centred on the township of Halls Gap, this picturesque region is renowned for its spectacular walks, mountain peak views, waterfalls and cool forests. Nearby **Lake Fyans** is also a place for leisure activities.

BUNJIL ROCK ART SITE

The Stawell-Grampians region is an area of significance for the Aboriginals. In the Black Range State Forest about 14 km south of Stawell on the Stawell-Pomonal Rd is the Bunjil Rock Art Site. Bunjil, the creator, was known and venerated by all the tribes of north-western and central Victoria. He was the principal legendary hero who created the world and gave the tribes their law and culture. The site was engulfed by bushfires late in 2005, but fortunately the artwork in a protected rock overhang is undamaged. It is the only known art site in Victoria containing figures painted in more than one colour.

STAWELL DEEP LEAD NATURE CONSERVATION RESERVE

The reserve is about three km north-west of town. It comprises 1823 ha of bushland containing a large number of threatened plant species. It is an ideal place to take self-guided nature walks. There are also vehicle tracks, although some are designated 4WD only. Visitors are advised to stay on the tracks to avoid the numerous mineshafts in the region.

Sisters' Rocks, KS
Bunjil's Cave, EG
Near Stawell, JM

Bendigo, PW
Daylesford, PW
Opposite: Near Anderson's Mill, PW

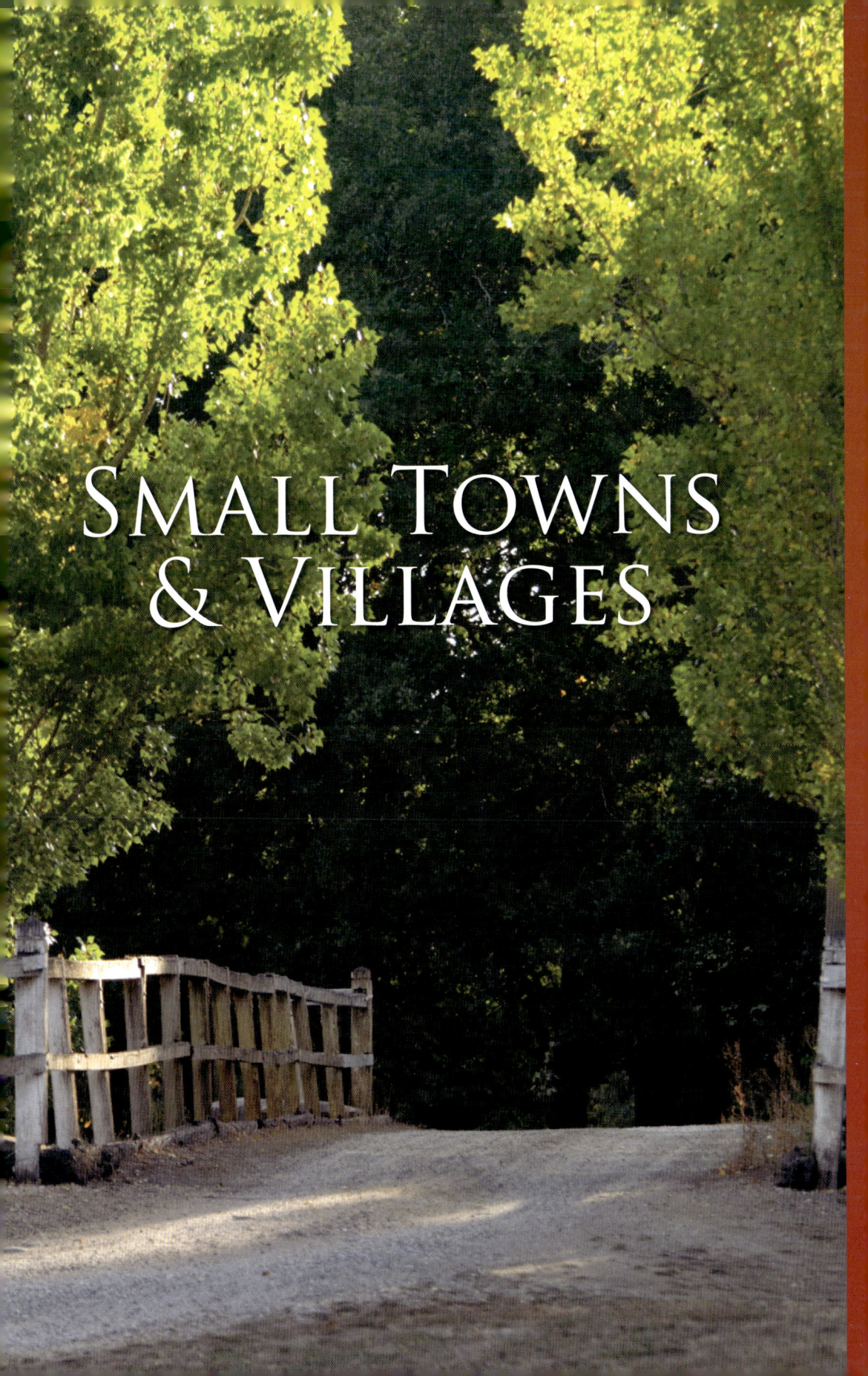

Small Towns & Villages

cooked by the obliging baker in his oven once the daily bread had been baked.

At the junction of the Glenlogie Creek and Avoca River there are brick foundations of the pump house, which once served the Amphitheatre Hotel, established in 1863, and still operating today.

AMPHITHEATRE TO ARARAT

Mt Cole and Ben Nevis, the latter a spectacular mountain, dominate the nearby small township of Elmhurst. Closer to Ararat the Pyrenees Mountains recede northward with some outstanding picture postcard landholdings at their foot.

AMPHITHEATRE

LOCATION: PYRENEES HWY, SOUTH-WEST OF AVOCA
POPULATION 1853: 6,000
POPULATION 2005: 249

As the Pyrenees Highway follows the south-eastern edge of the Pyrenees Mountains, the peaceful and almost abandoned settlement of Amphitheatre was a once thriving township on the fertile flats of the Glenlogie/Amphitheatre Creek and the Avoca River.

Gold was discovered in 1853 on the Glenlogie run, near the Amphitheatre, so named by Major Mitchell in 1836. The rush peaked in 1858-59 and miners were still making a reasonable living on gold dredging works as late as the 1890s.

Whilst there is little evidence today of the many businesses which once lined a busy Bailey St, in 1868 there were three butchers, two storekeepers, a draper, a shoemaker, two blacksmiths, a printer, a publican, a contractor, a policeman and a schoolteacher.

Bakery Park, was named after a once well established bakery on this site, where the locals could bring meat and vegetables for their roast dinner to be

BARKER'S CREEK

LOCATION: 9 KM NORTH-EAST OF CASTLEMAINE (MIDLAND HWY)

Barker's Creek takes its name from the first white settler in the area, a Dr William Barker who established the *Ravenswood No.1 Run* soon after Major Mitchell's exploratory journey of 1836/37.

The failed digger, turned notorious bushranger, Dan 'Mad Dog' Morgan, was arrested for the first time at Barker's Creek where he ran a slaughterhouse. Sentenced to 12 months hard labour for the robbery of four shepherds near Castlemaine, he was paroled in 1860 and promptly went on a robbery and killing spree in the district before the police shot him outside Wangaratta in 1865.

Near Amphitheatre, JM
Shop, Elmhurst, KS
Farm Shed, Barkers Creek, KS

On the road to Castlemaine, near Specimen Gully Rd, is the splendid building known as the Old England Hotel, which dates back to the early 1850s. It is now a private residence, but group tours can be organised with the owners. There is a museum at the old general store.

SPECIMEN GULLY RD

Specimen Gully Rd is a fascinating detour, just off the Midland Hwy on the outskirts of Castlemaine that gives a fascinating picture of the Goldfields. Turning off the highway, you pass under one of the many fine railway bridges common on the line between Melbourne and Echuca. Soon after you'll note a magnificent stone water-channel built to carry the creek under the line. The railway line was built to standards used in British India and this channel is a prime example of over-engineering!

Further along the road, there are some classic miners' cottages on the five acre (2 ha) lots that the miners were allowed to take up under the terms of their mining rights. There are signs of mining all along the road which eventually begins to climb the ridge that overlooks the Harcourt Valley.

Site of First Gold Discovery

Clearly visible from the road are the remnants of an old stone cottage once occupied by a shepherd named John

Worley (some records say George Wily) who worked on Dr Barker's original sheep station. A cairn and plaque indicates that Worley discovered gold nearby in 1851. One of Victoria's first finds, it was very rich and resulted in thousands of diggers from around the world descending on the region.

Specimen Gully Rd continues on to the ridge overlooking the Harcourt Valley. The ridge marks the dividing line between the gold bearing sandstone country to the west and the granite country around Mt Alexander which had no gold. There's a great view overlooking the valley to the east and back over the gold country to the west, with the remains of another ruined cottage lying just over the brow of the hill.

BEALIBA

LOCATION: 21 KM NORTH-WEST OF DUNOLLY; 37 KM NORTH OF AVOCA
POPULATION 1856: 12,000
POPULATION 2006: 250

Bealiba is a delightful small country town, full of character and charm, and with a history steeped in gold, timber and farming. Today Bealiba is very quiet, but for many years gold mines and timber mills operated here in abundance.

The first Europeans in the area, around 1840, were drovers. Squatters, George and Mary Coutts settled here in 1845. In 1856 gold was discovered at the foot of Mt Bealiba on Cochran's Run, but

Old England Hotel, Barkers Creek, PW
Bealiba Hotel, KS
Specimen Gully Rd, RE
Bealiba Shop, KS
Near Bealiba, KS

mining only lasted 18 months. Local agriculture boomed when the railway arrived from Dunolly in 1878. Bealiba's railway station began transporting timber to Melbourne, mainly to be used in railroad construction.

Evans Hotel was established in 1857 and is still operating today, as are the town hall (1879), the primary school (late 1870s), cemetery and Uniting church. The Anglican and Catholic churches also date back to the earliest days of settlement.

Pennington's Store (1860) was one of the first brick buildings in the main street. Originally used as a hay and corn store, it changed to a general store in 1891. The original International Hotel burnt down in 1867 and the Commercial Hotel was built in its place a year later. This once busy pub was last used as a butcher's shop and a peek through the front window today reveals a huge chopping block and meat hooks. Other notable buildings include the mechanics institute (library), railway station and post office.

There are walking tracks through the red ironbark forests of the neighbouring Bealiba Ranges.

Blackwood

LOCATION: 100 KM NORTH-WEST OF MELBOURNE, NORTH OF WESTERN HWY
POPULATION 1855: 13,000
POPULATION 2006: 300

Characteristic of Blackwood are well-kept timber houses, their verandahs at the road-edge with little picket fences. The Blackwood Hotel is known as an atmospheric watering-hole, sheltered beneath its immense oak tree.

Numerous streets, dotted with miners' cottages and newer weekenders, wind down to the river. These houses are

Blackwood Verandah, JK
Blackwood Cemetery, RE
Blackwood Hotel, RE

either particularly well-cared-for or spectacularly decrepit. In the river valley next to the caravan park are the Blackwood Springs, two of them only 30 m apart and yet with quite different mineral characteristics.

The cemetery, on the brow of a hill, has fascinating headstones and the grandiose Rogers family vault (1896). Matthew Rogers made his fortune from a mine at Simmons Reef. Being a stonemason, he built his own house and called it *St Erth*. Today it is famous as the spectacular Garden St Erth – two hectares of beautiful gardens, plus a café and horticultural centre selling the famous Digger's Seeds.

MT BLACKWOOD

No visit to Blackwood is complete without seeing Mt Blackwood. To get there, you travel south on the Myrniong Rd, then on dirt for about six km. After traversing densely-wooded slopes, the track arrives at open pasture-land and ascends what was clearly an old volcano.

The summit is occupied by a communications tower, but from the road there is a spectacular panorama, including the You Yangs, Corio Bay and Geelong, Werribee Gorge, Ballan, Warrenheip and Bunninyong and all the wild country of Lerderderg Gorge and the Great Dividing Range.

Bung Bong

Bung Bong is between Avoca and Maryborough and is a historic crossing place over Bet Bet Creek on the main route between South Australia and the Mt Alexander Diggings. The 1871 bridge was a prototype for a new series of locally constructed wrought-iron lattice-girder deck-truss main road bridges and it represents an important stage in the evolution of iron bridges. It was manufactured at the Ballarat Iron Works.

Buninyong

LOCATION: 125 KM NORTH-WEST OF MELBOURNE; 13 KM SOUTH OF BALLARAT ON THE GEELONG RD.
POPULATION 1851: 5000
POPULATION 2006: 1800

Buninyong is nestled at the base of Mt Buninyong, a dramatic extinct volcano that is a classic scoria cone. The main streetscapes of the town have changed little since the miners left for busier goldfields making it an attractive location from which to explore the region. The Aboriginal word, *Buninyouang*, means 'man reclining with bended knee' and is an apt metaphor.

Brief History

Buninyong claims to be the first inland town in Victoria. In 1838 the Learmonth brothers established a sheep station south of Mt Buninyong. Two of their employees established an eating house in 1841 and soon there was a small village servicing travellers on the route between Melbourne and Portland. In 1842 the first hotel license for an inland hotel was issued to the Crown Hotel.

On 8 August 1851 the local blacksmith

Thomas Hiscock struck gold in a gully three km west of the town. This discovery did not produce much gold, and most of the miners soon deserted to Ballarat, but in 1871 there were still 2281 people and 20 hotels at Buninyong and the quartz fields around the town produced payable gold well into the 20th century.

Visitor Information

Ballarat Visitor Information Centre
39 Sturt St, Ballarat 3350
www.visitballarat.com.au
tel: 5320 5741
freecall: 1800 44 66 33.

Sights

A walk in the town centre will reveal historic buildings dating back to the 1840s. Look for early shops with fine verandahs, the Eagle Store, Crown Hotel, former post office, former National Bank, and the former court house.

At the town's main intersection where Warrenheip St (the Geelong Rd) meets Learmonth St (the Midland Highway), the National Bank on the south-

Mt Buninyong, GS
Town Hall, BaT

eastern corner dates from 1867. The store on its eastern side was established in 1855. The commercial complex of the north-eastern corner incorporates an 1857 general store.

In Warrenheip St, there is the former Eagle Hotel (1858). The **Crown Hotel** at the corner of De Soza Park, claims to be the oldest continually licensed premises in the state, although the present building dates from 1885.

Built as the Presbyterian Church in 1860, the **Uniting Church** on Learmonth St. replaced an earlier church built in 1847. The **Church of England and Ireland** began in 1857 when the surviving Sunday School was built. **St Peter & Paul's Catholic Church**, Fisken St., dates from 1853.

Buninyong Town Hall

The Italianate Buninyong Town Hall, Council Chamber and court house complex was built in 1886. In 1994 the Shire of Buninyong became part of the Greater City of Ballarat, but the town hall has been carefully maintained and retains many original features, including a decorative stained-timber ceiling with metal ceiling roses for gasoliers.

The court house now has historic displays with a large collection of photographs and local records. Open Sundays from 1.30 pm to 4.30 pm.

Buninyong Library

The 1857 library is most unusual for its elaborate timber façade. It is now a tourist information centre and second-hand bookshop. The former reading room contains pictorial interpretation and text of Buninyong's early history in a series of panels.

Buninyong Botanic Gardens

Established in the 1860s, Ferdinand Von Mueller designed the Buninyong Botanic Gardens. The gardens feature a spring-fed lake (the Gong), barbecue facilities, a children's playground, 1870s swimming baths, converted into a courtyard garden, and the Queen Victoria rotunda (1901). On the southern edge of the gardens is the old miners' court, built in 1859.

De Soza Park

Originally the Buninyong Creek, this was an early mining site. There is a pleasant walking track, and an example of Chilean Mill, which was used to crush rock.

Buninyong Football Ground

Originally known as Royal Park, this was the site for Buninyong's annual highland games. The Duke of Edinburgh, visiting Buninyong on 13 December 1867, was so impressed by a highland fling danced by one of the locals that he consented to become the Patron of the Highland Society, and the ground was allowed to style itself on the Royal Highland Ground. Today it is known more simply as the Buninyong Football Ground.

Mt Innes - Birdwood Park

In 1918 trees were planted as a WWI memorial and there are excellent views of the township. The area has been incorporated into the Buninyong Walking Trail.

Eucalyptus Viminalis

The enormous manna gum in the middle of the unmade part of Nolan St, Buninyong, near Winter St, is by local tradition, the site of Aboriginal meetings. The tree is hundreds of years old and pre-dates white settlement.

AROUND BUNINYONG

Mt Buninyong

Mt Buninyong is an extinct volcano 1054 metres above sea level, offering magnificent 360 degree views of the entire goldfields region. The Ballarat district was first surveyed from this peak in 1837. Mt Buninyong is one of the most distinctive volcanic features of the Central Highlands group of newer volcanoes and is listed on the Register of the National Estate. There are a number of distinct craters, and a beautiful shady picnic area in the main crater. On a clear day you can see over nearby Ballarat and beyond to the Grampians in one direction and the ocean in another.

Travelling from Buninyong towards Geelong on the Midland Highway, take Lookout Rd on your left as you leave Buninyong. You can drive to the top, or park the car and take the pleasant walking trail, through the forest and the main crater, to the summit.

Cemetery & Site of Gold Discovery

On the Midland Highway, between Buninyong and Sebastopol, is Victoria's second oldest cemetery. Hiscock's Gold Monument is opposite Buninyong Cemetery. The Imperial Gold Mine site shows remains of large-scale gold mining activity.

Lal Lal, Historic Blast Furnace & Bungal Dam

Head south-east from Buninyong along the Midland Highway for 12 km to Clarendon and turn left onto the Clarendon-Lal Lal Rd (signposted for Lal Lal). It is four km to the hamlet of Lal Lal where the Lal Lal Falls Hotel, bluestone railway station and railway water tower are of interest.

After you cross the railway line take the first bitumen road on the right (the furnace is signposted). Continue straight ahead (the road becomes gravel) until you reach a picnic barbecue area (with toilets). Continue straight ahead, once again, for views over the dam and spillway. There are interpreted walks from the picnic area to the ruins of the blast furnace, which overlook the Moorabool River valley.

The National Trust considers the Lal Lal Blast Furnace an industrial site of great historical significance, and it is the only example of its type from the colonial era. The iron ore quarry and the smelting works were established by the Lal Lal Iron Mining Company in 1874. Remnants include the furnace, a Cornish flue, a tramway bed, mines, machinery sites, stone quarries and charcoal sites situated in five hand-built terraces cut into the side of the hill.

Lal Lal Falls

Lal Lal Falls Rd leads straight to the falls, situated on a tributary of the Moorabool River. They drop 34 metres down a gorge, created by the collapse of a lava tunnel, into a tranquil pool below. The local Aboriginals are said to have believed that Bunjil, their creator, resided at this place. The name is thought to mean 'dashing of waters'. There are picnic facilities and a small playground.

View from Mt Buninyong, AP
Lal Lal Blast Furnace, RE
Bungal Dam, RE
Lal Lal Falls, RE

CARISBROOK

LOCATION: 159 KM NORTH-WEST OF MELBOURNE; 7 KM EAST OF MARYBOROUGH
POPULATION 1863: 15000
POPULATION 2006: 600

Sprawled along a long stretch of flat highway, the former goldmining town of Carisbrook has some interesting treasures tucked away. Just minutes from Maryborough, Carisbrook is the centre of a rich agricultural area and possesses some fine old bluestone buildings.

At the junction of Tullaroop/Deep Creek and McCallums Creek, Carisbrook was surveyed in 1851 when Camp Carisbrook was set up as a police settlement to oversee convicts working on local farms. With the availability of water, Carisbrook developed industries such as a brewery, a cordial factory and a flour mill. At one stage there were 17 hotels in the town.

The Simson brothers made a great deal of money in the 1850s by collecting tolls from diggers travelling between Castlemaine and Maryborough. The township also grew as it met the need of through-traffic seeking facilities and supplies.

THE OLD LOG GAOL

Built in 1852, the gaol is made of interlocking horizontally-laid log walls. The original shingle roof has been replaced by corrugated iron.

Carisbrook Machinery Bed, DB
Junction Lodge, KS
Old Log Gaol, RE

JUNCTION LODGE

You can get an idea of how the local squattocracy lived by visiting *Junction Lodge*, a complex of farm buildings made of bluestone around 1873. Comprising a two-storey homestead, complete with stables, a barn, a blacksmith's, a kitchen plus workers' quarters on Camp St. Nearby are the ruins of a wall from Chalk's No.1 Mine.

WALKING TOUR

An hour-long walk through the centre of town includes such sites as the **Tilly Aston Memorial**, commemorating Carisbrook's most famous daughter, Matilda Aston, born 1873, who initiated the Braille Library and the Association for the Advancement of the Blind in Victoria.

AROUND CARISBROOK

The vast open landscape to the east of Carisbrook provides magnificent views to the south, taking in Mt Franklin and a seemingly endless vista of old volcanoes. Near **Moolort**, the highway is dotted with the bluestone remains of gracious homes, evidence of the rich pastoral history of the area.

Only minutes from Carisbrook toward Castlemaine is **Cairn Curran Reservoir** where superb sunsets can be viewed. Camping and caravans facilities are available.

Near town, there is a well-preserved

Aboriginal archaeological site – a stone arrangement measuring 5 x 60 m. It is one of only four stone sites in Victoria and the only one of a boomerang design. There are also a number of stone circles and a cairn. Unfortunately these are currently inaccessible, even to their traditional owners.

Most creek-side relics were destroyed by mining, but the local squatters did not want the creek ruined by gold washing and put their influence behind the development of Maryborough – which may have saved this site.

CHEWTON

LOCATION: 115 KM FROM
MELBOURNE; 5 KM EAST OF
CASTLEMAINE
POPULATION 1851: 30,000
POPULATION 1861: 3,353
POPULATION 2006: 400

The discovery of gold near Chewton at Specimen Gully in July 1851 pre-dated the discovery of gold at Ballarat, but it was kept secret until early September 1851. Such was the wealth of the new field, however, that Ballarat was soon virtually deserted.

The government camp on the new goldfield was relocated several times as the goldfield was developed, eventually coming to rest at the junction of Barkers and Forest creeks in Castlemaine. Outside the immediate control of officialdom, Chewton took shape with less regimentation than Castlemaine. Unlike Castlemaine, which was shaped by bureaucrats, Chewton's streets wander like the tracks they once were. In its shape and its names Chewton still feels like a diggings settlement.

Although Chewton is basically contiguous with Castlemaine, it has always jealously guarded its independence. It was large enough to operate its own affairs as a borough

until 1916 when a declining population finally forced the council to face the fact it was no longer viable. Rather than amalgamate with the next-door rival Castlemaine, the Borough of Chewton amalgamated with the Shire of Metcalfe (which took in Harcourt to the north, Taradale to the south, and Redesdale to the east).

The centre of Chewton is the group of buildings comprising the Red Hill Hotel, the post office (1879), adjacent town hall (1861) and the Primitive Methodist Church (1861) opposite. The church's twin flying buttresses are quite a spectacle. Next door, Trewartha House (1859), is a terrace noted for its zinc coated iron roof tiles imported from England before corrugated iron was used. Similar tiles cover the roof of the Theatre Royal in Castlemaine.

Post Office Hill to the south of Main Rd was one of the richest alluvial sites; the evidence of later sluicing is dramatic. Originally the gravel was so rich in gold that the first miners didn't even need to wash it. Many early diggers made their fortunes where Chewton Primary School now stands.

1850s Shops, KS
Cottage, KS
Cottage Window, KS
Primitive Methodist Church, KS

THE MONSTER MEETING

On the 15th December 1851, a Monster Meeting of 12,000 diggers met to protest against the mining license. This rally predated both the Red Ribbon protests in Bendigo (1853) and Eureka in Ballarat (1854) and was a key step in the move towards democracy. A plaque stands where the Monster Meeting was held at the Shepherd's Hut, just east of the junction of Forest and Wattle creeks.

MOUNT ALEXANDER HOTEL

On the bend in the highway before it descends into the centre of town, the sandstone building on the left was part of the Mount Alexander Hotel. Once the grandest hotel on the Goldfields, it was gutted by fire in 1863. Today the surviving half of the façade can be seen as a private dwelling. An old hut in its back garden, now incorporated into the main building, is believed to be a shepherd's hut dating back to the squatting era.

WESLEYAN CHURCH

The Wesleyan church (1861) & Denominational school, which became the Methodist church, then the Uniting church and Sunday school is now the community centre. The typical tiny 1860's shop next door is said to have been a baker, a boot maker's and stationery store. The 1850's sandstone building across from the store was the home of the Wesleyan church school's head teacher until 1895.

RED HILL HOTEL

The Red Hill Hotel, the first licensed hotel in the Chewton area, is still operating more than 150 years later. The original building was destroyed by fire and rebuilt in 1858 with an adjoining assembly hall. The gold taken from the soil excavated for the cellars is said to have paid for the construction of the whole building.

The cellars, with 60 cm thick sandstone walls, have been used for gold storage, as a temporary lock-up, as a cold storage prior to post mortem and they were, apparently, an excellent place for drinking after official closing time.

The assembly hall, known as the Adelphi Theatre, could seat 400 and over the years has been used for concerts, dances and parties.

On the northern side, a row of 1850's shops include a jewellery shop, bakery and butcher's shop with its carriage-

Wesleyan Church, KS
Red Hill Hotel, KS
Aerial View, BS

way through to stables at the rear. The Francis Ormond Mine was at the rear of the service station. The constant noise of its batteries and those of the Argus Hill mine over the creek must have been deafening.

THE POST OFFICE

The Chewton Post Office and residence was built in 1879. In 1997 ownership was transferred to the non-government Chewton Domain Society. Services do not include house-delivery, so the people of Chewton visit the building daily to collect their mail, making the post office the town's chief meeting place.

The old square derelict building opposite was once the Duke of Northumberland Bank.

CHEWTON TOWN HALL

Although this building has a reputation for being the smallest town hall in Victoria, it was not built as one. It was built in 1858 as a meeting room, and it was used by local organisations like the council and the Freemasons. It also housed the meetings and library of the Chewton Mechanics Institute.

The Chewton Borough Council used it for its meetings from 1861 to 1916. When Chewton was declared a municipality 28 councillors were elected, one of whom was James (later Sir James) Patterson. Patterson became mayor, was a prominent federationist, and Premier of Victoria

1893 to 1894. His portrait still hangs above the fireplace, and the building now houses other historic information and photographs; it's open from open from 1 to 4 pm on weekends.

Just west of the town hall in Ellery Park is the 1850's portable police lock-up, also owned by the community.

WATTLE GULLY MINE

Just off Fryers Rd to the south of Chewton is the Wattle Gully Mine. With its intact poppet head and machinery, Wattle Gully was one of the richest and longest running gold mines in Australia. The Wattle Gully Mining Company was formed in 1859 and mining operations continued until recently. It is currently a centre for re-exploration of the Castlemaine gold fields and it is hoped the mine may be revived.

CASTLEMAINE DIGGINGS NATIONAL HERITAGE PARK

The Castlemaine Diggings National Heritage Park is inspirational as a forest that has remade itself around the relics that the gold-diggers left behind. The park encapsulates and tells the story of the defining years (the early 1850s) of the Victorian gold rushes. It possesses story-rich landscapes that contain one of the greatest collections of authentic mining sites in Australia. Comparison with similar places overseas shows that the park comprises one of the last extensive mid-19th century mining landscapes left in the world. The full significance of the park can only be appreciated by visiting the sites.

The land and its heritage features have long been recognised as important touchstones for imagining and commemorating the gold rush and its consequences. As early as 1862 a

The park comprises one of the last extensive mid-19th century mining landscapes left in the world.

Post Office, KS
Town Hall, KS
Wattle Gully Mine, KS
Slate Mine, Castlemaine Diggings
National Heritage Park, KS

correspondent to a local paper was moved to write on the physical traces of the gold rush he observed in the gullies around Castlemaine: *'although only 10 years old… [it is]… an antiquity as real as any other ther world has reproduced.'*

The park is the sixth property to be placed on the National Heritage List. The goldfields it encompasses played a substantial part in all the changes gold brought to Australia: increased population, increased wealth, growth in manufacturing, improvement in transport, development of regional centres, development of a middle class, democratisation of political institutions, reform of land laws, and the genesis of an Australian Chinese community. Its impact was felt beyond Australia as well.

Chewton is an integral part of the Castlemaine Diggings National Heritage Park. The park itself is doubly unique because it brings together natural heritage (Box-Ironbark forests) and cultural heritage (Aboriginal sites, and the ruins and relics left behind by 19th century gold rushes).

The park includes the Garfield Water Wheel, Pennyweight Flat Cemetery, the Eureka Reef Walk and Spring Gully Junction Mine, as well as much more. The Visitor Information Centre can provide detailed information, including maps. Also, check the sections on Fryerstown and Vaughan.

There are many stunning walks and drives. Walks include:

Garfield Water Wheel to Expedition Pass

The eight km (return) walk begins at the Garfield Water Wheel and takes in Sailors Gully, Golden Gully, Donkey Gully, the remains of a blacksmith's shop, a puddling machine, ruins of the Welsh Village, slate quarries, a walled garden and Quartz Hill.

Chinaman's Point to Expedition Pass

The four km (return) walk begins from Chinaman's Point Rd car park and follows a track running parallel to Golden Point Rd. There is a derelict bridge at Donkey Gully and many gravels mounds and areas that become swampy after heavy rain.   Near the reservoir there are cascades, which look spectacular after rain, and sections of the water race to the Garfield Water Wheel.

Commuter Track

Castlemaine is linked to Chewton by a newly-created commuter track that allows walking or cycling along Forest Creek.

Garfield Water Wheel

Massive stone abutments and brick foundations are all that remain of the largest water wheel in the southern hemisphere. The water wheel, erected in 1887, was 22 m in diameter, and drove the 15-stamper quartz crushing battery of the Madame Garfield Mine until 1904.

Except for its bearings and iron

Welsh Village, Castlemaine Diggings
   National Heritage Park, KS
Garfield Water Wheel, Castlemaine
   Diggings National Heritage Park, KS
Water Race, Castlemaine Diggings
   National Heritage Park, KS

buckets, the wheel was constructed entirely of wood. The water that turned the wheel flowed three km from the Expedition Pass Reservoir, the final 250 m by an elevated wooden aqueduct. This flume delivered water to the highest buckets which turned the wheel. The Madame Garfield Mine's main shaft was sunk to 346 m.

To get there take North St, on the northern side of Pyrenees Hwy, opposite Fryers Rd (to Fryerstown). North St leads across Forest Creek past Bryce Ross Gully and into the hills where many deep mines extracted gold bearing quartz. Travel 200 m straight on then veer to the right to the Garfield Water Wheel ruin.

Returning toward Chewton, travel east along Walker St to see the site of the Argus Mine, a ruin, the stone walls in Forest Creek built to protect Chewton from flooding, and onto Shield's Tannery (1880).

Eureka Reef

The remains at Eureka Reef bring together many of the park's most important and interesting elements. They're found three km south of the intersection of Eureka St and the Pyrenees Hwy (the intersection is one km west of the Chewton Town Hall). The 1.8 km walk at the Eureka Reef includes the remains of some of the earliest quartz mining relics in Victoria, including the chasm where there was once an exposed quartz reef (see Big Reef, near Amherst), foundations of primitive crushing batteries, the ruins of a mining village and Aboriginal rock wells.

Forest Creek Historic Diggings

The Forest Creek Historic Diggings occupy the remains of White Hill and Red Hill where there have been nearly

150 years of continuous mining.

An interpretive trail (allow 30 minutes) shows how miners won gold through various phases of mining from the early 1850s to the mid 1950s. Try gold panning at the dam by purchasing a gold pan with instructions at the Castlemaine Visitor Information Centre.

The Forest Creek Historic Diggings are midway between Chewton and Castlemaine on the south side of the Pyrenees Hwy. In addition to the mining relics there is an indigenous garden and a maze designed to illustrate the region's geology.

Pennyweight Flat Cemetery

A wander through any of the local cemeteries illustrates how tough life was on the diggings. Typhoid and diphtheria were rampant. The few surviving headstones at the 1850s Pennyweight Flat Cemetery are a reminder of how vulnerable children were. The small cemetery has a secret feel even though it's so close to the town. To get there, take Farran St (between Chewton and Castlemaine opposite the Forest Creek Gold Mine) north, then turn right into Colles Rd. The cemetery is just over the bridge on the left.

Golden Point

At Chinaman's Flat, near the site of the Monster Meeting, Forest Creek takes a turn northward, past the place where gold was first found at Golden Point, and

Eureka Reef, Castlemaine Diggings National Heritage Park, RE
Forest Creek Historic Diggings, Castlemaine Diggings National Heritage Park, DB
Pennyweight Flat Cemetery, Castlemaine Diggings National Heritage Park, KS
Pennyweight Flat Cemetery, Castlemaine Diggings National Heritage Park, KS

on to the Expedition Pass Reservoir.

Take picturesque Golden Point Rd north towards Faraday. Golden Point still contains many stone ruins of homes, churches, schools, hotels and stores. There was a Welsh Church, also run as a school, on Chapel Street where the manse is now a private residence. Coy's wine and spirit store is also a private home.

Opposite the old bridge at Donkey Gully, a number of stone terraces can be seen on the eastern side of Golden Point Rd. They were part of the Faraday Tea Gardens where one could pick cherries, grapes or strawberries for a small charge.

Commissioner's Gully Rd is off Golden Point Rd. The police camp was established 200 m up the gully and some of the remnants of the house are still visible. A dam was built here to supply fresh water to the eastern part of Chewton, but during the flood of 1889 the wall was breached and has never been repaired.

Expedition Pass Reservoir

Expedition Pass was the route Major Mitchell found through the hills on his return journey to Sydney in 1836. A cairn beside the reservoir records the details. The reservoir was built in 1868 to supply water for the mines and for Chewton. The reservoir is filled by rain from surrounding hills and topped up via the Coliban Channel from Malmsbury. It is a popular local swimming place during summer.

Spring Gully

Spring Gully is a couple of kilometres west of Fryers Rd (to Fryerstown) and is now a beautiful, peaceful spot set in a heavily-mined landscape. There's a 1.5 km walk that takes in the ruins of

the rich Spring Gully Junction Mine and the old spring (now buried under five m of soil and rubble).

On the way to Spring Gully you pass the Wattle Gully Mine, a water race from Taradale (which crosses Fryers Rd at the top of the hill just past the Wattle Gully Mine), and the grave of pioneer Elizabeth Escott and her 16-year-old daughter Fanny.

Clunes

LOCATION: 148 KM NORTH-WEST OF MELBOURNE; 36 KM NORTH OF BALLARAT
POPULATION 1875: 6.000
POPULATION 2006: 850

Nestled in a picturesque valley, Clunes was the site of Victoria's first official gold discovery. It is also one of the most intact gold towns in Australia. Nestled in a valley, and surrounded by hills that are actually extinct volcanoes, it is a living museum that seems to have emerged virtually unchanged from the 19th century.

The streets are wide and few cars clutter the long stretch of verandahs or interrupt the quiet. It is easy to walk down the middle of the main street and imagine horses and carriages parked outside shops, or ambling along at a slow trot, dust trailing gently behind.

Not surprisingly, Clunes has provided a film location for such movies as *Ned Kelly, Mad Max, On the Beach* and ABC

Castlemaine Diggings National Heritage Park, KS
Expedition Pass Reservoir, Castlemaine Diggings National Heritage Park, KS
Spring Gully, Castlemaine Diggings National Heritage Park, KS
Elizabeth and Fanny Escott Grave, Castlemaine Diggings National Heritage Park, RE
Library, Clunes, JM

television series *Queen Kat, Carmel & St Jude* and *Something in the Air*.

James Esmond discovered gold on July 7, 1851 and triggered one of the world's greatest series of goldrushes. It is said that Esmond was looking for worms near the local creek when the glitter of gold took his mind off fishing! Esmond later fought at the Eureka Stockade with his friend Peter Lalor.

The initial excitement at Clunes only lasted a month, before discoveries at Ballarat took centre stage. In 1856 the London-based Port Phillip and Colonial Gold Mining erected a large stamping battery to exploit the quartz reefs beneath land that had been farmed by the region's original pastoralist, Donald Cameron.

In the late 1850s Clunes was the site for a bitter and acrimonious underground war between competing mines when their tunnels, following the same lead, met. Pitched battles required police reinforcements from Creswick and Ballarat.

VISITOR INFORMATION

Clunes Visitor Information
Bailey St, Clunes 3370
www.hepburnshire.com.au
tel: 5345 3896.

Clunes is within an easy drive to over 20 wineries.

SIGHTS

With some of the best-preserved 19th century architecture in the state, Clunes has over 50 buildings of historical significance and the only School of Mines outside a regional city.

Notably impressive are the Mannerist-style town hall & court house (1872) the Italianate post and telegraph office

(1879), the churches, the sandstone, bluestone and brick buildings in Fraser St and several old bank buildings.

The Wesleyans had the largest congregation due to the many Cornish miners in Clunes. Wesley College, Australia's largest co-educational private school, now has a campus for Year Nine students in the town centre.

Fraser St is wide and elegant, full of 19th century shops (most dating from the 1870s) with original storefronts and distinctive verandahs. In 1872 there were a staggering 23 hotels plying their trade along the length of Fraser St.

The Clunes landscape changed radically in the period 1880-1930 when bare hills ravaged by mining gave way to a tree planting program – both public and private – especially in Queens Park and along Creswick Creek.

Following the collapse of the mining industry in the mid 1890's, knitting factories reused earlier buildings and several factories were erected along the creek. Walking paths now lead along the banks of Creswick Creek and there's a fine view of Clunes from the scenic Rd just north of town.

Autumn, JM
Shop, JM
Guesthouse, JM

km west of Clunes. A huge old pine tree, planted in 1916, and magnificent views of the surrounding volcanoes are features of this picturesque granite outcrop. Legend has it that Captain Moonlight, a famous bushranger, used a cave on the mountain as a hideout – and that he was protected by locals. Today, despite being grazed, quarried and logged, Mt Beckworth has an excellent display of spring wildflowers. There are 250 species of plants, including 35 orchid species.

COLBINABBIN

LOCATION: 65 KM NORTH-EAST OF BENDIGO (BENDIGO-MURCHISON RD)
POPULATION 2006: 115

Colbinabbin is a small service town in an important sheep and grain growing area and an emerging viticulture industry. It has good access to the regional centres of Bendigo, Echuca and Shepparton. Crops include wheat, barley, oats, canola, tomatoes and Asian vegetables.

Port Phillip Mine

The remains of the **Port Phillip Mine** site form an imposing escarpment on the north edge of town, and a lookout provides a view of the creek and buildings in the tranquil valley. The mine sites are within easy walking distance of the town centre.

Clunes Museum & Bottle Museum

The **Clunes Museum & Bottle Museum** is housed in an old school (1881) and contains an intriguing collection with more than 6000 items dating back to the 1850s, including ginger beer bottles that were transported as ballast in sailing ships. It gives a fascinating insight into how food and drink was managed

AROUND CLUNES

Mt Beckworth Scenic Reserve

Mt Beckworth Scenic Reserve is eight

COLBINABBIN COMMUNITY WELL

The community well is a rare example of a once plentiful type of water supply infrastructure when wells, soaks, tanks and dams were in common use in rural settlements. At Colbinabbin, the earliest form of water technology (the whip) has survived from the 1860s and is one of only three extant whips in Victoria.

Port Phillip Mine, JM
Town Hall, PW
Detail, PW
Streetscape, KS
Colbinabbin Community Well, DB

COSTERFIELD & GRAYTOWN

LOCATION: 12-29 KM NORTH-EAST
OF HEATHCOTE, 120 KM FROM
MELBOURNE
POPULATION 1868: 30,000
POPULATION 2006: 0

Along the road from Heathcote to Nagambie you drive through the settlements of Costerfield and Graytown, two old gold mining towns offering a wealth of history, but few visible remains of their glory days. Refreshingly small and abandoned with some visible remnants of goldrush times, Costerfield is a delight for those who love real ghost towns. Travel a bit further on to Graytown and once again you'll find more ghosts than town.

COSTERFIELD

Costerfield's antimony mines once employed 700 men, but they closed in 1925. At one time local mining produced 92% of Victoria's antimony, a metal that is used in munitions and that was a prized contributor to Britain's wars until after the end of WWI. Just north of Costerfield,

Redcastle (previously Balmoral), once had 17,000 people, but ceased to exist when mining ended around 1910.

GRAYTOWN

Graytown, formerly known as Spring Creek, was surveyed in 1848. When gold was discovered in 1868 more than 30,000 miners arrived from nearby Heathcote, Whroo and Rushworth, significantly depleting the population of those towns.

To visit Graytown today it is almost impossible to believe that 90 hotels, several banks, two newspapers, a post office, a police station, courts, most church denominations, a concert hall, school, stores, all kinds of businesses and even a cricket team once existed.

When proclaimed a borough between 1868 and 1870, Graytown was compared in size to the Ballarat of goldrush days. In 1871 there were 511 buildings in the town. Today, the Heathcote-Graytown National Park forest has reclaimed the site of one of Victoria's biggest and shortest-lived goldrushes. All that remains are some white ant-riddled stumps of old poppet legs and fence posts, water-filled holes that were once cellars of hotels, empty streets and a chimney from a once busy boarding house. There are also water-filled mine shafts, some as deep as 500 ft (165 m).

At one stage at the height of the rush, water was so scarce that puddlers washed their gold in beer, as it was so plentiful and cheap – compared to threepence for a bucket of water! As is often the way in Australia, in 1870 there was too much water – when a freak storm and heavy rain caused the town to be abandoned. The mines filled with water and the miners moved on, leaving only a few families in the area.

Costerfield Cottage, KS
Costerfield Mine, KS
Graytown Water Course, KS
Graytown Mining Relics, KS
Graytown Cemetery, KS

The remarkably intact old cemetery, including a Chinese section, contains a marble headstone with text written in Welsh. Set 700 m deep into the forest off the highway, it can be reached by a dirt track on the northern side of the Heathcote-Nagambie Rd, near the Graytown sign.

Graytown is home to one of Victoria's oldest wineries, Osicka Wines.

VISITOR INFORMATION

Heathcote Visitor Information Centre
cnr High and Barracks Sts
Heathcote 3523
www.heathcote.org.au
tel: 5433 3121.

AROUND GRAYTOWN

The **Heathcote-Graytown National Park** includes some of the most significant environmental, cultural and recreational values in the largest remaining Box-Ironbark forest in Victoria. Significant Aboriginal cultural sites and places are evident. **Mt Black**, east of Graytown, is the highest point with fine views and picnic area. **Melville's Lookout** is a prominent ridge where bushranger, Captain Melville, waited to ambush miners during the goldrush. See separate section.

A 27 km dirt track to Rushworth takes you through the National Park via Whroo.

Dunolly Collectibles, KS
Broadway, RE

DUNOLLY

LOCATION: 23 KM NORTH OF MARYBOROUGH
POPULATION 1856: 45,000
POPULATION 2006: 750

Dunolly is a sunny place, with a strong connection to the past. This historical connection extends well past the main street, Broadway, which is itself lined with red brick buildings dating from the goldrushes.

More gold nuggets have been found in and around Dunolly than anywhere else in the world. Together with Moliagul and Tarnagulla, Dunolly forms a region known as the Golden Triangle.

The first European settler was a Scot, Archibald McDougall, who arrived with his sheep in 1845 and settled at Goldsborough, five km from the present township. He named his pastoral run and home after Dunolly Castle, the seat of the McDougall clan near Oban in Scotland.

McDougall was apparently unimpressed by the sudden influx of mining neighbours in 1852 because after gold was discovered and a town called Dunolly began to emerge near his home, he promptly sold up!

When a major strike occurred downstream in 1856 a new rush began and a new township, known initially as New Dunolly, emerged. The new townsite – the Dunolly of today – was surveyed in 1857.

The town was laid out in three distinct sections, with the commercial premises on Broadway, then a buffer zone of gardens, then government and administrative buildings on Market St (parallel to Broadway and two streets to the east). The town's centre of gravity settled on Broadway after the arrival of the railway line on the west side – but visitors should not miss Market St.

Broadway was flanked by stores, lodging houses, hotels, sly grog shops, theatres, booking booths, bowling and skittle alleys and brothels. The theatres offered a choice of Shakespeare, popular plays of the time, vaudeville and 'poses plastique' (nude posing).

As the alluvial gold was stripped from the surrounding creeks the population dwindled, only to revive again with a new strike in the 1860s. Nuggets have been found ever since.

The *Welcome Stranger*, at 2380 ounces (67.5 kg) the largest nugget ever discovered in the world, was found 14 km north at Moliagul in 1869 (see Moliagul). In 1980 the 875 ounce (24.8 kg) *Hand of Faith* nugget was found by a prospector using a metal detector, at Kingower, 30 km north of Dunolly.

There are many stories of smaller finds in recent times and prospecting in the attractive bush plains around the region remains a continuing passion for many people.

VISITOR INFORMATION

The Welcome Stranger Café offers tourist information, or contact Central Goldfields Visitor Information Centre Cnr Alma & Nolan Sts, Maryborough 3465, www.centralgoldfields.com.au tel 1800 356 511.

BROADWAY

A walk along Broadway's tree-lined footpaths will take in the post office (1890), the old Bendigo Hotel with its original interior and Cobb & Co stables (a huge red brick building at the rear of Daly's General Store), the Royal Hotel (first erected in 1856, but rebuilt in 1894), the Railway Hotel (built in 1858 as the Criterion Hotel), and the old London Chartered Bank (a two-storey classical revival building with Roman arches on the ground floor built in 1857). Finders Prospecting Supplies would once have sold picks and shovels. Today it sells and hires metal detectors!

Goldfields Historical Museum

Housed in an 1862 building on Broadway, the museum displays include a replica of the *Welcome Stranger* gold nugget. The anvil on which the nugget was cut is displayed outside. There are also relics of early European settlement and a four-million-year-old fossilised wombat jaw.

Town Hall

The town hall was originally built as a court house in 1884; the first town hall was built in the administrative section of the town, but because of acoustic and other problems the buildings were swapped in 1887. The court house became the town hall in Broadway, and the town hall became the court house on the corner of Bull and Market Sts.

Shop, KS
Broadway, KS
London Chartered Bank, RE
Ironmongers, KS
Goldfields Historical Museum, KS
Town Hall, KS

Court House

The restored court house is a magnificent Italianate building. Built in 1862, and originally the town hall, it became the court house in 1887, and operated until 1979. Next door to the court house, on the police station property, is the town's original brick lock-up (1859). Opposite is James Bell's mansion (1869). Bell was a successful local businessman with interests in mining and grain exporting.

Post Office

Designed by Henry Bastow in a Renaissance Revival style and built in 1891, this is a classic public building that reflects the town's strong civic pride. Few country post offices are as magnificent.

MARKET STREET

Old Post & Telegraph Office

Part of the original group of administrative buildings on Market St, the old post and telegraph office at the corner with Thompson St, is a fine example of the early standard designs the government used. The building was only used for its original purpose for 19 years before the grander new post office was built closer to the railway station on Broadway.

Anglican Vicarage

South of the Thompson St intersection, the Anglican vicarage was built in 1866 and is an unusual Gothic revival building classified by the National Trust. The vicarage is a single-storey building with steep roofs and decorative gables, erected in 1864-65. It is now a private residence.

CHURCHES

There are a number of fine churches that have survived – unlike many of the 14 hotels that once lined Broadway.

St John's Anglican Church

St John's Anglican Church, on the corner of Thompson and Barkly Sts, is a freestone structure erected between 1866 and 1869. Deason and Oates, the discoverers of the *Welcome Stranger*, contributed £1000 to its construction.

A little further down Barkly St is St John's Hall (1857) which was used as the first common school.

St Mary's Catholic Church

St Mary's Catholic Church, on the corner of Hardy and Alice Sts, was opened in 1871 – which meant the parishioners no longer had to meet in a local hotel. It is built of sandstone quarried in the nearby hills and granite from Mt Hoogly. The steeple was added in 1980.

Presbyterian & Wesleyan Churches

The Presbyterian Church, now the local RSL, was opened in 1864. The simple red brick Wesley Church, on Tweedale St, was opened in 1863 and the parsonage dates from 1880.

St Mary's Catholic Church, KS
St John's Anglican Church, KS
Laanecoorie Reservoir, KSt

CHAUNCY COTTAGE

Chauncy's Cottage on Havelock St (1859) is an attractive brick and stone cottage that has changed little since Phillip Chauncy lived there in the period 1860-1866. Chauncy was the local land surveyor. The house was originally intended to be an inn, but Chauncy bought it half finished and converted it into a family home.

AROUND DUNOLLY

Laanecoorie Reservoir

About 12 km east of Dunolly along the road to Eddington, the reservoir was built in 1889, and was one of Victoria's first reservoirs. It is a popular spot for picnicking, swimming, boating, fishing, and walking.

Wild Dog Diggings

East of Dunolly off the Eddington Rd, the Wild Dog Diggings was rushed in 1856. There are many shallow shafts overlooking Bet Bet Creek. The round shafts were dug by Chinese, the square by Europeans.

Dunolly Cycle Tracks

The Dunolly Cycle Tracks are a series of signed bush roads and tracks varying between 14 and 47 km in length. The area is rich in Box-Ironbark fauna and flora, and fascinating Goldfields history. There are cemeteries, old pubs, racecourses, abandoned town sites, and goldfields. Routes pass through the old gold towns of Dunolly, Tarnagulla, Bealiba, Waanyarra, Betley and Goldsborough. Maps are available from Dunolly Rural Transaction Centre and the Central Goldfields Visitor Information Centre in Maryborough.

FRANKLINFORD

LOCATION: 10 KM NORTH OF DAYLESFORD (OFF MIDLAND HWY)
POPULATION 1855: 100

On the rolling slopes below Mt Franklin, Franklinford is a quiet settlement open to a vast western vista of rich volcanic farmland and mine-ravaged bush. A dominant group of red brick churches and outbuildings, a corner hotel (now private residences) and a sandstone ruin are juxtaposed with a handful of contemporary homes, all peacefully converging on the five-way intersection that leads to Daylesford, Newstead and Castlemaine.

Franklinford is mostly remembered as part of the Aboriginal Protectorate that was established in 1841 by Edward Stone Parker to protect what was left of the original inhabitants of the region, the Dja Dja Wurrung, after white squatters and their sheep ended the traditional way of life and displaced them from their country. The Franklinford Protectorate began with 170 inhabitants. By 1855, 100 Aboriginals were working the cornfields of a self-sufficient run, but by 1864 only 10 remained and were moved to *Coranderrik*, a station at Healesville.

Mt Franklin, PW
Franklinford Mist, PW
Franklinford Cottage, PW

Mt Franklin was an active volcano and a significant ceremonial site during the time of Aboriginal occupation.

MT FRANKLIN

Mt Franklin, located eight km north of Daylesford, two km off the Midland Highway (Daylesford-Castlemaine Rd), was an active volcano and a significant ceremonial site during the time of Aboriginal occupation. Known as *Lalgambook* by Aboriginal people, numerous tools and implements used by them have been discovered on surrounding farmland.

Now extinct, the steep cone of the old volcano is covered in pine forest, making it a distinctive and somewhat forebidding landmark. It is considered a fine example of a breached scoria cone. A road takes you into the crater, via the breach, which is believed to have been caused by a flow of lava breaking through the crater rim. The crater now has a shady picnic ground. From the lookout there is a spectacular view of many distant volcanic peaks. There is a scenic walking track along the rim of the crater.

FRYERSTOWN

LOCATION: 120 KM NORTH-WEST OF MELBOURNE; 10 KM SOUTH OF CASTLEMAINE
POPULATION 1858: 15,000
POPULATION 1975: 56
POPULATION 2006: 120

The tranquil hamlet of Fryerstown is on the narrow, hilly road from Chewton to Vaughan Springs that snakes through deserted diggings and Box-Ironbark forests. The colour and shade of large old European trees lining the streets soften the harshness of the bush and the mining scars that surround the town.

Mullock heaps and the remains of the **Duke of Cornwall Mine**'s engine house and powder magazine announce the northern entrance to the quiet cluster of miners' cottages and once-public buildings nestled by Fryer's Creek. The Duke of Cornwall machinery, an impressive relic still remarkably intact, is classified by the National Trust.

Settlement began in early November 1851 with a few tents and stores around

Mt Franklin, PV (AP)

a shepherd's hut on Fryer's Creek where the quantity and accessibility of alluvial gold produced riches for many years. Nuggets up to 100 ounces (2.8 kg) were common. In 1853, the gold escort to Castlemaine transported 559,308 ounces (15.9 tonnes) of gold. The *Heron* nugget, found near Golden Gully in 1855, weighed 1,023 ounces (29 kg) and was described as: '… a Monster…completely eclipsing all former nuggets.'

Fryers Creek was proclaimed Fryer's Town (now Fryerstown) in 1854. In 1861 there were 213 homes and 30 hotels. However, 30 years later only 91 houses remained, and by 1975 many of those were deserted.

George Levi Carter established a boot factory at Fryerstown in 1853. A stand of large old box trees near the local hall were used to by police troopers to chain prisoners before a gaol of rough logs was built in 1858. A cottage built in 1854, representative of Goldfields administration quarters, survives as the township's oldest building. The mechanics institute, constructed in 1863 as a memorial to Burke and Wills who died at Coopers Creek 22 months earlier, was the social centre of Fryerstown and is still used today.

A few of the other early buildings, such as churches, hotels and stores, the post office and the court house, are now maintained as private homes.

There were more than 25 reefs discovered at Fryerstown after 1855. Heron's Reef, one of the longest lasting, produced gold for 30 years. The rich Mosquito Mine, which yielded 43,581 ounces (1.24 tonnes) in 1871 alone, was owned by the Rowe brothers – five miners from Cornwall known as 'the Quartz Kings of Fryerstown'. The Rowe brothers also invested in

the neighbouring Duke of Cornwall mine which housed a 75-horse-power beam steam engine transported from Cornwall. Edward Rowe's large sandstone home, *Lambruk*, is on the old road to Taradale.

The Fryerstown Antique Fair held annually on the Australia Day weekend at the end of January, is one of the longest running and largest in Australia. Held in and around the Burke and Wills Mechanics Institute and sprawled out beneath the massive old eucalypts, this huge country-style market, reminiscent of English fairs, attracts about 30,000 people.

FRYERSTOWN METHODIST CHURCH HALL

Next to the Fryerstown Methodist Church in Heron Street, the hall is a rare, intact example of the timber prefabricated buildings often used as court houses, schools and gold wardens' offices during the late 1850s. It is thought that it probably moved here after use in the surrounding goldfields as a court house. The walls are made of thick boards that are unlined inside. It is likely to have been used as a multi-purpose church and school until the adjacent brick and sandstone church was built in 1861, and the national school next door absorbed its pupils.

Court House and Main St, KS
Burke & Wills Mechanics Institute, Fryerstown, KS
Fryerstown Methodist Hall, DB
Cottage, Fryerstown, KS

AROUND FRYERSTOWN

The Fryer's goldfield quickly became overcrowded and diggers were forced to explore nearby Nuggety Gully, German Gully, Mopoke Gully and New Year's Flat, where they were successful in finding more gold.

The remains of many prosperous mines line the road from Chewton to Fryerstown. Spring Gully, the Eureka Mine and Heron's Reef are all part of the **Castlemaine Diggings National Heritage Park.** See separate section under Chewton. Many relics of the different types of mining activities scattered throughout the reserve include old shafts, tunnels and open-cut mines, water-races, the ruins of buildings, foundations and abandoned machinery. In contrast, during spring, many beautiful wildflowers adorn the park.

Spring Gully became a goldfield in its own right and, at its peak it had a population of several hundred Europeans along with a large number of Chinese. In addition, there was a major concentration of Irish at Irishtown between Fryerstown and Vaughan.

A digger named John Batey arrived in 1857 and spent much of his time digging miles of small water channels around the hills of Fryerstown and Spring Gully to bring water to his claim. Some of Batey's races still exist. In 1866 a 21 km aqueduct from the Loddon River was constructed to supply water to the puddling machines on the fields.

Heron's Reef is a gold mining site encompassing 45 ha which has been kept as it was when the miners departed. The site is classified by the National Trust as 'The Anglo-Australian Gold Mine Site Fryerstown' and is registered with Heritage Victoria. Relics span the three eras of gold mining activity – alluvial,

deep lead and hydraulic sluicing.

Heading south on Fryers Rd, immediately after passing the Wattle Gully Mine at the crest of the hill, the road passes over the water-race from Taradale. During summer, water can be seen cascading down steps on the left-hand side of the road.

After leaving Fryerstown there is a fork in the road to either Glenluce Springs or Vaughan Springs. On the way to Vaughan Springs you can see the extensive damage wrought on the landscape by sluicing and dredging at New Year's Flat and Red Knob. In the valley is a huge stand of poplars – a favourite spot for artists during autumn when the leaves are changing colour.

At a hill near The Monk, north of Fryerstown, is the **Eureka Reef**, discovered in 1854. A village was established and mining continued there into the 20th century.

GREAT WESTERN

LOCATION: 220 KM NORTH-WEST OF MELBOURNE: 17 KM NORTH-WEST OF ARARAT (WESTERN HWY)
POPULATION 2006: 250

Although gold discoveries in 1858 brought a swarm of diggers through the Great Western region, the main centres of mining activity were Stawell to the north-west and Ararat to the south-east. Great Western itself became better known as the home of liquid gold – the birthplace of

Duke of Cornwall Mine, KS
Town Hall, Fryerstown, KS
Shop, Great Western, KS

vineyards and wineries that are famous for their sparkling wines as well as fine dry white and red table wines.

The township allotments were surveyed in 1860 and the soil and climate attracted the attention of two Frenchmen who had met on the diggings at Daylesford. The area around Great Western reminded them of wine-growing areas at home and they planted their first grape vines in 1863. The experiment was a success and they quickly gained a reputation for their high-quality wines, even winning awards at international shows.

Much like diggers after a successful strike, two brothers – Joseph and Henry Best – followed the Frenchmen's lead. Each established separate vineyards in the region during 1865/1866. Both are still operating today.

A point of interest in the town of Great Western itself is an old **toll gate** that was erected in 1863 on the road to Melbourne to help pay for road construction costs. The structure has been restored and re-located at the entrance of the Great Western Mechanics Hall where it serves as a novel ticket office for any functions held in the hall.

GREAT WESTERN WINERY

Joseph Best established his vineyard in 1865 with cuttings from France. In a fascinating link with the district's gold mining era, the winery cellars are actually tunnels dug by miners in the 1860s and 1870s. But they weren't looking for gold. Joseph Best employed the miners specifically to dig the cellars out of the decomposing granite beneath his property.

Joseph Best died in 1887 and his property was purchased by Hans

Irvine, a man who set himself the task of producing champagne-style wine comparable to the best in France. To accomplish this he travelled to France to study the technique and even brought back French agronomists and vineyard employees to tend his vines. By the time Irvine sold the property and business in 1918, his sparkling wines had gained a wide renown and were regarded as a model for the district.

Irvine sold to Benno Seppelt and the vineyard, now known as Seppelt's Great Western, has continued its reputation for producing excellent champagne-style wines. The complex is now the town's major employer with up to 300 permanent and casual employees, depending on the season. Visitors today can wander past thousands of bottles of sparkling and table wines maturing in the tunnels that run for three km under the property.

BEST'S WINES (CONCONGELLA)

Joseph Best's brother Henry also left a fine tradition in wine making. He took up his property on the banks of Concongella Creek just north of Great Western in 1866 and planted his first vines two years later. After a success equalling his brother, the property was sold in 1920 to another established winemaking family named Thomson. Nevertheless the winery retained Henry's name and today is known as Best's Concongella Winery & Vineyards. The property's National Trust-listed cellar door is housed in the original stables built in 1866.

GARDEN GULLY WINERY

The vineyard is located on the site of the original Salinger Hockheim Winery established about 1890. The present vineyard began in 1985.

Great Western Cellars, BaT
Cottage, Great Western, JM
Best's Wines, Concongella, RE

GUILDFORD

LOCATION: 130 KM NORTH-WEST
OF MELBOURNE (MIDLAND HWY);
7 KM SOUTH OF CASTLEMAINE
POPULATION 1855: 6000
POPULATION 2006: 300

Guildford is a small settlement on the banks of the Loddon River. The road from Newstead in the west follows the Loddon River valley and becomes quite dramatic as it encounters the lava flows that poured out of Mt Franklin.

Level surfaces created by the basalt flow are evident on the hilltops and there is a softer sub-stratum of sandy shales and clay on the steep slopes down to the river. These lighter soils were rich in gold, so miners tunnelled underneath the basalt flows from the sides. In Guildford there is a steep track up to a lookout tower offering wonderful views over the plateau and showing the extent of the basalt plains, with the Loddon River and Campbell's Creek cutting through them.

The township began with a refreshment tent where the main track from Ballarat and Daylesford to Castlemaine crossed the Loddon River. Described as 'a lively little place', it served as a coach depot for Cobb & Co for many years.

With the discovery of gold in 1852, shafts and tunnels were excavated beneath the basalt on the Guildford Plateau – the tunnels were large enough for horse-drawn trucks to remove the wash dirt. However, the population fell dramatically as the alluvial gold declined and when the Borough of Guildford was proclaimed in 1866, there were just 250 ratepayers. Deep-lead mining was abandoned in 1876 and by 1888 the town had only three hotels and 200 residents.

A brief revival took place during the 1930's Depression. Between 1933 and 1939, the Guildford Plateau Company obtained 1,922 ounces (54.5 kg) of gold from a mine near Strangways.

Fine buildings of brick and stone, including old chapels, the police lock-up and some 19th century homes still survive. The **Guildford Family Hotel** (1856) is a major landmark at the intersection of the Midland Hwy and the Fryerstown Rd. Also known as Delmenico's Family Hotel, this marvellous old pub was one of 24 that operated at various times in the township and offered a music hall, billiard room, quoit matches and bocce court. The ruins of an old assembly hall are nearby and next door is London House (1856), which originally served as a store and post office.

On the west corner of the crossroad is the former Commercial Hotel, now a general store. On its northern side are the stables and outbuildings of a large department store that burned down in 1916. An Anglican church was built in 1861, while the present post office dates from 1901.

Guildford has been predominantly an Italian and Swiss-Italian town from its early days. Italians like Carlo Barassi produced wine in commercial quantities. Ron Barassi, one of the

Level surfaces created by the basalt flow are evident on the hilltops. In Guildford there is a steep track up to a lookout tower offering wonderful views over the plateau and showing the extent of the basalt plains.

Opposite: Aerial View of Guildford, Volcanic Plateau and Mt Alexander, BS Yapeen, Near Guildford, KS

all-time greats of Australian Rules' Football, grew up here. Summer sports for the town are cricket and bocce.

CHINESE TOWNSHIP

The largest Chinese camp in Victoria was on the road to Yapeen close to the junction of Campbell's Creek and the Loddon River near Guildford. Although no evidence remains today, up to 6000 Chinese miners lived there, largely in calico tents pitched in regular lines separated by narrow thoroughfares.

Every street was dotted with joss houses, restaurants, tea-houses, boarding houses, gambling establishments, opium dens, herbalists, barbers, cobblers and tailors. There were several two-storey timber buildings, literature and art shops, and scholars and interpreters wrote letters for illiterate diggers. The camp also had a theatre with circus performers, which later moved to Ballarat.

The Chinese worked as a co-operative group, using open-cut alluvial mining. Hostility toward them resulted in numerous local conflicts, but by 1865 most Chinese had departed for the Maryborough and McIvor diggings. Several years later the derelict camp was destroyed by fire. The few Chinese who remained – some well into the 20th century – turned to market gardening on the river flats.

THE BIG TREE

The largest specimen of river red gum in Victoria is at the intersection of Fryers St and Ballaarat St. This remarkable and beautifully preserved eucalypt, classified by the National Trust, measures 12.8 m at the base and stands 25.9 m high. A plaque recalls that Burke and Wills camped

beneath its shade on their journey from Melbourne to the Gulf of Carpentaria.

AROUND GUILDFORD

Yapeen

Two km north of Guildford is the small hamlet of Yapeen – the name thought to be an Aboriginal word meaning 'green valley'. The place has had several earlier names beginning with the Pennyweight Diggings in 1852 and Donkey Hill in 1856 when shaft mining averaged five ounces (142 g) of gold per load of wash dirt. It later became Strathloddon, after William Campbell's *Strathloddon* station. The ruins of an 1887 water wheel can still be seen in nearby Mopoke Gully.

Marsh House is a prefabricated two-storey building which was imported from England and erected in 1854 for William Mein, the son of a pioneer European settler.

Yapeen orchardists lay claim to the development of the Munro apple. They also grow pears, peaches, mulberries, olives, figs and walnuts.

Every street was dotted with joss houses, restaurants, tea-houses, boarding houses, gambling establishments, opium dens, herbalists, barbers, cobblers and tailors.

Mine on the Plateau, KS
Yapeen Store, KS
Yapeen Cottage, KS

The small town of Harcourt has a long history as a centre for the apple-growing industry in Australia.

HARCOURT

LOCATION: 130 KM NORTH-WEST OF MELBOURNE; 30 KM SOUTH OF BENDIGO
POPULATION 2006: 400

Set in a fertile valley at the foot of Mt Alexander, the small town of Harcourt has a long history as a centre for the apple-growing industry in Australia. Apart from its well-drained granite soils and temperate climate, since the 1870s the valley has had irrigation thanks to an ingenious system of gravity-fed water races carrying water from Malmsbury.

From 1853 market gardens, and from 1859 orchards, in the valley supplied fresh food to the nearby miners. Grapes were also planted early on. One of these first vines, a Shiraz, has survived at a house on Gaasch's Rd. It has been an important source of phylloxera-resistant root stock.

It is jokingly alleged that Harcourt was named when a fruit-grower caught some miners raiding his apple trees and called out, 'Hah! Caught Ya'!

The valley is a highly productive area, notable for apple and pear production and more recently for cherries and grapes, with several vineyards producing acclaimed wines, liqueurs and ciders. The Calder Freeway is lined with orchards and there are plenty of opportunities to sample local produce from roadside stalls – or to pick your own.

The Harcourt Valley runs north-south. On its western side a granite ridge, beginning at Faraday, marks the eastern edge of the goldfields that surround Castlemaine. The western slopes of the valley are covered with red gum and yellow-box trees. The eastern side of the valley rises to form the flank of Mt Alexander, the highest peak in the northern hills of the Dividing Range. The Dja Dja Wurrung knew it as *Lanjanuc*. A magnificent 'scar tree' (created by Aboriginals removing bark for various purposes) can be seen right beside the freeway, in front of the Shell service station, opposite the Harcourt general store.

The first European to traverse the area was Major Mitchell in 1836. He named Mt Alexander after Alexander the Great due to its proximity to Mt

Bowser, Harcourt, KS
Apple Orchard, Harcourt, KS

Macedon (Alexander was the son of Philip of Macedon). The mountain gave its European name to the goldfields that became the destination for thousands of miners from around the world.

The first white settler in the region was Dr Barker whose *Ravenswood No.1 Run* encompassed the present township. A roadside plaque on the eastern side of Eagles Rd, near Barker's Creek, shows where the homestead was located. It was on Barker's run that a shepherd named John Worley found gold in 1851.

The valley saw conflict between European settlers and the local Jaara Jaara people.   In January 1840, following the theft of a couple of sheep by the Aboriginals, a confrontation between three soldiers, eight armed men and eleven unarmed Jaara Jaara (two were boys) resulted in two Aboriginals being shot dead, another arrested and a number wounded. Edward Stone Parker, the Aboriginal Protector for the Loddon Region wrote: 'It seems evident, that in certain circles in the year 1840, sheep were regarded as more valuable than men, if they had the misfortune to be black men.'

The Heritage Centre, opposite the intersection of the Midland Highway and Calder Freeway, has a display featuring the history of the local apple industry. There is a large photographic collection.

MT ALEXANDER

Rising 350 m above the surrounding area, and standing 741 m above sea level, Mt Alexander is part of Victoria's mountainous backbone that includes the remains of twelve distinct mountain groups. .

Although it is now nothing more than a

worn stump compared to the lofty peak of earlier geologic time (it is, after all, 367 million years old), Mt Alexander remains a prominent landmark on the east side of the Loddon Valley. The mountain was important to local Aboriginal people as a high vantage point at the centre of their tribal area and as sacred ceremonial ground.

Major Thomas Mitchell was the first European to climb the mountain, in 1836. By the 1870s, its vegetation had been severely reduced to provide timber for the Goldfields. The slopes of the mountain are steep and heavily crowned with large granite rock outcrops. On the highest elevation is a stone trigonometry point from where much of Victoria was originally surveyed.

Mt Alexander Regional Park

The 1400 ha **Mt Alexander Regional Park** three km east of Harcourt includes the peak. The well-made **Mt Alexander Tourist Rd** traversing the ridgeline can be reached from the south by turning off the freeway at Faraday and from the north by turning onto the Harcourt North Rd. Just before the communications towers, there is a Rd on the left to **Lang's Lookout** from where there are excellent views.

The four-km-long **West Ridge Walking Track** links all the major lookouts and points of interest. The walk is best started at Lang's Lookout. The entire well-defined track can

Crooked Chimney, Harcourt, KS
Mt Alexander, KS
Dog Rocks, Mt Alexander, KS

be completed in half a day, taking in Shepherd's Flat Lookout and Dog Rocks, and concludes near the Leanganook Koala Park. **Leanganook Koala Park** is a koala enclosure with walking track circuits in the south of the park, but you are just as likely to see a koala in a manna gum anywhere on the mountain, usually in trees more than 30m high.

Near the koala park, set back in the bush amidst a pine plantation, is a little cottage built of granite blocks, now missing a roof and once surrounded by mulberry trees. Between 1873 and 1877, Mrs Bladen Neill attempted to establish a silk industry on the slopes of Mt Alexander. The venture, called the Ladies' Sericultural Company, was to offer employment to women, but it failed because the mulberry trees took too long to grow. The relics are not visible from the road.

Oak Forest

An extensive oak forest at the foot of Mt Alexander is a favourite picnic spot for locals. The Valonia oaks were planted at the turn of the century for use in hide tanning (tan bark production) and they are particularly magnificent in autumn. Even in summer they create a cool green escape, like a Scottish glen in the heart of Victoria – complete with intriguing cork bark trees, the delightful smell of pine cones, and a pleasant bubbling brook. To get there turn off the highway at the Shell Service Station into Market St. Continue straight on as the road becomes Picnic Gully Rd (unsealed) which leads into Mt Alexander Regional Park.

HARCOURT RAILWAY BRIDGE

With its classic features – Roman style arches, and superb granite masonry – the Harcourt Railway Bridge on Symes

Rd is a masterpiece that was built by German masons in 1859. This was one of a number of engineering works that stimulated the ongoing development of granite quarrying in the region.

HARCOURT GRANITE

Harcourt became known as a producer of high-quality granite, used all over Australia. Granite has been quarried on the slopes of Mt Alexander since the 1860s. Twenty-four of the old homesteads around Harcourt were built by the German masons (unemployed after finishing the railway bridge) from hand-cut granite blocks.

From the 1880s Harcourt granite began to appear extensively in Melbourne buildings, such as Flinders St Station, and in monumental and ornamental stonework. It was also used for Parliament House in Canberra. It took 26 horses to drag the large stone for the Melbourne Burke & Wills Memorial through bush tracks to the Harcourt Railway Station. Several quarries still operate today.

Lang's Lookout, Mt Alexander, KS
Harcourt Railway Bridge, KS
Oak Forest, KS

Heathcote

LOCATION: 110 KM NORTH OF
MELBOURNE, 47 KM SOUTH-EAST
OF BENDIGO
POPULATION 1853: 35,000
POPULATION 2006: 1800

Sitting at the foot of Mt Ida and beside McIvor Creek, Heathcote is surrounded by the largest remaining Box-Ironbark forest in Victoria. The town itself has an interesting and diverse streetscape of old shopfronts, hotels, churches and some distinguished homes.

There's now a vibrant café culture on the well-laid out highway frontage. It is worth stopping at Heathcote – at the very least for a bite to eat and to sample some of the region's excellent wine.

Although a relatively new wine region, Heathcote is fast developing a reputation for producing some of the finest Shiraz in the world, as well as high quality varieties such as Malbec, Merlot, Sangiovese and Cabernet Sauvignon.

Wine is the industry that flourishes today, but Heathcote was built largely on gold. The town has an excellent Visitor Information Centre, and from there you will be directed to some amazing nearby sights.

A number of prehistoric quarry sites were found at Mt Camel Range to the north of Heathcote, revealing an ancient past. The Wuy Wurrung Aboriginals previously inhabited the district. Pastoral settlement began in 1837. Heathcote was then just a couple of roadside inns beside a track along which flocks of sheep were driven down from northern sheep stations.

The town developed following a series of goldrushes along McIvor Creek, beginning in 1851. Behind Heathcote's old court house is the site of the major 1852 strike at Golden Gully. As the alluvial gold ran out in 1855, reef mining commenced. Nonetheless, by 1860 many ex-miners had taken up farming or joined the timber industry. Timber was used to fuel steam engines, provide sleepers for the construction of railways, for timber bridges, and to meet the needs of fast-developing Melbourne.

At the peak of the goldrush there were up to 35,000 people, largely housed in tents and shanties. There were three breweries, 22 hotels, two flourmills, a bacon factory, a hospital and several wineries.

The town's original gaol was built in 1853, but a break-out in 1859 led the commissioner of police to order the building of Camp Hill Gaol (1861), opposite the corner of Barrack and Hospital Sts. The original section of the Heathcote Hospital was built of sandstone in 1859. The Heathcote Winery, at the northern end of town, is housed in an old Cobb & Co coaching house and miners' store (1850s). The Union Hotel was opened in 1856 and was completed in its present form in 1871.

The trees along the main street and the plantation at Queen's Meadow were planted in the 19th century on the advice of Baron von Mueller.

Visitor Information

The excellent Heathcote Visitor Information Centre provides an extensive range of visitor information and maps, as well as souvenirs, gifts and local products.

Heathcote Visitor Information Centre
corner of High and Barracks Sts
Heathcote 3523
www.heathcote.org.au
tel 5433 3121.

Heathcote is fast developing a reputation for producing some of the finest Shiraz in the world.

Heathcote Winery, KS
Hospital, KS
Cottage, KS

HEATHCOTE COURT HOUSE & COUNCIL CHAMBER

Built in a style known as Victorian Mannerism, this is a rare survivor from the 1860s. It is unusual for a court house and a town hall to be incorporated in the one building.

MCIVOR RANGE RESERVE

From the information centre you can walk past Lions Park and the swimming pool, over McIvor Creek (a good fishing spot), turn right past Queen's Meadow and a caravan park in a forest setting, then through the Valley of the Liquidambars to the reserve.

Powder Magazine

The old 1864 powder magazine is situated within the reserve. It stored gunpowder used in gold mining and quarrying and is a fine example of a rare building type. Powder magazines played an important part in the development of underground gold mining activities in Victoria.

Viewing Rock

Past the powder magazine a track leads to Viewing Rock, a rocky outcrop that offers expansive views of the town and surrounding area. It was from here that the jail, powder magazine and gold assay house could easily be 'viewed' by the police guard during the gold era. This is an ideal lunch spot with extensive views to Mt. Alexander and surrounding countryside. From here, you can take a three to four km circuit around the top of the McIvor Range.

Devil's Cave

Also in the reserve is Devil's Cave, occupied by prospectors as far back as 1864. The track is signposted from the Valley of the Liquidambars. There are some good views of the town en route and visitors can see a profusion of wildflowers in spring. .

PINK CLIFFS

Heathcote's famous pink cliffs were created by sluice mining. The colours of the cliffs range from bright pink to yellow and golden ochre. Pink Cliffs Rd runs off Hospital St and there are a number of paths through the reserve which also has a picnic area.

AROUND HEATHCOTE

Heathcote-Graytown National Park

The Heathcote-Graytown National Park covers 12, 833 ha and incorporates much of the southern end

Court House & Council Chamber, KS
Powder Magazine, KS
View Over Town From Viewing Rock, KS
Heathcote-Graytown National Park, KS

of the extensive Rushworth-Heathcote State Forest, from the McIvor Range and Mt. Ida to Mt Black, Spring Creek and Graytown. There are areas for bushwalking, car touring, nature observation, horse riding and bike riding, picnicking and camping.

The park includes some of the most significant environmental, cultural and recreational values in a Box-Ironbark forest and supports 16 threatened species and many large old tree sites. There are also numerous significant Aboriginal cultural sites.

The best time to visit is when the wildflowers bloom in spring, but each section of the park has its own distinctive character and is laced with walking tracks that offer beautiful views.

You can explore goldrush and war-era historic features at Graytown, as well as other local mining communities at Balmoral (later known as Redcastle) and Costerfield.

Mt Ida Lookout

Head north to Elmore and a good gravel road will take you to the Mt Ida Lookout (450 m) which offers excellent views of the district.

Lake Eppalock

Lake Eppalock, 10 km out of town is not only one of Victoria's largest reservoirs, but is also popular for powerboat racing, swimming and fishing. Built for irrigation and flood-control purposes, Eppalock has been developed as a recreation area.

Duigan Memorial

On the Mia Mia-Lancefield Rd, west of town, is a memorial which commemorates the occasion in 1910 when the Duigan brothers built and flew the first Australian-made aircraft.

HUNTLY

LOCATION: 163 KM NORTH OF MELBOURNE (MIDLAND HWY); 14 KM NORTH OF BENDIGO POPULATION 1875: 3.590 POPULATION 2006: 2.247

Huntly is known as 'Home of the Whirakee Wattle' – a rare wattle species only found in this region that bursts out with vivid golden yellow flowers during late August and early September.

The town owes its existence to gold of another kind because miners struck it rich here in 1859. The initial alluvial rush was followed by intensive mining on a six-mile (9.5 km) deep lead at Huntly during the early 1860's. As many as 1,600 puddling machines worked along the lead and these, together with deep lead mining, extracted 500,000 ounces (14.2 tonnes) of gold.

The deep lead miners had serious problems with underground water, but mine tailings dumped into Bendigo Creek also caused severe flooding above ground. In the early 1900s a relief channel was dredged and a bank was built to confine floodwaters. Mining continued until the early 1930s and the Huntley Valley was left scarred

Huntly Council Chambers
& Post Office, KS

with old tailings, mine dumps and the sludge from battery crushings.

Mining was not the only attraction to the area. French, Italian, Swiss, Spanish and Chinese immigrants also saw the potential of the land for vineyards and market gardens. By 1878 a Monsieur Bladmire had established 42 acres (17 ha) of vines at 'Frenchman's Gardens' in the area now known as Sargeant's Lane. Huntly was also known for its high quality pipe clay, which was supplied to Epsom Pottery for the making of fine china.

The first Catholic Church in the area was established at Epsom in 1856 – a rudimentary affair with bark sides –but another 19 years passed before a permanent chapel was built. The gold rush brought an increased number of Cornish diggers and led to the establishment of the first Methodist church and Sunday school in 1860.

The Bird in Hand, built in 1859, was the first of at least 20 hotels trading in Huntly during its heydays. The shire hall, built in 1867, was used by council for 110 years and court cases were held in the council chambers before the court house was built in 1874. In 1994, the Shire of Huntly became part of the City of Greater Bendigo and the Huntly & District Historical Society now looks after the former council chambers, court of petty sessions, lock-up (a portable timber structure that could be moved around with the shifting goldfields population) and post office. These buildings comprise the **Huntly Heritage Centre** and are located on either side of the Midland Hwy. They are open on the first Sunday of the month (except January) from 2 pm till 4 pm and at other times by arrangement

Wattle, Melville Caves KS
Town Hall, Inglewood, KS

INGLEWOOD

LOCATION: 195 KM NORTH-WEST OF MELBOURNE; 45 KM WEST OF BENDIGO.
POPULATION 1860: 35,000
POPULATION 2005: 700

Inglewood is a classic goldrush town with a 'frontier' atmosphere. Time has stood still since the great days when the shops and hotels were full of diggers, but the empty shops with their shady verandahs are still standing.

Inglewood was the site of Victoria's last great 19th century goldrush. Known as the 'City in the Scrub', the town was established following the discovery of some of the richest reefs in the colony. The Columbian Reef yielded over 360 kg of gold from the first 1000 tonnes of ore mined.

During the goldrush the town became divided into Old Inglewood and New Inglewood, linked by Commercial St. The present township is at New Inglewood with its imposing historic buildings in Brooke St. Many were built after a cataclismic fire in 1862 that claimed most of the earlier wooden buildings. There's a magnificent private residence, **Tivey's House**, opposite the opulent old town hall in Verdon St. The historic court house (1860) is the best example of a number of regional Victorian court houses that were built in the 'Victoria Free Classical' style.

Inglewood is famous as the birthplace of Sir Reginald Ansett (founder of

Ansett Airways), Fanny Hines (the first woman to die in active service), and Jack Donaldson (who, from 1910 to 1948, held the world record for the 100 yard dash). Julius Vogel, who later became the Prime Minister of New Zealand, established a newspaper, the *Inglewood Advertiser* in the 1860s.

EUCALYPTUS OIL

Known as the 'Blue Eucy Town', Inglewood became the centre for eucalyptus oil production in the early 1900s. The Blue Mallee eucalyptus still provides 'the best quality eucalyptus oil in the world' in this uniquely Australian industry. The area is dotted with remnants of the industry's past such as at the **Old Blue Eucy Distillery**.

CHARLIE NAPIER HOTEL

This is a substantial 19th century hotel with folk-art murals that give a unique and important interpretation of colonial life.

AROUND INGLEWOOD

Kingower

The little town of Kingower lies between the granite tors of Mt Kooyoora and the steep Bald Hills of the Rheola Range. Famous for huge nuggets, Kingower was given the nickname, 'The Potato Diggings' during the goldrush. The discovery of a 720 ounce (20.4 kg) nugget in 1980 triggered a modern mini-goldrush. The rush was short-lived, but it did produce the *Hand of Faith* – the biggest gold nugget found in the 20th century. Kevin Hillier found the nugget using a metal detector near the Kingower schoolhouse. It was sold to a Las Vegas casino for $1 million.

Kooyoora State Park & Melville Caves

The **Kooyoora State Park** has panoramic views from a cluster of huge granite rocks, rare and beautiful vegetation, some of the most significant Aboriginal sites in Victoria, and a romantic association with a famous bushranger.

When they were in the area, Aboriginal people depended on natural springs and wells for their water as many rock wells in the granite outcrops stored rainwater. There are also scarred trees (where bark was removed to make shields and dishes), rock holes, shelters, ochre, quartz and mica quarries, tool manufacturing sites and three stone arrangements.

The **Melville Caves**, named after bushranger Captain Melville, lie just to the south of Mt Kooyoora. The caves are formed by granite boulders and fissures.

The rocks give excellent views of the flat plains to the south, so Melville was able to spy on the gold-bearing coaches that became his prey. He operated throughout south-western Victoria before he was captured in a Geelong brothel. It's believed that some of his treasure is still buried here.

The **White Swan Crystal Mine** in the park produced quartz crystals for communications equipment during WWII, and crystals can still be recovered from mullock heaps around the mine.

Streetscape, KS
Tivey's House, KS
Streetscape, KS
Shop Window, KS
Melville's Caves, KS
Over page: Franklinford, PW

'Murder, robbery and lynch law is perpetrated with impunity.'

Joyce's Creek

LOCATION: 22 KM WEST OF CASTLEMAINE: 7 KM WEST OF NEWSTEAD
POPULATION 1855: 5.000
POPULATION 2006: 10

Alfred Joyce arrived in Port Phillip with his brother in 1843. A year later the pair selected a run of 10,000 acres (4000 ha) on the Moolort Plains of central Victoria. They named it *Plaistow* after the family's country home back in England and it proved to be a rich and productive property. In 1853 the government granted 640 acres (260 ha) at *Plaistow* to the Joyce family and they were soon caught up with a goldrush. Today the old Georgian-style homestead offers B&B.

The first discoveries of gold made at Joyce's Creek were in June 1854. The diggings meant profit for Plaistow as the main road from Adelaide to Mt Alexander passed by the head station. The stream of diggers coming overland created a demand for stores and meat. The station took in travel-weary horses, sold wheat and hay and later sold dairy products, fresh garden produce and candles on the goldfields.

In *A Homestead History – reminiscences and letters of Alfred Joyce, 1843-64* the early settler writes of the harshness of rural life in Australia, as well as the situation faced by the diggers. Many were too poor to buy necessities and equipment, and many struggled to cope with the hard labour of digging and washing. 'Murder, robbery and lynch law is perpetrated with impunity,' Joyce says, adding that there were only a dozen police amongst a population of up to 30,000 people.

Kyneton

LOCATION: 85 KM NORTH-WEST OF MELBOURNE
POPULATION 2006: 5000

The town of Kyneton is built on a gentle westward slope and sits within a right angle described by the Campaspe River. The town, unlike most settlements in this region, predates the goldrush.

It grew from the earliest run, *Carlsruhe*, taken up in 1837 by Charles Ebden. In late 1839 James Donnithorne took over the lease of the northern part and named it St Agnes. A station complex was built on the west bank of the Campaspe where Rock House stands today, next to the present homestead called *St Agnes*. By 1850 Kyneton was a growing rural centre, just in time to play host to the human flood after the discovery of gold at Mt Alexander the next year.

Because it took shape while Victoria was still a pastoral colony, Kyneton has a pre-Victorian character. Kyneton is a bluestone town. A high proportion of the 19th century buildings are of squared or rock-faced basalt and brick does not make an appearance until quite late in the century. The earliest house remaining is the Old Rectory (now a B&B with a Paul Bangay garden), on corner of Piper and Ebden Sts, built in 1851.

The old churches of Kyneton are best seen from Ebden St – all are bluestone and have lots of character, particularly Holy Rosary Roman Catholic Church. Many interesting and beautiful old houses are to be found in the streets of the town. Some are characteristic timber or brick buildings with bow-fronted windows of the inter-war period. One superb bluestone structure is the former Kyneton Hospital, well proportioned and with an excellent cast-iron verandah, located at the western end of Simpson St.

VISITOR INFORMATION

Kyneton Visitor Information Centre
High St, Kyneton
tel 5422 6110
freecall 1800 244 71

PIPER STREET

Piper St was the commercial centre during the 1850s and has not changed much. The whole street is a remarkable survivor and still has original bluestone paving, kerbs and gutters, and gaslights. Today there are antique shops, a restaurant or two, a bakery and some fine old pubs.

Museum

The first Bank of New South Wales (1855) now serves as the museum (corner of Powlett St) and has an excellent display of 19th century interiors complete with costumed figures, as well as an original squatter's house and stables.

Flour Mills

Two original flour mills remain. **Willis Brothers Mill** (1862),on the corner of Ebden St, has machinery intact and is still making flour. The other, now known as the **Old Butter Factory** (corner Wedge St), contains **Meskills Woolstore** and sells pure spun wool, knitting yarn and knitwear.

St Paul's Anglican Church

Halfway along Ebden St is **St Paul's Park** with a view up the hill to the church. St Paul's Anglican Church is best approached from Powlett St where the pathway rises up the hill with dramatic views of the handsome tower through mature stands of atlas cedars and a large Lebanon cedar close to the church.

Zetland Lodge

On the corner of Piper and Mollison Sts, Zetland Lodge (Freemasons) is a rather plain red brick building. However the lodge room inside is a well-preserved example of ritual furnishing used in Freemasonry. The highlight is the decorative scheme, including scenes of the Nile River and the pyramids beautifully painted in deep perspective by Thomas Fisher Levick during the 1920s. Levick was an art-teacher at Kyneton and Castlemaine, and a founding spirit of the Melbourne Workingman's College (later the Prahran Institute of Technology and now Swinburne University).

Ellim Eek

Ellim Eek, at the intersection of Mollison and Mair Sts, was built in 1890 then purchased for a surgery by a local doctor who added oriental terracotta ornament on the parapet, coloured ceramic panels, and an unusual verandah.

MOLLISON & HIGH STREET

The arrival of the railway from Melbourne and the siting of the Kyneton station on the other side of the Campaspe River shifted commerce away from Piper St. The large bluestone station is a rare example of an 1860's

Streetscape, Kyneton, MH
Mural, Zetland Lodge, KS
College House, Piper St, Kyneton
Streetscape, Kyneton, KS
Elim Eek, KS

station with all its appurtenances intact – level-crossing gates, signals and goods sheds. Trains began running in 1861 and the town grew out to meet them.

Consequently, the present commercial centre in Mollison and High Sts dates mainly from the 1860s to 1910s, with some older buildings located in High St. It is an impressive townscape, although there are a few major gaps on street corners.

The magnificence of the old bank buildings testifies to the financial strength of Kyneton through the late 19th century. The National Bank still operates from original premises designed in 1877 by Leonard Terry. Opposite is the art nouveau former Bank of New South Wales. The date 1817 refers to the founding of the banking company, not the building which is nearly 100 years younger.

Post Office

Kyneton Post Office is a superb civic building, mainly in Italianate style with an open arcade on cast-iron columns, although the clock tower is more Gothic with a clock face on the diagonal and spiky pyramid roof. The town hall offices were built in the late 19th century and extended in the same handsome style when the hall itself was added as a memorial to WWI. The excellent hall functions as the

town's civic centre and hosts musical and theatrical presentations touring Victoria's major country centres.

Mechanics Institute

Past the traffic lights is the old mechanics institute, perhaps the finest original building in town. The temple-like structure has a two-storey central façade with Doric pilasters and pediment. The symmetrical wings are single-storey. Inside is an atmospheric hall with stage and gallery, now used as a community centre. The town library on the right-hand side is due to be renovated and extended, while the rest of the building is to be restored – all this with the help of Kyneton Bowling Club which has a long-term lease of the rear of the building.

BOTANIC GARDENS

The Kyneton Botanic Gardens were commenced in 1863 and were given a boost in 1866 with a gift of plants from Baron Ferdinand von Mueller, the first Director of the Royal Botanic Gardens in Melbourne. As well as a fine collection of oaks, the gardens also contained Giane redwoods, but only a few remain today. Significant plants still growing include the rare Chilean Wine Palm and a fine row of Atlas cedars.

AROUND KYNETON

The Calder Fwy after Woodend gives clear vistas towards Hanging Rock and the northern face of Mt Macedon. Then Jim-Jim Hill at Newham shows its complex volcanic profile and the Cobaw Range further north grows distant as the road follows the valley of the Campaspe River past Carlsruhe towards Kyneton.

War Memorial & Mechanics Institute, KS
Post Office, KS
Catholic Church, KS

Mills

The old bluestone mill near the river is the **De Graves Mill**, and nearby stands **Skellsmergh Hall**, a grand bluestone house with wide verandahs. There were six flour mills in Kyneton during the first 20 years of the goldrush.

Huge mills were erected in the 1850s when demand for grain was heightened by the sudden influx of people on the Goldfields. It soon became clear, however, that the lighter soils of the north and west of the state were easier to plough and less boggy for reaping..

Mineral Springs

Kyneton Mineral Springs are two km west of the town on the Old Calder Freeway (Piper St). There are trees, picnic facilities, a hand pump for mineral water and a period rotunda.

Turpins Falls

Turpins Falls are set in beautiful rolling grazing country, with outcrops of granite boulders, red gums on the creek flats, and patches of good-quality bush. The gorge – worn through the basalt – and the falls themselves, come as a dramatic surprise. They have been a famous beauty spot since the 1880s.

To get to Turpins Falls from Kyneton, take the road towards Heathcote. At Langley head left down Manns Rd towards East Metcalfe for 2.5 km, then left for 2.6 km, then right for 1.3 km and finally turn off right to the carpark and picnic area. Contact the Visitors' Information Centre for more detailed directions and a map.

LEXTON

LOCATION: 49 KM NORTH-WEST OF BALLARAT (SUNRAYSIA HWY) 23 KM SOUTH OF AVOCA POPULATION 2006: 250

Squatters settled in the Lexton area from 1838, but the nucleus of a township (initially known as Burnbank) was not established until 1845 when two enterprising settlers, David Anderson and William Millar, built a pub, a general store/post office, a blacksmith's shop and wheelwright shop. Anderson built a separate post office in 1848 and gradually other businesses began to appear.

The first Anglican and Catholic services were held in 1850 and 1851 respectively, coinciding with the influx of diggers rushing to the first gold strikes in the district. A local residence was converted into a court house and police magistrate's residence. The Lexton Hotel was built in 1852 and a school was established in 1855.

The town grew quickly and by 1860 it had become a centre for government administration, remaining so until 1994 when the Shire of Lexton was amalgamated into the new Pyrenees Shire. When the gold production dwindled, Lexton once again became a service centre for the local pastoral industry and today the main industry is wool.

Turpins Falls, RE
Farm Near Lexton, KS

HISTORIC WALKING TRAIL

A stroll around Lexton's walking trail provides a number of reminders of the town's past. Today's **Toll Bar Park** was the site of the Lexton Hotel (1852) which was also used as a Cobb & Co staging post. The **Pyrenees Hotel** was built in 1859 and replaced the original 1845 Burbank Inn & Store.

The first post office, located in a section of the original timber general store, was moved into a separate building in 1848 and remained the town's official post office until 2001. In 2000 the local Progress Association purchased the post office license and opened the Co-operative Post Office in the former Lexton Shire Offices. It became a Rural Transaction Centre in 2004.

Another general store, built in 1854 and incorporating a butcher's shop, is close to Burbank Creek between Anderson and Clapperton Sts, while one of the oldest homes in Lexton – the Anderson family's *Sunnyside* – is opposite the public hall in Campbell St.

Lexton's second court house, in Williamson St, was built in 1874 and purchased by the shire in 1936. **St Mary's Anglican Church**, built in 1874, is in Skene St. Methodist services were held in a private home until the first church opened in 1863. The present Methodist chapel was brought from Clemenston near Creswick in 1912. Presbyterian church services were held in the St Andrew's School at the eastern end of Williamson St until the first **St Andrew's** church was built next door in 1856. It was rebuilt on the original foundations in 1876. St Joseph's Church wasn't built until 1894.

Just outside Lexton, on the road to Talbot, is a canoe tree – a survivor from the days when Aboriginals roamed the plains.

Ruin, Glengower, Near Majorca, KS

MAFEKING

LOCATION: 40 KM NORTH OF DUNKELD. JIMMY CREEK RD. OFF THE DUNKELD – HALLS GAP RD
POPULATION 1900: 10,000
POPULATION 2006: 0

Mafeking was the scene of Victoria's last gold rush. The diggings were named after the famous siege in South Africa where British troops led by Colonel Robert Baden-Powell (who later founded of the world scouting movement) held the town of Mafeking for 217 days from October 1899 and May 1900 during the second Boer War.

The Mafeking rush was a short-lived affair which began when the Emmett brothers made a strike on the east side of Mt William in 1900. The population peaked at 10,000 within a few months, but by 1902 most diggers had left the area. The last miners' cottages were burnt in a bushfire in 1960.

Despite the short duration of the rush, the miners devastated the area in their search for gold. Much of the gold was taken from open-cuts using a hydraulic sluicing technique. A jet of water was directed onto the face of a cutting. The earth was then shovelled into a contraption known as a 'Tom' that consisted of two boxes. Water was directed into the upper box where a grate trapped the coarser gravels, stones and rocks while the finer particles of gravel, sand and gold fell through to the second box. A series of bars or ripples at the bottom of this box helped trap fine gold particles, while the water and lighter material ran off as overflow.

Today there is an attractive picnic area, a camping ground and an information board. However, this area is definitely unsuitable for children as there are a number of dangerous mineshafts scattered through the bush.

BROWNING'S WALK

Browning's Walk (one hour return) takes in some remaining historic features. A pamphlet is available from the Grampian National Park Visitors' Centre at Halls Gap. It identifies various features of the walk, including an old-growth stringybark, a regenerated gully, the site of the first claim, tail races, old shafts, a dam embankment used for water storage and open-cut mine sites.

MAJORCA

LOCATION: 7 KM SOUTH
OF CARISBROOK
POPULATION 1863: 15,000
POPULATION 2006: 50

Majorca owes its existence to the last significant gold discovery in the Maryborough district, made in March 1863. The town name was chosen to maintain a Mediterranean theme started with diggings a little to the north-west, known as Gibraltar. Gold in the area was found in broken basalt. Some prospectors recovered an ounce of gold (28.3 g) per load and the discovery of an 80 ounce (2.3 kg) nugget has been recorded.

Majorca grew rapidly and, by May 1863, there were 250 stores and businesses as well as a police camp in the town. More substantial brick

buildings, including the post and telegraph office and a government building, were erected in 1864 and in the following year Majorca became a municipality.

Ten thousand diggers crammed their tents along a 2.5 km lead. In winter Majorca was described as: '…a sea of mud and also noise… (caused by) the sounds of dogs, cradles, axes, the calls of top-men, the cursing of draymen, and the crushers pounding away at cement all night.'

Just as the initial rush declined, another began when the Hanoverian Lead was discovered just north of the town in 1864. By 1865 many small companies with steam-driven machinery were established on the field. However, within five more years the Majorca rush had subsided and the town, even with 24 remaining hotels, became a quiet country place surrounded by farms. Businesses began to move to nearby Maryborough.

Today, few buildings remain, but not only because the gold ran out. Devastating bushfires swept through the town in January 1985, destroying many buildings and 51,000 ha of surrounding farmland.

Ruin, Majorca, KS
Trough, Majorca, KS

Main St, Maldon, GC

MALDON

LOCATION: 139 KM FROM
MELBOURNE: 19 KM NORTH-WEST
OF CASTLEMAINE
POPULATION 1854: 20,000
POPULATION 1932: 700
POPULATION 2006: 1500

*Every street yields
an interesting
view – whether
it is of an old
chapel on a
hillside, or a
tall industrial
chimney.*

Maldon is a picturesque 19th century town that remained remarkably unspoilt. Every street yields an interesting view – whether it is of an old chapel on a hillside, or a tall industrial chimney. The town has great charm. Although there are no dominating buildings, the streetscapes remain historically consistent. The elements that make up a 19th century gold mining town are all clustered within an area that can easily be covered by foot in an hour or so.

Trees, gardens, fences, stone gutters, buildings and open spaces all contribute to produce a unique and coherent visual composition. Look beyond streetscapes, however, for the fascinating variety of detail in picket fences, cast iron verandahs, windows and roofs.

BRIEF HISTORY

Maldon was built on the slopes of Mt Tarrangower, right where gold nuggets were picked up off the hillside. Tarrangower is a big straggly hill, prominent in the last low ranges of central Victoria at the edge of the vast plain spreading north to the Murray River. Alluvial gold was discovered in December 1853, but it soon became clear that Mt Tarrangower's golden treasure was principally held within quartz reefs.

Initially individual miners crushed their own quartz in primitive iron 'pots', but large stamping plants were soon set up and there was an inexorable drift towards control by larger companies. Digging into the mountain to follow the quartz required large amounts of capital and holdings were consolidated until there were around 40 major mines.

The yield of gold taken from the mountain is astonishing. Some of the richest seams in the country were mined in shafts which can be seen just west of the Newstead Rd. Here are the workings of the North British Mine, site-works and tunnels and the kilns where quartz was heated until it was brittle, before it was crushed to release the gold.

The town of Maldon is one of the many that grew with the great alluvial mining rushes of the 1850s, but survived beyond that golden decade thanks to quartz reef mining. Until the 1870s quartz and alluvial mining employed roughly equal numbers of men.

After the mid 1870s quartz reef mining continued to flourish and was profitable until the 1920s, although the population began to decline from the beginning of the 20th century. When the North British Mine closed in 1928 it had produced more than 242,000 ounces (6.9 tonnes) of gold.

The town survived as a service centre until the 1960s when tourism began to play an important role. The National Trust recognised the heritage values of the town when it classified Maldon as the first 'Notable Town' in Australia. Mining has recently started again, but Maldon's greatest treasure is its 19th century streetscapes.

Visitor Information

The Maldon Visitor Information Centre has a wide range of maps and brochures, plus a free accommodation service.

Maldon Visitor Information Centre
93 High St, Maldon 3463
tel: 5475 2569 email:
maldonvic@mtalexander.vic.gov.au.

The **Victorian Goldfields Railway** is a heritage railway that links to Castlemaine on Wednesdays and weekends (see the Castlemaine chapter for more details). There are regular bus links to Castlemaine and Bendigo.

The Township

Maldon has few spectacular buildings, but the overall impact of the town is memorable, because many interesting buildings come together to make up a coherent whole.

The main street – at one point called Main St – curves around in wayward fashion, just as a mining town's main street should. It is a town for walking and for browsing – there are antiques, books, excellent cafés, and plenty of atmosphere.

Maldon Historical Museum

The old market building is now used as the town museum. One of the prized exhibits is a union banner that celebrates the cooperation between the miners and mine owners.

Maldon Hospital

Maldon Hospital, built in 1860, is a striking and grand building standing on a hilltop with great presence – all pilastered and porticoed.

Cottage, Maldon, RE
Trough, Maldon, RE
Maldon, GS
Maldon, GS
Maldon Hospital, JK

Anglican Church

The Anglican church in High St (1861) is a prime example of local ragstone with dressings of grey granite. The interior is quite beautiful, with a scissors-truss roof and excellent stained glass. Many other little churches and chapels dot the hillsides.

Post Office

The current post office was built in 1870, and is famous as the home of Ethel Richardson, the famous author who wrote under the pen name Henry Handel Richardson. Ethel's mother was a postmistress and ran the Maldon Post Office from 1880 to 1886. Ethel was very happy in Maldon and she described her life in her books (*The Fortunes of Richard Mahoney*, and *The Getting of Wisdom*).

Royal Hotel & Theatre

The Royal Hotel dates to back to 1857 and is significant for its intact front façade, and its theatre, which is a rare survivor.

Beehive Gold Mine

At the far end of Main St is a tall chimney marking the remains of the Beehive Mine. Built in 1862s, it stands 30 m high, and was designed to minimise wood consumption (the top two metres were knocked off in a lightning strike in 1923). Chimneys like this were once common around the goldfields, but this is the only one to survive. Around the chimney there are many other signs of mining activity to explore, starting with the original open-cut mine. The Beehive Mine was one of Victoria's most productive mines, yielding 210,000 ounces (6.5 tonnes) of gold.

Brooks Store

Built in 1866, Brooks Store functioned as a general store for 120 years. The two-storey section built as a grain store was a later addition. The rest of the building is generally unchanged, and is considered to be a rare example of a 19th century country store complex.

Maldon State Battery

The Maldon State Battery was one of 60 built by the government. This one dates from WWI and was built to encourage small operators to keep searching for new reefs. It was finally decommissioned in 1995.

AROUND THE TOWNSHIP

Mt Tarrangower

Mt Tarrangower rises 571 m above sea level and is at least half that height above the surrounding countryside. On top there is a lookout tower made from an old poppet head from a Bendigo mine. The view is wonderful: across plains and ranges from the Dividing Range and the potato-shaped peaks around Daylesford to the Moolort Plains and the endless northern horizon.

The tower is illuminated during the annual Easter Fair (held annually since 1877) and can be seen from 50 km away. The mountain is the focus for a hillclimb event, held annually since 1928 (with a few breaks) – the longest-running motor sport event in Australia.

Lisles Reef Walk from the carpark at the top of the mountain takes about one hour. In addition to great views, it has interpretive signage and a number of significant mining relics (including a whim platform) and ruins (including blacksmiths' shops).

Cemetery

Beyond the Grand Union Mine, is the picturesque cemetery – with a gate lodge (1866) and sentinel cypresses. Relatives still paint the railings of family tombs. A brick funerary oven stands in the Chinese section for the ritual roasting of food to honour the dead; unfortunately the Chinese grave markers, being timber, were all burnt in a 1969 bushfire.

AROUND MALDON

Carman's Tunnel

Carman's Tunnel was dug in an unsuccessful search for gold between 1882 and 1884. It was dug into the side of a hill, through solid rock. Today it is possible to experience guided tours by candlelight. The 570 m tunnel is dry,

spacious and level – but the experience of going into the heart of Mount Tarrengower is unforgettable.

The tunnel is about two km south of the town. Each guided tour takes 30 minutes, and the tours run in the afternoon on weekends and during school holidays.

North British Mine

Virtually opposite Carman's Tunnel (which found very little gold) the North British Mine was once one of the richest in the world. It was owned by one very rich man: Robert Dent Oswald. The site contains a 500 m deep shaft, mine dams, mullock heaps, the foundations for various machines and – most dramatically – three large quartz roasting kilns.

Porcupine Flat Dredge

About 1.5 km from Maldon on the Bendigo Rd, a huge dredging machine sits incongruously beside the road. These huge machines were once common throughout the goldfields and were used to process alluvial gold. The dredge floated on water and the large buckets scooped up soil ahead of the machine. The soil was washed on board, the gold captured and the residue dumped behind. The dredge moved along digging its own canal and filling it in as it passed.

The process was extremely efficient (especially when compared to individual men with pans and cradles!) but the environmental impact was horrendous. This particular dredge was used between 1958 and 1984.

Lisles Reef, DB
Porcupine Flat Dredge, PV
Carman's Tunnel, RE
North British Mine, KS
Maldon Cemetery, DB

*Malmsbury
is the heart
of bluestone
country and
the entry to
the Goldfields.*

MALMSBURY

LOCATION: 95 KM NORTH-WEST
OF MELBOURNE. CALDER FWY
POPULATION 2006: 500

The tiny town of Malmsbury clings to the sides of the Calder Freeway, which rolls up and down like a big dipper crossing the steep valley formed by the Coliban River. Malmsbury is the heart of bluestone country, with bluestone bridges, viaducts, mills, public buildings, churches and cottages. Most of them date from the 1860s and 1870s.

This is also the entry to the Goldfields. The dam wall of the Coliban Reservoir is directly up-river from the railway viaduct and can be seen from trains approaching the station. The water from this reservoir flows under the Calder Freeway just past the Coliban River Bridge and, in an amazing 19th century engineering feat, flows from here to Harcourt, Castlemaine and Bendigo.

Malmsbury expected to be bigger than it is. There are several pieces of evidence that give this away. The most obvious is the layout of the town, which is

clearly designed to accommodate a large population. The grand avenue through the centre, Mollison St, is used as actual roadway for less than half its width and shops are only found on the southern side. The design is the work of the government surveyor, WS Urquart, who is also responsible for Sturt St in Ballarat – although in this latter case the city has actually grown to match the street's scale.

MECHANICS INSTITUTE

Don't miss the tiny mechanics institute (1862) at the top end of the hill on the Kyneton side of town. In 1855 on her return from England, social and family rights advocate Caroline Chisholm threw herself into the task of providing safe, decent accommodation for people and families travelling to the Goldfields. The existing inns were expensive and often little more than dangerous sly-grog shops.

In response to her pressure, the government funded 10 shelter sheds. In fact, each location had a number of sheds built to accommodate single men, single women, married couples

Aerial View of Malmsbury Viaduct, BS
Mechanics Institute, JK

and families. The sheds had a station keeper who sold water, wood, candles and provisions.

None survive, but one stood on the site of the Malmsbury Mechanics Institute. The others were at Essendon, Keilor, Keilor Plans, The Gap, Gisborne, Black Forest, Woodend, Carlsruhe and Elphinstone.

THE MALMSBURY MILL

At the Bendigo end of town is an imposing bluestone flour mill dating from 1861. An explosion in 1862 destroyed the building and killed two workers. It was rebuilt that year and operated until 1895. Then it was used as a chaff-cutting mill until the early 1900s. The building now serves as a restaurant and gallery and, unlike most central-Victorian mills, it retains its towering chimney.

ST JOHN'S ANGLICAN CHURCH (1861)

The Anglican church is set back from the main road on the north-eastern slope near the town-centre. It was built to seat over 450 people, commensurate with a parish three times the size of the town in its heyday. The church, designed in 1861 by Melbourne architects Purchas and Swyer, comprises a double nave divided by arches. The unusual façade has a star-shaped rose window and an octagonal belfry.

BOTANIC GARDENS

When Malmsbury was surveyed in 1863, 22 acres (about 9 ha) were set aside for botanic gardens, making them the third oldest in regional Victoria (behind Geelong and Portland). The town received thousands of plants

and seedlings recommended by the famous botanist, Baron Ferdinand von Mueller. A feature today is a rare hybrid strawberry tree. The gardens, which also include an ornamental lake, run along the side of the Coliban River and give access to the railway viaduct.

The **Malmsbury Town Hall** (1867) is next to the Botanic Gardens.

RAILWAY VIADUCT

The magnificent railway viaduct over the Coliban River was constructed in 1861-1862 as part of the nationally significant Murray River Railway that linked Melbourne to the Goldfields and Echuca. The 25 m-high viaduct, with five 18.3 m spans, is an outstanding engineering work and is considered one of Australia's finest early bridges. Constructed of finely dressed bluestone, it is the largest stone bridge in Victoria, and when completed it was Australia's longest masonry bridge. The structure can be viewed from the southern end of Ellesmere Place, on the left side of the Calder Freeway coming from Melbourne, or by walking through the park to the reservoir.

COLIBAN RESERVOIR

The dam wall of the reservoir is directly up-river from the railway viaduct. The channel flows under the Calder Freeway just past the Coliban River Bridge. Water is reticulated from here via Harcourt to Bendigo, Castlemaine and Maldon.

Mill, KS
Church, PW
Botanic Gardens, KS
Aerial View of Town & Coliban
    Reservoir, BS

Moliagul is an inviting out-of-the-way spot for a picnic.

Moliagul Hotel, KS
Moliagul, KS
Fence, Moliagul, KS

MANDURANG

LOCATION: 10 KM SOUTH-EAST OF BENDIGO

Mandurang is the geographical centre of Victoria – to be precise, the front steps of the disused Mandurang Uniting Church are the spot. It is also home to a number of wineries, including Chateau Dore, an early vineyard where the first grapes were planted in 1866. It was established by Jean Theodore de Ravin, one of the international fortune-seekers attracted to the Victorian goldfields. De Ravin was born on the French island of Martinique in the West Indies. His property on Sheepyard Creek also contained a market garden. The vineyard survived until 1901 and was re-established in 1975 by Ivan Grose, de Ravin's great grandson.

MOLIAGUL

LOCATION: 225 KM NORTH-WEST OF MELBOURNE; 15 KM NORTH-WEST OF DUNOLLY
POPULATION 1855: 16,000
POPULATION 2006: 200

Timeless Moliagul, the place where the world's largest gold nugget was discovered, is now virtually a ghost town. Small and lush-green in springtime, it's an inviting out-of-the-way spot for a picnic. Moliagul has a pub, a church, a cemetery, a school and a few obvious houses – although there are more houses scattered through the bush.

At the corner of Bealiba and Murray Sts there is a monument to Moliagul's most famous son, John Flynn, the founder of the Royal Flying Doctor Service who was born in 1880 on the site. Opposite Flynn's memorial is an old, abandoned timber cottage well-hidden beneath old peppercorn trees. The house is on the corner of a road

that leads to a very different landscape – red, raw and ravaged, the pure heart of gold country.

Gold was first found at Moliagul in late 1852 in Queen's Gully. 4000 miners rushed to the area and a police camp was established. By January 1853 there was a store, blacksmith's and butcher's shops. A new rush began when gold was found at Little Hill in July 1855.

The **Moliagul State School** (1872) and the **Anglican church**, built in 1864, are still in use. The **Mt Moliagul Hotel,** built in 1856, has budget accommodation and an interesting collection of farming and mining memorabilia.

THE WELCOME STRANGER NUGGET

Two Cornish miners, John Deason and Richard Oates, who were mates from childhood, arrived in Moliagul in 1862 after eight years of moderate success on the Bendigo goldfields.

Aware of large nuggets found in Black Gully, Deason and Oates pegged a puddling claim on the side of the hill. They also selected nearby land, which they farmed while working the puddling claim. In 1866 they found a 36 ounce (1 kg) nugget. On the morning of 5 February 1869, Deason was breaking up soil on the claim at Bulldog Gully when he hit what seemed to be a rock an inch (2.5cm) below the surface. Before he realised what he had discovered, he broke his pick handle and resorted to using a crowbar.

Deason's son called Oates who was ploughing nearby, but not wanting to alert nearby neighbours, the pair covered their find and carried on as if nothing had happened. They had just found the largest gold nugget in the world.

Later that day they carted the nugget by dray to Deason's house where it was placed in the fire. Fifty seven pounds weight (26 kg) of quartz was prised away and later crushed. On the following Monday they revealed their find to friends who then escorted it by convoy to the Bank of London in Dunolly. The nugget was too large to be placed on the bank's scales and had to be broken into pieces to be weighed. The anvil on which it was broken up can be seen in Dunolly.

The 2380 ounce (67.5 kg) nugget, including the crushed quartz and bits of gold broken off and given to friends, was worth £10,000 at the time. It's difficult to calculate its relative value today, but the gold alone would still be worth close to $2 million – which makes no allowance for its rarity value, or the relative purchasing power of money. In real terms something like $5 million is probably a better valuation.

Ironically, Deason and Oates were refused a bag of flour on credit a week before the strike. Soon after, Oates returned to Cornwall to marry. He then brought his wife back to Moliagul and worked the claim with Deason until it was bare in 1875. He lived out his life at Dunolly. Deason continued to mine and invested in a quartz-crushing battery and a further property known as *The Springs*, where he and his family lived. Deason died in 1915 Moliagul aged 85. His descendants still farm in the area.

MOLIAGUL HISTORIC RESERVE – WELCOME STRANGER SITE

A granite obelisk erected in 1897 by the Mines Department stands on the site of the nugget's discovery, two km south-west of the town. The **Welcome Stranger Discovery Walk** starts from

the monument and explores the reserve, which has a picnic area. John Deason had a small two-roomed shack near the start of the track. A little further on is the puddler where Deason and Oates treated the wash from their claim. The pile of stones further on is all that remains of Richard Oates' house.

A wooden headstone along the fence line is thought to mark a Chinese grave dating from the 1860s. There were many Chinese in the area at that time, but stones from fireplaces and raised dirt floors are all that remain of their camps. The walk continues past an old puddling machine, then crosses Black Gully, so named because the gold tended to be stained with black ironstone, as was the *Welcome Stranger* itself.

AROUND MOLIAGUL

Mt Moliagul

To the north of Moliagul is Mt Moliagul, which offers excellent views from the summit.

Kooyoora State Park & Melville Caves

Just north of Moliagul on the Rd to Inglewood is the 11,646 ha Kooyoora State Park. See the Inglewood section for more details.

They had found the largest gold nugget in the world.

Cottage Near Moliagul, KS
Welcome Stranger Monument, KS
Mining Landscape, KS

MOONAMBEL

LOCATION – 201 KM NORTH-WEST
OF MELBOURNE; 20 KM NORTH-
WEST OF AVOCA
POPULATION 1860: 30,000
POPULATION 2006: 100

On the Pyrenees Highway between Avoca and Stawell, Moonambel is a quiet and charming village. Moonambel, the Aboriginal name for 'hollow in the hills', was part of the *Mountain Creek Run* in the 1840s. Gold was discovered in the area in 1860.

When tens of thousands of goldminers arrived, the town flourished, with breweries, a flourmill, soap factory and newspaper, all of which have long disappeared. Of the many hotels that once existed, only the 1866 Commercial Hotel remains. The local general store is worth a visit to see the old photographs on display. Town tours can be arranged.

Some hauntingly-interesting deserted wood-shingled cottages, with outbuildings made from with wide hand-cut timbers and earth roofs, are set back from the highway and are dotted around the side streets.

Most mining was initially alluvial, and shallow alluvial shafts are still visible on the banks of the creek. A track crosses the creek opposite the hotel (to a beautiful picnic area) and continues into state forest where mullock heaps mark reef mines, some of which continued operations into the early 20th century.

Orchards and vineyards were established in the 1860s. The wine industry was quiet from 1948 to 1969, but has re-emerged since then and the area now boasts a number of Victoria's finest wineries. In the heart of the flourishing wine district there are many fossicking areas.

Main St, Moonambel, KS
Cottages, Moonambel, KS
Commercial Hotel, Moonambel, KS
Government Battery, Mt Egerton, DB

MT EGERTON

LOCATION: 25 KM EAST
OF BALLARAT: 8 KM SOUTH
OF WESTERN HWY
POPULATION 2001: 198

The Mt Egerton goldfield was discovered at what became known as 'All Nation's Gully' in 1853 by a party of Ballarat miners. The field was rushed in 1854 with miners concentrating their efforts on extracting gold from one long line of quartz reefs. By 1856, the field had a population of 600 and a flourishing township was taking shape.

By 1867 the town was dependent on two mining companies: the Egerton Company and the Black Horse. These two companies continued to carry the field for many years. In the early 1870s, a third local company, Parker's United, also prospered. In 1877 there were 78 men were working at the Black Horse, 50 men at Parker's United, while the Egerton Company was employing an average of 175 men.

The riches began to decline during the 1890s and one by one the town's principle mines suspended or cut back operations as known gold reserves were exhausted. The government installed a small crushing battery at Mt Egerton to assist the locals and it survives today as a testament to the times when the bulk of the township's population were involved in gold mining in some way.

Today Mt Egerton is a centre for the surrounding farming community.

MOYSTON

LOCATION: 239 KM NORTH-WEST
OF MELBOURNE;
15 KM WEST OF ARARAT
POPULATION: 200

In the centre of Moyston, a monument proudly stands to commemorate this small town as the birthplace of Australian Football, which was inspired by an Aboriginal game played in the region.

Gold was discovered in Moyston in 1857 on Campbells Reef and there are numerous reminders of the gold-rush era in and around Moyston. The solitary pine tree in the town's Avenue of Honour marks the site of the Campbell's Reef graveyard. The Jallukar State Forest protects an abundance of wildlife and old mine shafts.

AUSTRALIAN FOOTBALL

Four men are credited with first codifying Australian Football's rules: Tom Wills, his cousin Henry Harrison, WJ Hammersley and JB Thompson. Together, on 17 May 1859, at the Parade Hotel on Wellington Pde, they drew up 10 rules, making Australian Rules Football the oldest officially codified football game in the world.

There is growing consensus that these men drew initially on at least three sources: the informal ball games played by convicts and immigrants, rugby, and a Aboriginal game, known as marngrook which was played in Western Victoria. It is also clear that Gaelic football has had a long-term impact on the development of the game, but Gaelic football was not codified until 1884.

Harrison's family had a property near Moyston and Wills' family also had a property in the Western District.

Harrison and Wills were both descended from emancipated convicts who were forced to fight bitterly for their rights within the colony and who were strongly pro-Australian. A unique Australian game, free of overt British influence would have appealed to them; it is also highly likely that both men would have at the very least seen, if not played, marngrook, which was described in 1841:

'The men and boys joyfully assemble when this game is to be played. One makes a ball of possum skin, somewhat elastic, but firm and strong. The players of this game do not throw the ball as a white man might do, but drops it and at the same time kicks it with his foot. The tallest men have the best chances in this game. Some of them will leap as high as five feet from the ground to catch the ball. The person who secures the ball kicks it. This continues for hours and the natives never seem to tire of the exercise.'

Grampians Near Moyston, RE

Some surveyed towns, such as South Muckleford, did not develop as expected and were known as 'paper townships'.

MUCKLEFORD

LOCATION: 129 KM NORTH-WEST OF MELBOURNE; 10 KM WEST OF CASTLEMAINE

A mixture of farmland, historic goldfields, and Box-Ironbark forest covers much of the area between Castlemaine and Maldon, known as Muckleford.

The earliest alluvial diggings date back to the 1850s with the most recent operations dating from 1958. In the summer of 1858-59, Joseph Day built a dam across the Loddon River near the junction with Muckleford Creek and installed extensive sluicing works that recovered gold 'in every dish'.

Some surveyed towns, such as **South Muckleford**, did not develop as expected and were known as 'paper townships'. At South Muckleford, the brick school that opened in October 1871 closed in 1927 due to declining numbers. The local progress association converted the building to a hall that is still used today by the small rural community.

GOWAR STATE SCHOOL

Until 1880 Gowar was known as North Muckleford. The school, built in 1874 of stone and shingle, was unlined and would have been bitterly cold in the winter. Thirty-four children attended the school in 1882, but it closed in 1908. The circular brickwork on the west side of the school ruins is the top of a well which supplied the children with drinking water.

There is an eight and a half km walk from the school through Smith's Reef Forest. The walk starts at the school ruin. Evidence of small-scale alluvial mining can be seen on the gully flats. Sawpit Gully, on the east side of the school, was rushed in 1857. Further on, Smith's Reef, was home to a 25-horsepower engine that drove stampers to crush quartz.

The track is close to the Muckleford Fault. Extending many km in a north-south direction, movement along the fault line over millions of years raised the land to the west.

RED, WHITE & BLUE MINE

Red, White & Blue Mine lies in the Muckleford State Forest and has been worked on and off since 1871. Between 1909 and 1915 a total of 2369 ounces (67.2 kg) of gold was recovered from 4,452 tons (4523 tonnes) of ore. In 1958 the current poppet head was brought to the site from the Deborah United Mine in Bendigo. The mine was reopened and renamed 'The Golden Age', but only a small quantity of gold was extracted.

The walking track from the car park near the poppet head displays a cross-section of Goldfields history – alluvial diggings, puddling machine sites, battery tailings, mining dams and campsites. The scars of shallow alluvial mining are still clear at the gully near the start of the walk. Some of these mining works date back to the Great Depression in the 1930s.

From Castlemaine take the Maldon Rd. Turn south onto Muckleford Rd, then right into Muckleford School Rd to the start of the Muckleford Forest. The walk starts at Bells Lane track. Picnic tables are near the mine.

Muckleford South
Muckleford1.tif

NEWSTEAD

LOCATION: 134 KM NORTH-WEST
OF MELBOURNE; 16 KM SOUTH-EAST
OF CASTLEMAINE.
POPULATION 1855: 16,000
POPULATION 2006: 200

Newstead is a peaceful rural town quite separate from the gully settlements of Castlemaine. The long elm-lined main street curves down onto the Loddon River's flats between the verandah fronts of old stores and workshops.

The bridge over the Loddon River brings into view immense river red gums that were seen by Major Mitchell in 1836 when deep mud slowed the exhausted expedition. The bridge deck is high – for good reason – dramatic floods have threatened the town on a number of occasions.

A primitive punt connected Lyons St with the 'Junction', a settlement on the west bank. The crossing made Newstead an important staging post for miners making their way from Adelaide and Ballarat to the Mt Alexander diggings.

In 1860 the punt sank and a toll bridge attracted coach traffic between Maryborough and Castlemaine. However, in the winter of 1861 the crossing still had mud that 'came up to a man's middle'.

Newstead grew as the service centre for quartz mines at Welshman's Reef, Mia Mia and Green Gully, and for small farms that had their first harvests in 1855. Local industries included two flour mills, a ginger beer and soda water factory, and a biscuit and confectionery factory.

For 135 years from 1860, Newstead was a local government headquarters. It is now part of the Shire of Mt Alexander. Newstead Shire included

Campbells Creek, Clydesdale, Fryerstown, Green Gully, Guildford, Joyce's Creek, Sandon, South Muckleford, Strangways, Strathlea, Tarilta, Vaughan, Welshman's Reef, Yandoit, and Yapeen.

There is a group of tiny buildings near the river, some still occupied, some rather neglected. One is called **The Shambles**, a traditional name for a butcher shop.

Government and religious buildings dominated the higher, drier, ground of Lyons St, including a police station, court house, churches, a school and shire offices.

The first building to appear in Lyons St, the National Hotel, was on the present site of **Newstead Park**, next to the Loddon River. This modest park has an interesting arched iron gateway. The surviving **Crown Hotel** was originally a bakery and store built in 1857.

On the south-west corner of Wyndham and Lyons Sts, where the Shamrock Hotel once stood, large numbers of Aboriginals used to camp and hold corroborees, as late as 1856.

The **General Store** began in 1868 and was further developed over the years

Newstead Avenue, KS
Window, Newstead, KS
The Shambles, Newstead, GS

CHURCHES

The **Primitive Methodist Church** opened in 1860 and was sold to the current owners, the Masonic Lodge, in 1907. The 1868 Gothic style **Anglican church**, built of red brick with white pressed brick laid on blue mortar, has an open-framed interior made of oak and a very large slab from the Barker's Creek slate quarries. Four Baptist Churches were built in the 1850's and 60's, two of which survive. The **Uniting church** still stands, as does a splendid classical bank building (now a dwelling).

OLD MILL

The Sheehan family built the Old Mill in Layard St in 1869. One of several mills processing grains in the area, it became waterlogged during floods in 1909 and closed in 1914. It is now a private residence. Next door was the millowner's home, also built in 1869, as a four-room cottage of hand-made bricks and shingle roof.

BUTTER FACTORY

In 1905 the Newstead Co-operative Butter and Cheese Factory opened and for many years its famous 'super-fine' product underwrote Newstead's prosperity. Changes in the dairy industry led to its closure in 1975. It is now home to Oz Candles and Newstead Winery.

into what is now the supermarket and bakery. On the north-east corner of Lyons and Panmure Sts is the site where Richard Rowe established a successful blacksmith, wheelwright and coach building business.

COURT HOUSE

The 1863 court house is a fine example of the small court houses designed by the Public Works Department. It is considered to be the most accomplished example of the Free Classical style. In 1987, the Newstead and District Historical Society converted the court house into an archive and museum.

MECHANIC'S INSTITUTE

In 1868 an impressive brick mechanics institute was erected, to replace an earlier one built in 1854. Newstead's institute was important to the town's social, political and economic history and today the 146-year-old building

Main St, Newstead, KS
Old Mill, KS

still functions as a public hall. A recently opened community centre has been sensitively designed to complement the mechanics institute.

NATIONAL SCHOOL

The first National School opened its doors in 1859. The building on Campbell St, which replaced tents that were used for the first three years, still stands next to the Newstead Pottery on the road to Daylesford overlooking the old Newstead Racecourse.

RAILWAY STATION

The red brick and bluestone station and matching goods shed make a handsome pair, beautiful in their decrepitude. Trains between Castlemaine and Maryborough no longer stop here and the reserve near the station is a quiet spot to pause for a rest. Nearby, the **Railway Hotel** was built shortly after the arrival of the railway in 1874.

AROUND NEWSTEAD

Castlemaine – Newstead

The low, lightly-wooded hills between Castlemaine and Newstead were devastated by gold mining. The soil on the slopes looks depleted, and it is – the regrowth forest is not yet a century old. Some pleasant farms are found on the creek flats alongside the road, where soil is clearly deeper and more fertile.

Daylesford – Newstead

The drive between Newstead and Daylesford is particularly beautiful, and follows the shallow valley of the Wombat Creek (a tributary of the Loddon) towards its source on the Great Dividing Range beyond Daylesford. The remains of small townships, a cluster of cottages, an abandoned pub, are evident every 10 km or so along the road.

Solid old red gums and yellow box trees line the creeks, while stringybarks cover the scourged hillsides where miners sluiced away the topsoil. As the road climbs through Franklinford toward Mt Franklin, open pasture on the rich soil of volcanic hills contrasts with thick forest regrowth on the mine-ravaged slopes of rubble and clay.

The drive between Newstead and Daylesford is particularly beautiful, and follows the shallow valley of the Wombat Creek.

Avenue, KS
Mechanics Institute, DB

REDESDALE & MIA MIA

LOCATION: 91 KM NORTH
OF MELBOURNE;
20 KM SOUTH-WEST OF HEATHCOTE
POPULATION 2006: 500

Set amidst picturesque rolling hills and the Campaspe River's steep ravine, Redesdale and Mia Mia are two small settlements midway between Heathcote and Kyneton.

Major Mitchell passed through the district in 1836 and crossed the Campaspe (which he named) near the site of present-day Redesdale. Within a year, overlanders and squatters were moving into the area and there was much conflict with the local Aboriginals. Burke and Wills camped near the present-day Mia Mia Bridge crossing the Campaspe River on their way north from Lancefield.

Over the years Mia Mia's role declined and Redesdale became the principal town. The atmospheric old Redesdale Pub was rebuilt in bluestone in 1856 and is the town's social centre. The Redesdale Presbyterian and Catholic Churches, still in use today, were built in the 1870's.

Towards Mia Mia, past the handsome St Laurence's Church, the road descends suddenly into the narrow valley of the Campaspe River and crosses on a wonderful iron-framed double bridge – one structure for each lane of traffic.

REDESDALE BRIDGE

Redesdale's famous 'basket handle' bridge, has an eventful history. The bridge was built in England and transported to Australia, originally intended to span the Yarra River at Hawthorn. However, the ship caught fire and sank and the 200-ton bridge was raised and brought to Redesdale by bullock dray. The bridge was officially

Redesdale Bridge, JM
Turpins Falls, RE

opened in 1868 with a banquet and ball. Locals celebrated with champagne and danced on the bridge until the early hours of the following morning.

Sadly the unusual design has in the past had tragic consequences. A woman from Mia Mia was killed when her horses in a buggy bolted down the Redesdale Hill and one horse went each side of the central stone abutment.

AROUND REDESDALE & MIA MIA

Aviation Memorial

In 1910 the brothers John and Reginald Duigan made Australian aviation history when they built and flew the first Australian-made aircraft at Mia Mia. The pilot, John Duigan, perched on the lower wing of the flimsy self-constructed biplane, took to the air on the family property at *Springs Plains Station* for a seven-metre hop. About three km from Mia Mia on the Lancefield Rd there is a memorial. The machine itself is now in the Melbourne Museum.

Lake Eppalock

Lake Eppalock is seven km north of Redesdale. Constructed in 1964 to provide irrigation supplies from the Campaspe River and town water to Bendigo, water skiing, swimming, fishing and boating are popular recreational activities. Facilities around the lake include caravan parks, camping grounds, picnic areas and public boat ramps.

Turpins Falls

South of Redesdale, the beautiful Turpins Falls are a short drive off Redesdale Rd. There is a large lagoon-like pool at the bottom of the falls. See the Kyneton section for more information. If coming from Redesdale turn right just after Barfold (nothing but a town hall) down a dirt road where the highway curves around a bend and some letter boxes appear. Follow the road signs after that. There's about three or four turns to make over a few km of dirt track. Contact the Heathcote VIC for a map.

The Cascades

The Cascades, an interesting set of rocks, pools and water paths on the Coliban River, are just east of Metcalfe about 20 minutes drive from Redesdale. The Cascades are just past Metcalfe, down a short dirt road.

Death Valley, Redesdale, KS
The Cascades, PW
Redesdale Hotel, KS

Rushworth retains much of its original character, many early buildings and intact streetscapes.

St Paul's Church, Rushworth, KS
Commercial Bank, KS
Shire Hall, KS

RUSHWORTH

LOCATION: 166 KM NORTH OF MELBOURNE
POPULATION 1853: 40,000
POPULATION 2006: 1000

Rushworth, an old goldmining and timber town, is one of two National Trust classified historic precincts in Victoria. It retains much of its original character, many early buildings and intact streetscapes. The Rushworth State Forest, surrounding the town, is the largest surviving ironbark forest.

Known as the 'Gold & Ironbark town', Rushworth began in the 1850s as a stopping place for those travelling between the Bendigo and Beechworth diggings. Gold was discovered about a mile (1.5 km) east of present-day Rushworth in August 1853 when local Aboriginals showed some gold-bearing stones to a group of diggers who had camped overnight. Alluvial gold was abundant throughout the area and underground shafts were later sunk to a depth of 886 ft (270 m).

The present township was surveyed in 1854 and named by poet Richard Horne, a friend of Charles Dickens, one of the two gold commissioners overseeing the rush. By 1858 there was a police camp, a timber court house, five hotels, two breweries, a school, seven large stores, 20 tradesmen's shops and two banks – all at the southern end of High St. The first local newspaper, the *Waranga Echo*, started in 1868.

The town continued to prosper as timber became a major local industry with seven sawmills in operation. Only Risstrom's Sawmill remains. The last steam traction engine to haul timber from the surrounding forest is classified by the National Trust and can be seen at Rushworth & District Historical and Preservation Society.

An old steam traction engine whistle post (1906) stands on the south-eastern corner of High St. This signalled the engines, which had absolute right of way, to sound a warning whistle.

HIGH ST HERITAGE WALK

Rushworth's main thoroughfare, High St, leads south off the highway to Whroo, seven km into the Rushworth State Forest, then to Graytown, a further 20 km down a dirt track. Until the railway arrived in 1890, the southern end of the street was the hub of local business, so this is where many of the older buildings are found.

On the hill at the southern-most end is **St Paul's Anglican Church**, of Gothic Revival style built in 1869-70. In the middle of High St, and for its entire length, the central plantation is clearly the town's centrepiece with wide immaculate green lawns, palm trees, and the band rotunda (1888). The Rushworth Brass Band has been playing since 1874. The imposing fire tower dates from 1900.

Diagonally opposite the whistle post is the court house (1870s). The Waranga Shire Hall (1869) is at the corner of Horne St. Continuing northwards along High St, on the right-hand side is the town's oldest building – a private residence built in 1854 as the Imperial Hotel. Just beyond are the first and second CBC Banks (1883) and Cracknell's Bakery – all now private homes. Opposite is the Criterion Hotel (1856).

The first school was established on the south side of High St in 1858. Continue along the left side to see the Glasgow Building (1858) and between Wigg and Parker Sts, a series of old shops. The former *Chronicle* newspaper (1888) office is at Parker St.

At High St and Moora Rd the post office (1885) is on one side of the Rd and on the other is the old Presbyterian church (1858-59). The Rushworth Hotel (built in 1878 as the Cricketers Arms) and the first Catholic church (1861) are west along Moora Rd.

RUSHWORTH MUSEUM

At High and Parker Sts the Rushworth Museum is in the former mechanics institute, built in 1913 to replace the original structure of 1861. A wealth of items pertaining to local history and assistance with family genealogy is available. Open Tuesdays from 2 pm to 5 pm.

GROWLERS HILL LOOKOUT TOWER

The fire authority's lookout tower at Growlers Hill offers views over the town, the Waranga Reservoir, the Rushworth Forest and the Goulburn Valley. Head west along Parker St then turn right into Reed St.

AROUND RUSHWORTH

Rushworth State Forest

Immediately south of town via Whroo Rd, is Rushworth State Forest, part of Victoria's Box-Ironbark forests. It is home to red ironbark, yellow gum, grey box and a profusion of wildflowers and orchids following autumn and spring rains. Fauna includes 100 bird species, echidnae, possums, kangaroos, wallabies, wallaroos and the rat-sized marsupial known as the Tuan.

For many years the forest provided employment for timber cutters and carters. Electricity poles, railway sleepers, road markers, fence posts and firewood were cut from the ironbark trees.

Whroo Historic Reserve

See the separate section on Whroo, once a once thriving goldmining town with 1000 people and 139 buildings. Visitors can walk to an Aboriginal waterhole from the Whroo Cemetery, but the highlight is the extremely dramatic Balaclava Mine, a 25 m deep open-cut mine with an extensive network of tunnels and shafts.

Jones's Eucalyptus Distillery

Just south of Rushworth is a distillery to extract eucalyptus oil from blue mallee gum. Visitors should call first, as opening hours are irregular. Head south on the Whroo Rd and turn off into Parramatta Gully Rd.

The Gold & Ironbark Trail

The Gold & Ironbark Trail is the original route taken by gold miners travelling between Bendigo, Heathcote, Rushworth/Whroo and Beechworth goldfields. The trail begins at Toolleen on the Northern Highway, but you can join at any point along the way. The Colbinabbin Range, overlooking the beautiful vineyards of Colbinabbin 21 km west of Rushworth, is one of Victoria's oldest landforms. Also known as the Mt Camel Range it is formed of greenstone, some of the oldest rock in Victoria.

Main St, KS
Building, KS
Rushworth State Forest, KS
Criterion Hotel & Shops, KS
Over page: Amherst Big Reef, JM

Lavandula has a cluster of 1850's stone farmhouse buildings, an old-world garden and a beautiful lavender farm.

SCARSDALE

LOCATION: 23 KM SOUTH-WEST OF BALLARAT
POPULATION 2001: 114

The first squatters settled in the Scarsdale region, south-west of Ballarat in 1837. The town origins lie in the development of two groups of deep lead gold mines that operated during the 1860s and the 1870s, while the name is believed to honour the English Baron Scarsdale, of Scarsdale in Derbyshire.

The first group numbered 30 mines, led by the Croesus Prospecting Company. They operated on the leads around Black Hill and Sugarloaf Hill to the north and east. The second group comprised nine mines, including the Bute, Golden Stream, Scarsdale Great Extended/Galatea, and Wheal Kitty, that worked along the main trunk lead running south from Scarsdale to Piggoreet. However, none were very successful. By the end of the 1870s only machinery foundations, tailings and mullock heaps remained and Scarsdale returned to agricultural pursuits.

Of historic interest in the township today is the Scarsdale General Store, the 1883 weatherboard post office, the 1861 Scarsdale Hotel (previously known as the Royal Exchange) and the former town hall , now a community hall and kindergarten.

A prominent geological feature in the form of a volcanic plug, known locally as 'Rocky', is visible from Community Park. The park itself provides parking, picnic, BBQ and toilet facilities and access to a rail trail that passes through the area.

SHEPHERDS FLAT

LOCATION: 36 KM NORTH-EAST OF BALLARAT; 10 KM NORTH OF DAYLESFORD

The hamlet of Shepherd's Flat is just 10 minutes drive from Daylesford. It is home to the **Cricket Willow** factory, the birthplace of the Australian cricket bat. The story goes that during the English cricket team's tour of Australia in 1902 it was revealed that no cricket bats were made in the country. Cuttings of willow were subsequently sent to Australia by ship and the Crockett Family began crafting bats in its small workshop at Shepherd's Flat.

Cricket Willow is now owned by a multi-national sports goods manufacturer and has been greatly expanded. Visitors can see all the stages in bat manufacture. There is also a picturesque cricket oval and pavilion, and a cricket museum housing an extensive collection of memorabilia.

A little further along the Hepburn-Newstead Rd, on a bend of the Jim Crow Creek, is **Lavandula**, the Swiss-Italian Lavender Farm. This is a traditional European-style settlement with a cluster of 1850's stone farmhouse buildings and an old-world garden. The original methods of planting and harvesting are still used.

Previous page: Amherst Big Reef, JM
Sculpture, Scarsdale, GS
Undertaker, Scarsdale, GS
Shepherds Flat, JM

SMEATON

LOCATION: 142 KM NORTH-WEST OF MELBOURNE; 31 KM NORTH OF CRESWICK

Smeaton is a small town surrounded by extinct volcanoes, fertile volcanic soils, and relics from a rich gold mining past. There's a pub and a handful of closed-down shops at a quiet crossroads.

Captain John Hepburn, one of the first European settlers in the area, arrived in the district in 1838 and took up a squatting run in the area, which he named *Smeaton*.

During the 1860's, the entire landscape north of Creswick developed into a prosperous agricultural community. Disenchanted miners purchased small freehold plots or tilled the fields of Captain Hepburn's estate.

THE BURIED RIVERS OF GOLD HERITAGE TRAIL

Throughout this area, beneath the rolling hills, are long buried underground streams – once full of gold. A deep lead could be described as a 'buried river of gold'. Between Smeaton and Creswick five different deep leads, each part of this ancient river system, were chased by approximately 120 mines. The Buried Rivers of Gold Heritage Trail will take you to the sites and tell of the area's gold mining history and settlement.

In 1872 a small prospecting party struck a rich lead of gold in the area that was to become the township of Bloomfield. After 2,790 ounces (79.1 kg) of gold were extracted, mining here finished. Following this success, however, another landowner opened his fields to mining and 14,000 ounces (396.9 kg) were found between 1873 and 1875. By 1878 there were over 20 mining companies working in the

Between Smeaton and Creswick five different deep leads were chased by approximately 120 mines.

Bury No 1, RE
Berry Consolidated Tailings, DB

region. Successful mines on the *Seven Hills Estate* included the Madame Berry, the Ristori Freehold and the Lone Hand. In the early 1880's, Madame Berry produced the most with 387,314 ounces (11 tonnes).

South of Smeaton, the towns of Allendale and Bloomfield grew to accommodate the miners, mostly immigrants from Cornwall. Miners' homes were typically three to four room timber dwellings. Cornish traditions including Methodism, temperance and friendly societies were a part of the culture. Cornish technology, essential in the deep leads, included Cornish beam pumps that drained the buried rivers. Ruins of the Cornish-style engine houses can be seen on the site of the Hepburn Estate and Berry No.1 mines (where gold was not found until 1890).

At the peak of their productive days extensive plants, including winding and pumping engines, boilers and puddling machines, surrounded mineshafts. In 1883 the Madame Berry No.2 mine was dominated by 90 ft (27m) sawn timber poppet legs and 90,000 bricks were used to construct buildings. Today all that can be seen at most sites are the brick footings, huge mullock heaps and sand dumps left beside the abandoned shafts. An example of the quantity of material extracted can be seen at Clover Hill.

By the late 1880s the Berry Deep Leads began to decline and by 1900 the annual yield had dropped and pumping out water had become a critical problem.

Booklets are available at the Creswick Museum, and the visitor information centres at Creswick, Daylesford, Ballarat and Castlemaine.

Anderson's Mill, PW
Anderson's Mill, PW
Anderson's Mill, DB

Anderson's Mill

Tucked into a sheltered valley, Anderson's Mill, a magnificent four - storey bluestone building, survives as a reminder of a flourishing industry that followed the goldrushes. The architecture – Georgian-inspired colonial industrial – has a classic simplicity. The flour mill has a huge chimney and a 25-tonne water wheel which is eight an a half metres across.

The Anderson brothers arrived from Scotland in 1851. After success as diggers they supplied timber for the Goldfields from their Wombat Forest sawmills. In 1861, they built the mill in response to the local agricultural boom.

The mill was sited near Birch Creek so that water could be utilised as a power source, and the height of the building reflected the need to use gravity in the milling process. Water was released into Birch Creek from Hepburn Lagoon, about five km away, and channelled into a water race to turn the wheel. The operator of the release gates at Hepburn Lagoon would release 'half oats water' or 'full flour water' depending on the product being processed.

The Anderson family owned and operated the mill for almost 100 years, but prosperity was short lived. The railway bypassed Smeaton and the wheat industry gradually shifted north and west towards the Mallee and Wimmera. After the mill closed in 1959, most of the machinery was sold for scrap. The building was unused until 1974 when it became one of the first buildings to be included on the Historic Buildings Register. A restoration program was established and the mill was purchased by the State Government in 1987.

Anderson's Mill is open on the first Sunday of each month from 12 noon to 4 pm. The beautiful grounds around the mill may be enjoyed at other times. For more information contact the Parks Victoria Information Centre (tel 13 19 63). To get there, head south out of town on the Creswick Rd and the mill can be seen after crossing a bluestone bridge built in 1892.

KINGSTON AVENUE OF HONOUR

One of the most impressive and moving of the Goldfields' many Avenues of Honour is centred on nearby Kingston, a small town that was once an administrative centre for

the Shire of Creswick. The planting of the 286 elms started in 1918, and the Avenue was officially dedicated in 1927. The trees were planted to honour the men and women who enlisted from the Shire of Creswick to serve in WW1. Each tree has a cast-iron nameplate and the names are also recorded in the Roll of Honour.

SMYTHESDALE

LOCATION: 19 KM SOUTH-WEST OF BALLARAT (GLENELG HWY)
POPULATION 1859: 17.500
POPULATION 2006: 800

Smythesdale was named after Captain John James Barlow Smythe, a squatter who took up a run in the area in 1838 and called it *Nintingbool*. Gold was discovered in the nearby Woady Yaloak Creek in April 1853 and more than 1000 diggers soon flocked to the area, shattering the peace. Bouts of drunkenness and rioting were common until law and order was established towards the end of 1854. Even so, the township did not get a court house until 1861, two years after it was officially named Smythesdale. A police camp was established by 1863.

At the end of 1861 the town boasted 100 dwellings, stores and shops,

Looking West From Kingston, KS
Kingston Avenue of Honour, KS

including five butchers' premises and a few boarding houses. By the mid-1860's, the town had grown to 209 buildings, including three theatres and a newspaper office. In addition, more than 100 tents and bark huts, half of them occupied by Chinese, were erected on the outskirts.

Nevertheless, the Smythesdale gold began to peter out around this time. It had been lucrative while it lasted – more than a million ounces (28.3 tonnes) of gold were won in just over eight years. The number of miners quickly dwindled, with only about 130 still at work in 1871. By 1882 the number had decreased to 70. Nevertheless, the town survived to serve as the law administration centre for the nearby Ballarat Goldfields up until the turn of the 19th century.

One of Smythesdale's famous sons was Arthur Alfred Lynch – poet, novelist, journalist, soldier, parliamentarian and rebel – the son of an Irish goldminer briefly imprisoned for a role in the Eureka Stockade. Inheriting something of his father's 'rebel' nature, Lynch formed and led an Irish contingent against the British in the Boer War. Later, when elected to represent Galway in the British Parliament, he was arrested and sentenced to life imprisonment for treason over his South African activities. He was pardoned by Edward VII and served for the Allies in WWI.

SMYTHESDALE TOWNSHIP WALKING TRAIL

Court House Hotel, Smythesdale, GS
Court House, Smythesdale, GS
Sculpture & Lock-up, Smythesdale, GS

The history of Smythesdale can be best appreciated with a walk around the town. The **Court House Hotel** (originally a wooden building erected in 1859, later expanded to the present two-storey brick) is now the only operating hotel in town.

The **Police Camp Historic Precinct** includes the court house built in 1861 and now home to the Woady Yallock Historical Society, some brick stables and the single-storey bluestone lock-up built in 1869 with a gabled roof and an iron grille door. The precinct also contains the post and telegraph office – a two story brick building built in 1867 and now a private residence, the mechanics institute and library, the former Union Bank building and the home of the multi-functional Matthew Veal – a coach builder and undertaker of the 1860s.

Visitors can also stroll through the Smythesdale Botanic Gardens, past the war memorial and old brewery dam.

AROUND SMYTHESDALE

Enfield Forest

A state park occupies the southern half of Enfield Forest which lies to the south of Smythesdale. A number of walks take in the historic mining area of Surface Point.

Surface Hill

These old diggings represent one of the best examples of gold sluicing in the district. A walk across the Argyle Dam wall and up the hill reveals the scarring produced by sluicing. Access is off the Smythesdale/Ross Creek Rd just out of Smythesdale (Surface Hill Rd). The large Chinese population of this area worked Misery Creek, leaving behind substantial evidence of their labours. Access is via Misery Creek Rd at Staffordshire Reef, or at Dereel off the Ballarat/Colac Rd.

Boden's Water Race Walks

Water races were constructed by Thomas Boden to direct water to mining sites in the 1860's. The race network, hand dug by miners over the summer periods, transported water from various catchment areas via dams, around the land contours from valley to valley, finishing at gold sluicing areas. There are three easy walking trails two to six km long that follow the water races and take between 75 minutes and three hours to complete.

Jubilee Mine Historic Reserve

Located at Italian Gully, this is one of Victoria's best-preserved 19th century gold mining sites. Between 1887 and 1913 the mine employed around 300 men and produced today's equivalent of $80 million worth of gold. The remains of a deep lead mine, water races, old cyanide vats, mullock heaps and machinery foundations, as well as derelict houses and other buildings can still be seen. Access is via Browns Rd at Smythesdale, or via the Newton/Berringa Rd at Newtown. Picnic, toilet and wood BBQ facilities are provided and there is a walking track with interpretative signage.

Linton

Linton is close to the Jubilee mine site about 13 km south-west of Smythesdale via the Glenelg Hwy. The steep main road is lined with weatherboard shops and small houses with cottage gardens. The public library was built in 1874.

Ballarat to Skipton Rail Trail

The trail consists of 53 km of the old railway line from Ballarat to Skipton through Haddon, Smythesdale, Scarsdale, Newtown and Linton. It can be accessed wherever it crosses the highway.

Brown's Diggings Township site

In 1861 Brown's Diggings was a town with a population of 2351 and 734 dwellings. It lies between the Glenelg Hwy and Brown's Rd.

Nimon's Bridge

This is the highest wooden trestle bridge in Victoria, built in 1899. Access is via Galatea Rd.

Devil's Kitchen Geological Reserve

This is a good chance to see a volcanic lava flow which has been exposed by water erosion. Travel south from Scarsdale and turn off on the road to Cape Clear.

Jubilee Mine Ruins, GS
Nimon's Bridge, GS
Linton, GS
Linton, GS
Devil's Kitchen, GS

STEIGLITZ

LOCATION: 86 KM WEST
OF MELBOURNE;
37 KM NORTH-WEST OF GEELONG
POPULATION 1855: 2,000
POPULATION 2006: 0

Steiglitz is an abandoned goldmining town on the edge of Brisbane Ranges National Park, between Anakie and Meredith. Now under the control of Parks Victoria, the old town and its surroundings is part of the 469 ha Steiglitz Historic Park. Declared in 1979, the park contains preserved remnants of the town as an example of what happens when gold runs out and there is no other reason for people to stay.

Squatters Charles and Robert von Steiglitz were the first European settlers to establish a run in 1847. Alluvial gold was found on their property in 1853. A cairn by the bridge is made of stone from the original Von Steiglitz homestead. It was near this site in 1855 that the discovery of a gold reef on Sutherland Creek sparked a rush of 2000 people and led to the establishment of the town.

By 1856 alluvial mining had finished and the population temporarily declined. A second wave of mining in the 1860s, using quartz-crushing batteries, renewed interest. Steiglitz became a busy township with four hotels, a variety of shops, churches, a newspaper, and an undertaker. But profits diminished in the late 1870s and the population again dwindled to a few hundred.

Mining found another lease on life in the 1890s, this time when the old tailings were treated with cyanide. The town boomed and the population once more surged to 2000. However, returns were disappointingly low and by 1896 commercial activity had slowed down considerably. Many buildings were sold in 1900 when the population dropped to 300. Some mining continued in the 20th century, but the last mine closed in 1941. By the 1950s, fewer than 100 people lived in Steiglitz. The school closed in 1958 and the post office in 1966, by which time the population was down to 13. When the last residents departed the town all the buildings and roads were in a state of decay.

Bush Near Steiglitz After Fire, RE
Bridge & Scott's Hotel, KS
Cottage, KS
Hotel Verandah, KS

Today Steiglitz is a ghost town. Preserved remnants in the park include Scott's Hotel, a post office, a weatherboard cottage, a brick shop, a school, some building foundations, some old streets, shafts and tailings, and two timber churches – St Paul's Anglican Church and St Thomas' Catholic Church. Interestingly, the latter had been moved to Geelong in the 1950s, but was returned to Steiglitz in 1982 as part of the preservation project.

THE COURT HOUSE

Built in 1875 when Steiglitz was a flourishing gold town, the slate-roofed red-brick court house closed in the late 1870s when gold became scarce. It reopened in 1895, but finally ceased operating in 1899. The National Parks Service took control of the building in 1977 as the first stage of preserving the town. The court house is now a museum containing a display of old photographs, maps and relics of the gold years. It's open Sundays and public holidays from 10.30 am to 4.30 pm. Information regarding 'Discovering Old Steiglitz', a 45-minute self-guided walk of the town, can be obtained at the court house.

Much of the surrounding park can be explored on foot and is particularly rewarding when a variety of wildflowers bloom in spring. During some holiday periods, the rangers organise guided walks and other activities. Walks along bush tracks reveal evidence of mining

and gold panning is permitted one km downstream from the Meredith Road Bridge on Sutherland's Creek. Pans, cradles and hand tools can be used within the creek bed. A miner's right is required. Bert Boardman's recreation area provides information and picnic facilities, including toilets, water and fireplaces.

STUART MILL

LOCATION – 23 KM SOUTH OF ST ARNAUD (SUNRAYSIA HWY)
POPULATION 1861: 700
POPULATION 1869: 7,500
POPULATION 2006: 50

Stuart Mill is a quaint hamlet nestled in an amphitheatre of fertile grassy flats surrounded by attractive hills. A steep spur of the Pyrenees watches over the settlement and Strathfillan Creek, lined with magnificent red gums, meanders through the village. Stuart Mill is also remembered as the place where department store founder GJ Coles once lived.

Squatters took up land in the area during the 1840s and Strathfillan Creek is named after one of them. Stuart Mill was originally known as Alberton (Albert Town), but was renamed in April 1863 after John Stuart Mill, the English political economist.

A party of Italians were the original discoverers of alluvial gold in the hills around Stuart Mill itself and they worked in secret in 1859 and 1860. The secret was blown in June 1861 when Andrew Stranger found gold just behind where the town's Methodist church now stands. The resultant stampede of 700 miners required the presence of a mounted trooper. The diggings were rich and Stranger and his party averaged 12 ounces (340 g) to the load. They also found a 12 ounce nugget. Within a month, 4000 diggers had swarmed over seven alluvial gullies in the immediate area.

Court House, Steiglitz, KS

The Stuart Mill township was surveyed in 1865, ironically just before a lull in the mid-1860s as the initial alluvial gold ran out. But several new discoveries in 1869 brought large numbers of diggers back and Stuart Mill became known as the 'City of the North', eclipsing St Arnaud in importance. As one report stated: (Tents went up) '…as thick as mushrooms after a heavy rain.'

Businesses boomed. The town soon boasted 10 hotels and dancing saloons, along with a telegraph office, bank, and shops for bakers, grocers, butchers, chemists, ironmongers, a blacksmith and wheelwright, brick makers, a flourmill and sawmill, boot shops, drapers, jewellers, a post office and a police station. The doors of a mechanics hall and library were open every night. Stuart Mill also had a rifle club and a Light Horse Brigade, comprising 60 mounted men. Several private schools operated before the first state school was constructed in 1870.

Some of the old buildings are still standing today. Old Malcolm's Inn – a general store and hotel, complete with original kitchen, laundry and maids quarters, built in 1868 – is now a B&B and licensed tea rooms. The

rebuilt Methodist (1894) and Catholic (1912) churches have also survived.

When the gold finally ran out and the diggers had gone, the town turned back to agriculture. The deep, fertile soil along the creek maintained life in the district and the region became known for its fine merino wool. .

AROUND STUART MILL

St Arnaud Range National Park

The park lies east of Stuart Mill and can be reached by following the signs to Teddington Reservoir. It contains one of the most intact large areas of Box-Ironbark vegetation in Victoria. Significant Aboriginal sites within the park include scarred trees, mounds and stone artefact scatters.

The rocky ridge tops offer fine views for bushwalkers, mountain bike riders and four-wheel drivers. Hiking in the steep and rugged terrain is popular. A number of old mining sites can be found in the park and prospecting is permitted in designated areas. A current miner's right is necessary.

Teddington Reservoir

The Teddington Water Scheme supplied water flowing by gravity through wooden pipes to St Arnaud. Two reservoirs (upper and lower) were completed in 1900 and 1929 respectively. The system was downgraded in 1947 and now only supplies nearby Stuart Mill. The banks of the Upper Teddington Reservoir provide a pleasant area for camping and picnics. Toilets and fireplaces are provided.

Teddington Reservoir, KS
Old Malcolm's Inn, KS

TALBOT & AMHERST

LOCATION: 159 KM NORTH-WEST
OF MELBOURNE; 18 KM NORTH
OF CLUNES
POPULATION 1854: 15,000
POPULATION 2006: 300

A little off the beaten track, midway between Clunes and Maryborough, Talbot is one of Victoria's best preserved gold towns and rewards exploration. For decades Talbot was virtually a ghost town, its streets deserted, but today it is slowly reinventing itself as a quiet and attractive place to live.

Talbot grew from diggings that were known as Back Creek where gold was first found in 1854. The name was changed from Back Creek to honour Lord Amherst, the Governor of India, in 1861.

The town was surveyed in 1859 following the 'Scandinavian' rush, still remembered by Scandinavian Crescent, one of the town streets. Initially Talbot consisted of six streets, including Scandinavian Cresent, Bond, Russell, Chapman Oxford and Ballaarat (the original spelling is still used) Sts. They have changed little, but it is still hard to imagine 15,000 diggers, the noise, the dust and the general tumult.

By the mid-1860s the population had dwindled to around 3000 and more substantial brick and bluestone structures began to replace the tents and timber huts. Other industries such as soap and candle factories, flour mills, a theatre and a gas works provided alternative employment. A soft drink factory, owned by the Cohn Brothers, later of Bendigo, was founded in 1861.

By the end of the century, scaled-down mining resulted in the population declining to 1300. A brief revival from 1934 to 1940 saw one of the old mines reopened.

Described by some as a living ghost town, Talbot has some wonderful buildings in its fascinating historic precinct. The magnificent former court house (1866), is located opposite the Court House Hotel (1860), still operating, one of the 100 pubs and sly grog shops that once existed. The former Union Hotel also survives. Opposite Talbot's imposing Town Hall is the former Bull and Mouth Hotel, in Ballaarat St,an 1860s bluestone structure which has been converted into a restaurant. Long term publican Mrs Chesterfield's ghost is said to reside there.

Talbot's Post Office is the oldest functioning post office in Victoria. Next door is the original police station and behind is the old bluestone lock-up. Now privately owned, inspection is possible by appointment. Of the five churches still remaining, Wardell, the architect of St Patrick's Cathedral in Melbourne, designed St Michael's Anglican Church. The old cast-iron key still opens the front door. Inside, servant's pews still exist. The Talbot Railway Station now houses a Gallery, Museum & Nursery.

Talbot Building & Sign, KS
Window, JM
Store, JM

VISITOR INFORMATION

Visitor information, including a walking tour, is available the restored Visitor Information Centre in London House (open Wed to Sun, tel 5463 2193), from the museum, the post office and the local store. The surrounding countryside includes two of the Goldfields region's most fascinating secrets: the Big Reef, and Stony Creek Elementary School. Ask for up-to-date directions.

ARTS & HISTORICAL MUSEUM

The former Primitive Methodist Church (1870), now a local history museum, is an excellent example of Talbot's historic architecture. The Museum contains an impressive collection of goldmining memorabilia and a vast array of photographs from the 1850s to the 1900s. The historical society is also in the process of converting a former school building into an educational museum. Open Sunday 1pm to 4pm or by appointment. There is a small admission fee.

WALKING TOUR OF TALBOT

The best way to explore the town of Talbot is on foot. A walking tour of the small township reveals heritage buildings largely untouched by renovator's hands. A brochure/guide is available from a number of outlets in town, including the VIC.

ABORIGINAL SITES

Not far from town are there some Aboriginal drinking wells and a hollowed-out Red Gum, birthing or 'shelter' tree where Aboriginal women gave birth to their children. The tree is just off the Maryborough Rd. Turn left after you cross Back Creek on London Bridge, into Pollocks Rd, and the tree is about 250 metres on your left.

AROUND TALBOT

Amherst

There was a tussle between Amherst and Talbot as to which town would dominate. Talbot won. Amherst is 5 km north-west of Talbot, and it was once the centre of a lively municipality founded on a population of 30,000 diggers. Surveyed in 1855, at one stage the town included seven general stores, an inn, various tradesmen's enterprises and a hospital (which survived until the 1920s).

Walking Tour of Amherst

There's a walking tour that shows the extent of development in the late 1850s. Although now virtually deserted, the layout – and the ambitions – of the town are evident. The superb bluestone culverts under the Talbot-Avoca road (just past the intersection with Bakers Rd if coming from Talbot) are the most striking and monumental remains.

Big Reef (Quartz Mountain)

One of the Goldfield region's most

astonishing secrets, the Big Reef (also known as Quartz Mountain) hides in the forest about three km from Amherst. It is a massive quartz outcrop (or blow), thc biggest left in Victoria. Some claim it is the largest single piece of pure quartz rock in the world.

Prior to the goldrushes, these quartz blows were relatively numerous, but all the others have been mined. Classic examples existed at Victoria Hill in Bendigo and at the Eureka Reef outside Chewton – but like every other example no quartz remains at these sites – all you will see is a hole in the ground. The quartz has been fed into crushers.

A shaft has been dug into the Big Reef, but the reef obviously owes its survival to the fact that no gold was found. The surrounding bush has been heavily mined, with shafts, dams and their associated puddlers, clearly visible.

The reef forms a north-south ridge about a km long. Massive pure quartz outcrops stand six metres above the ridgeline and in one section a magnificent balancing rock of at least 40 tonnes dominates the scene. It's an arresting site, with an aura of power; it surely played a role in Aboriginal life.

If possible get directions, because the tracks into the bush are poorly signposted. Take the Talbot-Avoca Rd from Amherst towards Avoca. After approx 1.5 km take the left-hand fork on the dirt Glenmona/Lillicur Rd. After .7 km turn left onto Tyler Track. After .6 km turn right onto Quartz Track; the reef is signposted and is a short, steep walk from the track after 2.4 km. Take the Mia Mia Track back to Talbot, past the Talbot Cemetery.

Stoney Creek Elementary School

The Stoney Creek Elementary School

is another secret spot, tucked away in a peaceful stretch of forest south-west of Talbot. All that remains of the school that occupied the site from 1865 to 1916 are the evocative, remnants of rock gardens, and in particular, a large map of Australia created out of quartz stones.

It could be said that these stones stand as a monument to Miss Elizabeth James, who was the head teacher from 1905 to 1912. A keen gardener she and her pupils created a garden that in addition to the map included a summer house and a sun dial. The school porch was hung with creepers.

The site is surrounded by attractive bush and the signs of intense mining activity. It's hard to imagine the hardships the community faced, and how important the school and its garden must have been. The modest scale of the ruins and the location are intensely moving.

To get there, take the Lexton Rd out of Talbot. Just before the golf course turn left into Nuggetty Gully Rd (which becomes Nuggetty Track) and drive south for 6 or 7 km until the road does a right-angle turn to the west, following a water race along the southern edge of the of the forest. The Stoney Creek School is just after the right-angle turn.

Culvert, Amherst, JM
Rock Map of Australia, Stoney Creek
Elementary School, DB
Stoney Creek Elementary School, RE

TARADALE

LOCATION: 100 KM NORTH-WEST
OF MELBOURNE ON THE CALDER
FREEWAY
POPULATION 1861: 5000
POPULATION 2006: 500

Taradale was once a thriving town on the way to the Goldfields. The basalt-flow ridge to the south of town has been mined by shafts and tunnels and huge mullock heaps are seen on either side of the road. As the road descends into the village of Taradale, the magnificent railway viaduct can be seen to the right. The rails are 40 m above Back Creek, and the viaduct was for many years the highest on the Victorian railway system.

The road that leads beside the creek should be followed as far as the base of these great piers, so that their true scale can be appreciated. There is also a path through the Fairy Dell (elm thicket) along the creek.

Gold was first discovered at Taradale in 1852, but it was not until 1855 that the first great rush occurred at Yankee Point and Liberty Flat. One of Taradale's earliest mines, the Phoenix, began working in 1858 and ore was still being processed at the quartz battery until 1957. The Quartz Block mine and Old Comet mine on the corner of Henry St and Old Drummond Rd were worked up until late 1930's.

By 1861 Taradale was a thriving community with four square miles of streets, 20 stores and 23 hotels, 40 liquor licensees and several banks to service the 2,000 Europeans and 3,000 Chinese that used the town. By 1910, when Taradale had a population of 500, five churches remained and so did five hotels. Taradale has lost many houses over the last century. Most were timber-built and were either transported to another place or simply crumbled away.

The former **Methodist Chapel** is now a charming residence and behind is the old court house. On a hill to the east stands **Holy Trinity Church** with a magnificent Wagnerian façade (1858). Along with the local school

Holy Trinity Church, Taradale, JK
Aerial View of Taradale, BS
Near Taradale, KS

and community hall, it is one of the only facilities in town still in use.

TARADALE WALKING TRACKS

There are seven walking tracks that take in all the interesting sites around town, including the viaduct, churches, historic houses, mines and the local cemetery. There are also fine walks along the tracks of the Fryers Ranges State Forest on the western outskirts of Taradale.

THE COLIBAN MAIN CHANNEL

A very significant feature of the Coliban River valley, the Coliban Main Channel passes close by historical features, including the Syphon outlet and inshoot, the brick abutments of an old flume, an arched brick-lined tunnel, brick and bluestone overshoots, native cherry trees and Taradale Reservoir. The channel traverses a wide range of scenic areas from pastoral plains to steep hillsides.

In 1864 Joseph Brady, described as the most accomplished civil engineer ever to have worked in Australia, won a prize for designing the best scheme for supplying water to Bendigo. A series of channels, tunnels, syphons and aqueducts following the ridges from Malmsbury brings water to Bendigo and Castlemaine, as well as to all the towns and farms along the way. The entire channel from Malmsbury to Bendigo is 65 km long. The scheme was called the Coliban Water Supply and the Coliban Main Channel System opened in 1877. It has flowed continuously since.

Coliban Main Channel Walk

A 13 km section of the historic Coliban Main Channel, accessible to walkers, dates from around 1870. Turn west into Roderick St opposite the Independent service station and

travel about one km to the channel. There is a cleared space suitable for parking just over the channel. The entrance to the walk itself is through a pedestrian gate on the Taradale side of the channel. Access to the southern section is directly from the Rd; access to the northern section is via a short detour through the bush. Follow the sign 'B' (Bendigo.)

TARADALE VIADUCT

The magnificent Taradale Viaduct is one of the longest and highest railway bridges in Victoria. Leave your car in the main St and walk towards the bridge. A footpath takes you through open ground with a water race on one side and a splendid 19th century house and barn on the other. The viaduct is 650 ft (198m) long and 120ft (36.5m) high – a remarkable achievement considering it was constructed without the use of modern machinery. In the 1930s, framed steel trestles were added between the bluestone piers to give extra support for the newer and heavier locomotives.

The Viaduct took four years to build. It is estimated that 2000 men and 1600 horses built the line between Gisborne and Castlemaine. One man from Taradale fell from the top of the viaduct during its construction, but luckily landed in the creek unharmed. Apparently a short time later the same man fell into a relatively small hole, broke his neck and died!

Viaduct, GC
Holy Trinity Church, KS
Cobb & Co Stables, KS
Viaduct & Moon, PW

TARNAGULLA

LOCATION: 45 KM WEST OF
BENDIGO (WIMMERA HWY)
POPULATION 1852: 5,000
POPULATION 1865: 20,000
POPULATION 2006. 200

The sleepy-town atmosphere of Tarnagulla today belies its boisterous, golden past. The story began in 1852 when a party of miners bound for the Korong goldfield camped on the *Tarnagulla* pastoral run (taken up in the 1840s) and found alluvial gold in nearby Sandy Creek. Thousands of miners swarmed to the spot. Originally known as Sandy Creek, the settlement was renamed after the surrounding property in 1860.

The area soon proved to be very rich and excitement grew in 1853 as a number of gold nuggets were found in Nuggety Gully, a bit less than one km south of the town. Amazingly, one nugget of 192 ounces (5.4 kg) was uncovered in a dray track while diggers were marking out the ground.

The action switched back to Tarnagulla in 1854 when a digger named David Hatt discovered a rich lode in a massive quartz band which he called Poverty Reef. There was no irony intended in the name. Rather, it was chosen to commemorate his rescue from drowning a little earlier in Poverty Bay, New Zealand. An even happier sequel occurred when Hatt married his rescuer, a young Maori girl, and brought her to live in Tarnagulla. They are buried together in the town's cemetery.

Poverty Reef was exceedingly rich, at one stage yielding 13.3 tons (13.5 tonnes) of gold in 13 months. In 1859 a single crushing at the Prince of Wales claim obtained two cakes of gold weighing 1,389 and 1,504 ounces (39 kg and 42.6 kg). In 1866 the newspaper *Dunolly & Bet Bet Shire Express* reported that an area of 90 sq m had yielded gold worth £1.25 million.

The mid-1860s proved to be the peak of the boom, although alluvial finds were still made fairly frequently over the next 40 years. The 953 ounce (27 kg) *Poseidon* nugget was discovered in 1906 and named after that year's Melbourne Cup winner. Incredibly, nuggets named *Leila* (675 ounces/19 kg) and *Hazel* (502 ounces/14 kg) were found a few days later.

During its boom years Tarnagulla was a bustling metropolis. It seems incredible now that there were nine general

Court House, Tarnagulla, KS
Grey House, DB
Ruin, KS

stores, four hotels, two breweries, two banks, a saw mill, a corn factory and four crushing machines. The list goes on to include five bakers, three butchers, three surgeons, two chemists, four blacksmiths, two boot makers, two drapers, a tailor, a fruiterer, a tobacconist, a watchmaker, a gold broker, a share broker, an ironmonger, a miller, two wheelwrights and a painter!

Cobb & Co's horse-drawn coaches left the Victoria Hotel daily for Melbourne with passengers and mail, often making the trip in three days – so long as bushrangers such as 'Black Douglas' (an American negro who had jumped ship in Melbourne and gone on a crime spree in the Goldfields) and his gang did not interrupt the schedule. Horse-drawn transport gave way to steam when the railway reached Tarnagulla in 1888.

Gradually, as the gold yields dwindled during the latter part of the 1800s and into the 1900s, settlers turned back to agricultural pursuits.

SIGHTS

The old Union Bank, built in 1859, is now known as *Locharron* and is privately owned. The Colonial Bank, opened in 1866, has a tall chimney stack used for the smelting of gold. It is now the Tarnagulla B&B.

The old Sandy Creek Post Office, opened in 1856, was originally intended to be the gold warden's office. The building was sold in 1887 and has been a private residence since that time. The replacement Tarnagulla Post Office and Residence, completed in 1886, is also a private residence today, as is the court house, which held its first sitting in 1863 and the Wesleyan Methodist church, which was built in 1865.

Tarnagulla Public Hall (Former Victoria Hotel & Theatre)

The Victoria Hotel and Theatre, built in 1862, is the oldest known purpose-built theatre attached to a hotel in Australia. It saw many big vaudeville names of the mid-1800s entertain the miners from its stage. The hotel license was withdrawn in 1916, after which the building became a private residence. However the town bought the building in 1924 to be used as a public hall.

Soldier's Memorial Park

This park features a cannon taken from the British flagship HMS *Nelson* and presented to the town in 1898. It has been fired five times. The first four were from its initial position in the Recreation Reserve – twice in 1900 associated with the relief of Mafeking during the Boer War, once for the Coronation of Edward VII in 1902, and once at the end of WWI. The cannon was moved to its present position after WWII and fired for the fifth time during the Australian Bicentennial celebrations in 1988.

Tarnagulla Public Park Reserve and Cricket Pavilion

The recreation reserve was opened in 1865. It is notable for the two-storey pavilion built in 1884 and a band rotunda built about 1886. Both have been recently restored.

Colonial Bank, KS
House, KS
Shop, KS

Timor General Store, KS
Lock-up, KS
Diggings, KS
Grand Duke Mine, KS
Grand Duke Tailings, KS
Water Stand, KS

TIMOR

LOCATION: 8 KMS NORTH
OF MARYBOROUGH
POPULATION 1857: 30,000
POPULATION 2006:

Timor has a general store (which no longer operates), one school, one church and a cemetery – and it is hard to imagine the need for eight daily Cobb & Co coaches to Maryborough, or the post office handling 32,543 outgoing letters in one year…. Nonetheless, Timor was once known as the Junction of the Deep Leads and it was indeed a bustling community of thousands of people.

Only the mounds of crushed quartz and the Grand Duke and North Duke arches remain from the great days of deep lead mining.

Gold was discovered in 1854, and within a month there were 18,000 people on the diggings. In 1857 the number of people camped at Chinaman's Flat grew to 30,000.

Timor was originally named Coxtown after a local butcher who built a bridge (Butcher's Bridge) at his own expense across Bet Bet Creek. It was not a coincidence that he also owned the Bridge Inn!

High St, Timor, was a mile long and at the height of the town's prosperity it had thirty-eight hotels, four churches, three butcher's shops, two banks, three bakeries, two large stores, several small stores, a police station and several lock-ups.

One of these lock-ups still stands adjacent to the general store (1870) – and it was well used. Drinking, gambling, hold-ups, robberies, pick pocketing and violence were rife.

A wander through the cemeteries around Timor leaves a lasting

impression of the fragility of life on the goldfields. There were several tragedies at the mines and the cemetery has many tombstones relating these stories.

The Timor primary school was established in 1873; before that classes were held at the Wesleyan Church at Chinaman's Flat and then at the Mechanic's Hall. It was reported that 203 children crowded into the one room. Tragically, in 1875 14 children died in a measles epidemic.

Although there were many successful mines in the Alma, Timor, Chinaman's Flat region, the Timor area was dominated by the Duke & Timor Gold Mining company which leased produced 216,054 ounces (6.1 tonnes) of gold over 27 years from 1874.

A massive Cornish steam engine was installed in 1874 at what was eventually known as the **Grand Duke Mine**, on the eastern side of the township. The huge engine had a cylinder that was 80 inches (2 metres) in diameter with a 10 foot (3 metre) stroke, and it was capable of pumping 2000 gallons (9000 litres) of water per minute from the mine.

The **North Duke Mine**, adjoining the Grand Duke also pumped out vast quantities of water, and also recovered large quantities of gold: 30,071 ounces (852 kg) in one five year period. It ceased operation in 1904. The Duke and Main Leads Consuls was the last mine on the lead and it stopped working in 1918.

TRENTHAM

LOCATION: 97 KM NORTH-EAST
OF MELBOURNE: 24 KM SOUTH-EAST
OF DAYLESFORD
POPULATION 2006: 700

Trentham is full of charm and character thanks to its gold and timber town beginnings. Near the end of the Great Dividing Range, just north of the Wombat State Forest, the town is surrounded by rolling hills and impressive stands of manna gums.

Picturesque High St includes a former guesthouse, the Old Bakery, and a row of timber-fronted shops (1866). Tragically the famous and historic Cosmopolitan Hotel (1865) was severely damaged by fire in 2005.

High St also features an imposing, slightly incongruous, square clock in memory to a much-loved local character, Dr Gweneth Wisewould, the town's doctor from 1938 to 1972.

On a shrubby rise on the edge of town is the Catholic church of St Mary Magdalen, a red brick building used as a setting for funerals or weddings in the long-running TV drama series, *Blue Heelers*.

Although gold was significant, Trentham has also been an important logging and milling town for most of its life. Until 1997 it was also home to the Trewhella engineering works, which was started by two brothers in 1887 and at one stage employed 60 workers. A branch was later established in Birmingham and is still trading today. Billy, one of the founding brothers, invented a jack later used in the timber industry.

The closing of the Woodend to Daylesford railway in 1978, and the recent reduction of timber harvesting in the surrounding forests have left the town a quieter place – although it can still be busy at the weekend. The old railway station is now the Visitor Information Centre.

Trentham has a beautiful little park with ornamental lake, sweeping lawns, seats, picnic tables, and an undercover rotunda.

Bowser & Sign, Trentham, RE
Trentham Railway Station, PW

Ball and Welsh set up a drapery shop and Mrs Ball imported from dresses London.

Dog & Streetscape, PW
Trentham Falls, PW

VISITOR INFORMATION

Trentham Visitor Information
Old Railway Station, Victoria St
Trentham 3458
www.hepburnshire.com.au
tel: 5424 1178.

The local historical society has brochures for some excellent walks.

TRENTHAM FALLS

Trentham Falls is Australia's highest single drop waterfall, a 35 m unbroken cascade over basaltic columns. The falls are created by the Coliban River pouring over the end of a six million-year-old lava flow, onto the quartz gravel of original river bed dating from 270 million years earlier.

The falls were once harnessed for power via a turbine, which drove the machinery at a nearby sawmill.

The Trentham Falls Reserve provides an excellent example of unspoilt local vegetation. The dominant tree species is manna gum with messmate stringybark, and an understory of blackwood and silver wattle. Before logging, the forests around Trentham would have averaged 30 to 40 m high with taller trees reaching 65 m.

Since the early days of settlement the falls have been a popular picnic spot.

AROUND TRENTHAM

Tylden

Just near Trentham, Tylden is a small township surrounded by magnificent countryside with sweeping views to Mt Macedon. The district was originally part of the massive *Mollison Run*, a squat of 256,000 acres (103,600 ha). Toll gates operated along the Tylden-Trentham Rd when, in 1853, a huge influx of people came into the district on their way to the diggings at Blackwood.

Newbury

A scatter of houses is all there is of Newbury, but it lies at the summit of the Great Dividing Range. Within a radius of 10 km, six major rivers flow from natural springs in these low hills. The Campaspe and Little Coliban Rivers rise near East Trentham, the Coliban just west of Newbury, and the Loddon a little further toward Bullarto.

VAUGHAN

LOCATION: 10 KM SOUTH
OF CASTLEMAINE
POPULATION 1857: 30,000
POPULATION 2006: 150

The quiet village of Vaughan is set in beautiful undulating countryside along the old road between Guildford and Fryerstown, in the valley of the Loddon River. In 1853 gold was discovered at the meeting of Fryer's Creek and the Loddon River, known as 'the Junction'. As many as 13,000 diggers quickly rushed to the spot and there are reports of twice that number spread out in neighbouring gullies.

A general store and a brewery began to service the district and a township sprang up virtually overnight. The Junction Hotel opened in a calico tent, later replaced by a weatherboard building. More substantial buildings, such as the Bank of Victoria, were built in brick. Gold buyers, C Ball and his nephew WH Welsh diversified to set up a drapery shop in Borgoyne Street, while Mrs Ball imported dresses from London. It was said that: '…women for miles around would drive into Vaughan to patronise the Store.' The business moved to Castlemaine in 1882 and from there grew to be an important retail outlet with stores in Melbourne and throughout Victoria.

The settlement was surveyed in

1856 and named after Charles Vaughan, a Melbourne businessman, parliamentarian and prominent Victorian Baptist. Perhaps to counter this local Wesleyans built a stone church in 1858 on the west hillside above town, where it stands today as 'a picturesque ruin'.

When the alluvial gold began to peter out in the early 1860s, Vaughan's population declined dramatically to fewer than 170 people. A revival activity in the mid-1860s followed the discovery of deep leads beneath the basalt of nearby Table Hill and Bald Hill. Homes built of logs with chimneys made of clay or bark and cottages of mud brick, with gardens, were dotted everywhere. There was a regular coach service to Castlemaine and Vaughan became headquarters for the surrounding shire. In 1871 the court house became the shire hall.

The origin of many diggers is clear from the names of the surrounding settlements which were called Italian Hill and Irishtown. Irishtown was about 800 metres to the east of the main township and was home to (the inevitable) Shamrock Hotel. The only Roman Catholic church in the district was built in 1865 to accommodate a congregation of 400.

Initially, Chinese worked only one quartz claim on New Year's Flat, but by 1861 the entire valley around Vaughan appeared to be dominated by Chinese miners. A number stayed in the area until the early 1900's, well after the gold had been worked out. They found seasonal agricultural work and established market gardens along the Loddon River between Glenluce and Vaughan, and sold their produce in Fryerstown. They also owned commercial premises, but tragically, in 1917, a fire destroyed all the Chinese

buildings, many of which were two-storey shops. Today, a small Chinese cemetery on a rise above the mineral springs, and occasional crops of wild onions along the river banks, are the only evidence of their presence.

THE VAUGHAN MINERAL SPRINGS RESERVE

The Castlemaine Diggings National Heritage Park includes the Vaughan and Glenluce Mineral Springs Reserves. The reserve is on the banks of the Loddon River just below the Chinese cemetery. Created in 1878, the small original reserve now covers 100 ha along the river.

The mineral springs were first found in 1910 by a young boy who wandered over slate exposed by hydraulic sluicing and noticed water bubbling to the surface. Mining activity had stripped the river valley of all its trees, but in the early 1900s the local community developed the reserve into an attractive park setting and Vaughan was a popular health and recreation centre until the 1950s.

Today, visitors can still see historic mining sites and relics throughout the reserve, as well as a band rotunda in the picnic area.

Bridge, Vaughan, KS
House, Vaughan, KS
Springs, Vaughan, KS
Loddon River, Vaughan Mineral Springs
Reserve, KS

AROUND VAUGHAN

Red Knob

Red Knob was once one of a string of hills between Chewton and Vaughan capped by ancient river sediments. However, most of it was sluiced away in the 1950s and the ochre colours of the remaining 'earth pillars' are typical of the eroded landscape created by mining. There is a viewing point near Red Knob that looks out across the broad valley once covered with natural bush. Today the landscape is filled with introduced poplars.

Red Knob, Near Vaughan, KS
Houses, Vaughan, KS
Ruined Church, Vaughan, KS

WEDDERBURN

LOCATION: 214 KM NORTH-WEST OF MELBOURNE; 75 KM NORTH-WEST OF BENDIGO (CALDER HWY)
POPULATION 2006: 900

Wedderburn was once a rich gold-mining settlement made famous because more than 80% of the world's largest nuggets have been discovered within a 40 km radius of the town. Today Wedderburn has one of Victoria's few active gold mines and fossickers continue to seek their fortune in this area, known as the Golden Triangle …with good reason.

Three nuggets worth £10,000 were found in a Wedderburn backyard in the 1950s. Then, in 1980, three schoolboys with metal detectors uncovered the *Beggary Lump* nugget weighing 2.4 kg. Within a week others had found another 3.7 kg of gold. To cap it off, later that same year the *Hand of Faith* nugget weighing 27.2 kg was discovered near Kingower to the south-east. It sold for US$1 million.

Wedderburn was first settled by squatter John Catto in 1840. A shepherd named Brady found nuggets on the run in 1852 and the rush began. Lack of water restricted alluvial mining and quartz reef mining took over when the initial rush faded. In 1859 a public crushing plant opened and continued operating on and off until the end of the century. Large-scale dredging was attempted in 1910, but again lack of water was a problem.

Fortunately the Victorian Government's Land Act of 1869 opened up the large squatting runs to small landowners which provided other means of employment when the gold production dwindled. Agriculture, including the distillation of eucalyptus oil, has been an important part of the local economy ever since.

SIGHTS

Today, Wedderburn is a mish-mash of old and new buildings typified by a quaint café and museum that adjoins the modern visitors centre. Buildings of note include the simple Gothic Revival brick Methodist church in High St with its small central tower built in 1866. Also in High St is the ANZ Bank building (1875), the old Commercial Hotel and the former Royal Hotel (1867) now a B&B with a gallery. In Wilson St you will find the primary school, the Presbyterian church (1868), the Church of Christ & Holy Trinity Anglican Church – the latter a Gothic Revival sandstone structure with lancet windows built in 1866. The chancel was added in 1909. Some old mud-brick houses and stone cottages are signposted off the highway in Tantalla St.

The Coach House Museum

The museum is located in an old store built in 1864. It contains original period items, a blacksmith's display, photographs and other memorabilia, while a collection of carts and buggies are housed in an outbuilding. Open every day except Monday. .

The Lonely Grave

At the northern end of town is the isolated bush grave of a Scottish miner, who died in 1856.

AROUND WEDDERBURN

Fossicker's Drive

This is an interesting circuit that incorporates many of the old gold sites and relics in the district.

Christmas Reef Gold Mine

The mine is four km east of town on the Borung Road. It is a small-scale working gold mine, complete with antiquated equipment that provides an opportunity to see some old-fashioned reef mining. The tunnels have been dug and blasted by hand and you can see a 100-year-old stamping battery at work. There is also a museum display in a nearby bark hut. The mine is open daily for guided tours through the workings. BBQ and picnic facilities are provided.

Hard Hill Tourist Area

Hard Hill is part of old diggings. There is an original ore-crushing battery and a puddler. There are picnic facilities, walks through the bush and old mining tunnels. To get there, turn west off the highway into Tantalla St, then into Wilson St and Hard Hill Crt.

Eucalyptus Distillery

A eucalyptus distillery, one of a number which used to operate in the district, is near Hart Hill.

Wedderburn General Store, KS
Wedderburn Police Station, KS
Shire Hall, KS

Mount Korong

The first European to climb Mt Korong was explorer Thomas Mitchell in 1836. The Major Mitchell Trail now passes through Wedderburn and on to the mountain. To intersect the trail follow the Calder Hwy for nine km south-east of town, then turn left onto Powlett Rd. After about six km you will come to an unsealed road on the left. It leads through a gate to a picnic area. Visitors can stroll along the trail through the open bush or scramble up over the granite boulders to reach the summit.

Wychitella Flora and Fauna Reserve

A 16 km drive north from town takes you to Wychitella Forest Reserve which was created to protect a remnant of mallee forest occupied by the rare lowan (mallee fowl) and the marsupial mouse. Mallee fowls mate for life. They build a mound up to five m in diameter and one m high within a large distinctive ring of leaf litter, bark and twigs and the hen lays its eggs inside. The pair's territorial 'patch', about 25 m in diameter, is maintained year round by the cock.

Colourful wattles bloom in the reserve during July and August and many orchids appear in spring.

Façade, Wedderburn, KS
Aboriginal Rock Wells, Whroo, KS
Balaclava Mine, Whroo, KS

WHROO

LOCATION: 7 KM SOUTH OF RUSHWORTH
POPULATION 1853: 1,000
POPULATION 2006: 0

Just a short drive south of Rushworth, the 500 ha Whroo Historic Reserve encompasses the remains of Whroo and its goldfield. Whroo was once a thriving goldmining town with 139 buildings; all that remains are peppercorn trees, wells, grassy clearings and the now silent Balaclava open-cut mine. The mine is huge – and made ghostly by its emptiness and the whistle of wind through its many tunnels and the surrounding trees.

The Whroo cemetery, on a lonely hillside south of the mine, contains about 340 graves and reflects the harshness of

life on these goldfields. A sign-posted walking track takes visitors through the old township to the cemetery, beyond which is an Aboriginal waterhole known as the Ngurai-illam-wurrung rock well. It is thought the town's name comes from the local Aboriginal word 'wooroo', meaning 'waterhole' and given to the area.

Alluvial gold was found in a gully near Balaclava Hill in 1853. The hill was named because the find was made on the same day as the Battle of Balaclava fought during the Crimean War. The discovery brought thousands of hopeful diggers to the area overnight. Balaclava Hill itself became the site of a gold strike in 1854. The resultant 25-m deep open-cut mine produced gold worth more than £1 million from the 15 cm wide gold veins crisscrossing through the quartz.

The town grew as stores and hotels appeared, along with butcher shops, dairies, a bakery, banks, a mechanics institute and library – one of the first on the Goldfields. Balls were held at the mechanics institute from 1859 to as recently as 1955, but only on moonlit nights owing to the lack of electricity. The site is marked today by a row of sugar gums.

Mining activity wound down in the 1890s and the town went into rapid decline. The buildings were soon removed or demolished. Today all that can be seen are relics of cyanide vats and a puddling machine facing each other to the south-west of the hill.

The Whroo Historic Reserve lies within Rushworth State Forest and the town remnants can also be accessed from Graytown, 27 km south along a dirt road through beautiful box-ironwood countryside.

It is a great area for bushwalking, cycling, horse riding and forest drives. Explore the reserve via a network of sign-posted walking tracks. The picnic area at Whroo has an information centre – a mud-brick building which sells Devonshire teas.   There are toilets, BBQ and picnic facilities, but no water. Camping is allowed, but few facilities are provided.

YANDOIT

LOCATION: 20 KM NORTH OF DAYLESFORD
POPULATION 1857: 5000
POPULATION 2006: 50

All that remains of the once rich gold town of Yandoit are a handful of stone homesteads and timber cottages dotted along the long Main St. A few small churches and a school are still in use, but stone ruins are scattered around, including a church standing roofless on a nearby hillside. .

Captain Hepburn was the first European to settle in the region. Other pastoralists followed and there was some conflict with authorities when Yandoit Creek and Yandoit Hill were included in the Mount Franklin Aboriginal Reserve in 1841.

Rushworth State Forest, KS
Balaclava Mine Tunnel, Whroo, KS
Whroo Cemetery, KS

The first discoveries of gold were made in the tributary gullies of Yandoit Creek. Alluvial gold was so rich that miners reported sinking shallow holes into beds of nuggets. In 1853, a party led by Captain Anderson (sometimes known as Atkinson) worked at Nuggety Gully and found 100 to 120 ounces (2.8 to 3.4 kg) per digger. One of his men, William King, discovered gold in King William Gully but the news was not made public until months later when a German blacksmith who sharpened the party's picks followed them back to the site. Some 5000 men rushed the area in March 1855, but by winter that year the population had dropped to 600.

Activity intensified again at New Nuggetty Gully in July 1858 when mining began at a deep lead called Forty Foot. By August 1860 Yandoit was busy with stores and hotels and the town was surveyed in 1861. New Nuggetty Gully continued to attract prospectors who enjoyed success for a number of years.

Agricultural settlement continued throughout this early period and by the late 1860s Yandoit supported a mixture of farming and mining. Many of the miners came from Italy and the Italian-speaking part of Switzerland, forced out of their homeland because of tensions between Hapsburg-controlled Switzerland and Garibaldi's Italy. In particular, Italian speakers who lived on the Swiss side of the border felt isolated and persecuted. As a result many left for the goldfields of Victoria and they stayed in the district when the gold petered out.

Most were originally tradesmen – millers, bakers, exporters of small goods, masons and hoteliers – and they reverted to their original skills and trades. The Italians built much of the surviving local stone architecture. They planted gardens and orchards and crops. They made cheese, wine and their spicy sausages (known as bull-boar sausages by the English miners because the main ingredients were beef and pork). Some grew vineyards on the Yandoit Hill slopes and, in 1883, 100,000 gallons (455,000 litres) of wine was being produced in the Yandoit area. Unfortunately phylloxera wiped out the industry a few years later.

Today, the orchards and old vineyards have gone. A few of the old homes still stand, but most are crumbling ruins. The sausage, however, remains an important regional speciality.

Horse, Yandoit, KS
Letterboxes, Yandoit, KS
Ruined Church, Yandoit, KS
Store, Yandoit, KS
Ruin, Yandoit, PW
Opposite: Yapeen, KS

Ballarat Railway Station, BaT
Maldon, RE
Opposite: Shamrock Hotel, RE

PRACTICAL INFORMATION

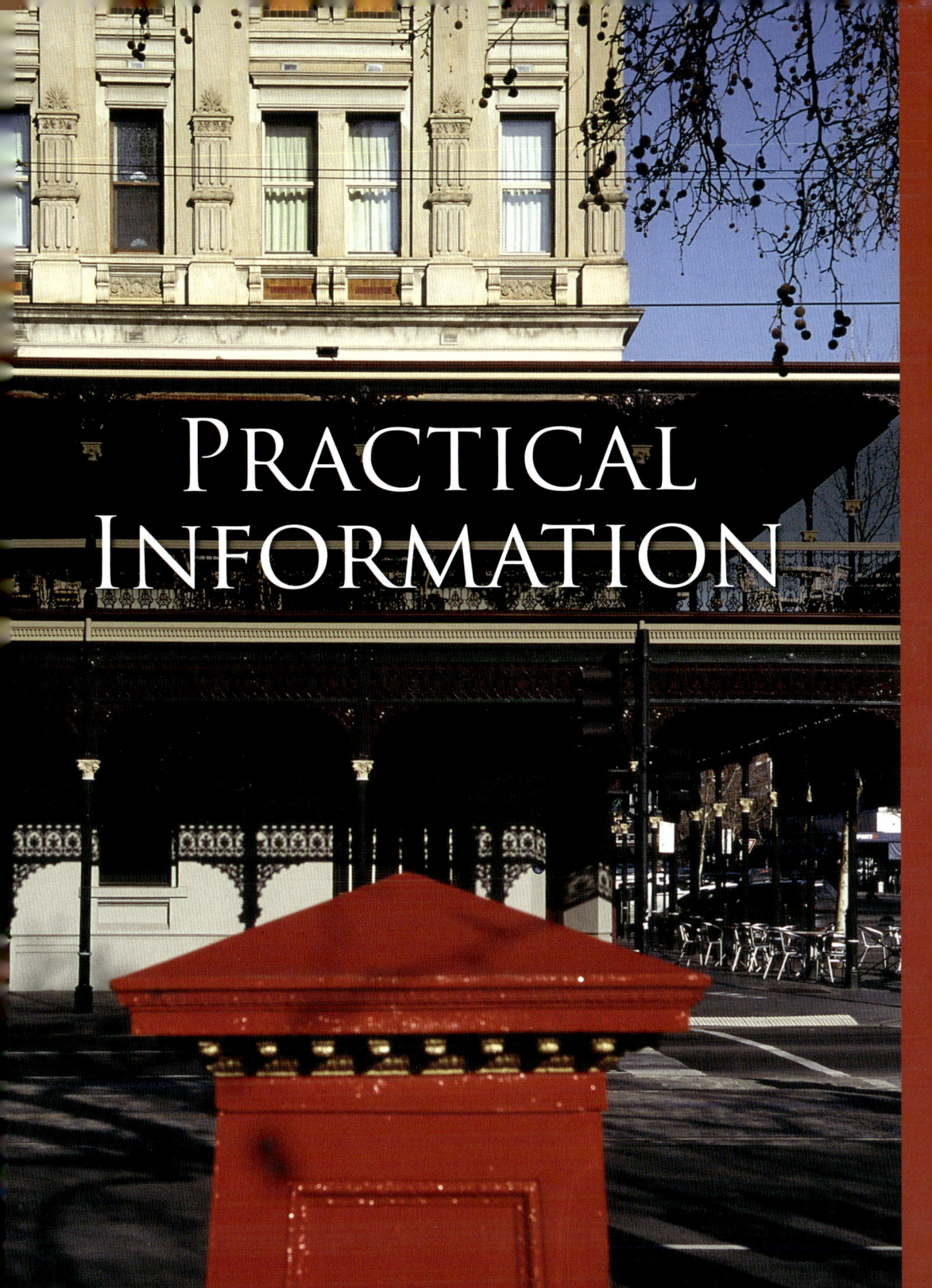

VISITOR INFORMATION CENTRES

ARARAT

Ararat Information Centre
Railway Station Complex,
91 Vincent St, Ararat 3377
Tel: +61 3 5355 0281
Freecall: 1800 657 158
(within Australia only)
Email: tourinfo@ararat.vic.gov.au
Web: www.ararat.vic.gov.au

AVOCA

Avoca Information and
Community Resource Centre
122 High St, Avoca 3467
Tel: +61 3 5465 3767
Freecall: 1800 206 622
(within Australia only)
Fax: +61 3 5465 3597
Email: avoca@netconnect.com.au
Web: www.pyreneestourism.com.au

BALLARAT

Ballarat Visitor Information Centre
The Eureka Centre,
cnr Eureka & Rodier Sts,
Ballarat 3350
Tel: 1800 44 66 33
Email: information@ballarat.vic.gov.au
Web: www.visitballarat.com.au

BEAUFORT

Beaufort Visitor Information Centre
72 Neill Street, Beaufort 3373
Tel: +61 3 5349 2604
Fax: +61 3 5349 2933
Email: visinfo@beaufortcrc.com.au
Web: www.pyreneestourism.com.au

BENDIGO

Bendigo Visitor Information Centre
Historic Post Office, 5167 Pall Mall,
Bendigo 3550
Tel: +61 3 5444 4445
Freecall: 1800 813 153
(within Australia)
Fax: +61 3 5444 4447
Email: tourism@bendigo.vic.gov.au
Web: www.bendigotourism.com

CASTLEMAINE

Castlemaine Visitor Information Centre
Historic Market Building,
44 Mostyn St, Castlemaine 3450
Tel: +61 3 5470 6200
Freecall: 1800 171 888
(within Australia only)
Fax: +61 3 5471 1746
Email:
visitors@mountalexander.vic.gov.au
Web: www.maldoncastlemaine.com

DAYLESFORD

Daylesford Visitor Information Centre
98 Vincent St, Daylesford 3460
Tel: +61 3 5321 6123
Fax: +61 3 5321 6193
Email: visitorinfo@hepburn.vic.gov.au
Web: www.visitdaylesford.com

HEATHCOTE

Heathcote Visitor Information Centre
Cnr High and Barrack St,
Heathcote 3523
Tel: +61 3 5433 3121
Fax: +61 3 5433 3901
Email: vicheathcote@hotkey.net.au
Web: www.heathcote.org.au

KYNETON

Kyneton Visitor Information Centre
High St, Kyneton 3444
Tel/Fax: +61 3 5422 6110
Freecall: 1800 244 711
(within Australia only)
Email: vic@macedon-ranges.vic.gov.au

MALDON

Maldon Visitor Information Centre
93 High St, Maldon 3463
Tel: +61 3 5475 2569
Fax: +61 3 5475 2007
Email:
maldonvic@mountalexander.vic.gov.au

MARYBOROUGH

Central Goldfields
Visitor Information Centre
Cnr Alma & Nolan Sts,
Maryborough 3465
Tel: +61 3 5460 4511
Freecall: 1800 356 511
(within Australia only)
Fax: +61 3 5460 5188
Email: visitorinfo@iinet.net.au
Web: www.centralgoldfields.com

ST ARNAUD

St Arnaud Tourism
4 Napier St, St Arnaud 3478
Tel: +61 3 5495 1268
Freecall: 1800 014 455
(within Australian only)
Fax: +16 3 5495 2059
Email: stainfo@ruralnet.com.au
Web: www.ngshire.vic.gov.au

STAWELL

Stawell & Grampians
Visitor Information Centre
52 Western Hwy
Stawell 3380
Tel: +61 3 5358 2314
Freecall: 1800 330 080 (within
Australia only)
Fax: +61 3 5358 4366
Email: stawell.info@ngshire.vic.gov.au
Web: www.ngshire.vic.gov.au

Old Convent, PW

GETTING AROUND

One of the best aspects of travelling in the Goldfields is the ease of getting around. For speed, independence, flexibility and the ability to get off the beaten track there's still nothing to beat private cars, but the Goldfields can definitely be enjoyably explored using other means. All the major towns are linked by bus or train. In particular Ballarat, Castlemaine and Bendigo have fast train connections to Melbourne. Combining public transport with walking or cycling requires a little bit of planning, but if time allows, this is definitely the best way to explore the region. There are some substantial hills and ranges that are great for walking, but these can easily be skirted by cyclists if they choose. There are some great (flat) cycling routes.

The region is now home to some exceptional accommodation options, ranging from grand hotels, boutique B&Bs, and luxury self-catering cottages to traditional options like motels and caravan parks.

The resurgent wine industry is evident throughout the Goldfields and there are scores of top quality vineyards, including some that are well-known internationally. The wine produced reflects regions that are radically different in climate and geology: from the big reds around Bendigo and Heathcote, to the lighter cool-climate varietals around Ballarat and Macedon.

The major centres, particularly Ballarat, Daylesford and Bendigo, have outstanding restaurants and cafés, most of which showcase the growing variety of regional produce (cheeses, olives, unusual meats like venison, trout, sausages and more). There are also a number of outstanding restaurants scattered further afield.

The toughest challenge for a visitor is putting all these elements – history and natural heritage, accommodation, food, wine (and perhaps a spa) into a package. Planning time will be time well spent. This book, plus internet research, plus advice from the appropriate visitor information centres will enable you to put together one of the greatest travel experiences in the world.

CAR

There is a network of high-quality highways and main roads that enable you to get between major towns quickly – if you choose. In fact the direct routes – and the time you can travel them – can be misleading. A drive that theoretically takes 30 minutes could easily absorb a day if you choose to stop and explore, and especially if you choose to take some of the side roads.

Petrol costs are increasingly an issue, but the Goldfields are compact, so even a grand tour won't break the bank. Most places in this book are less than two hours from Melbourne. If you choose one of these places as a temporary base you could, without exception, spend a fascinating weekend without travelling more than 30 minutes from your temporary base and you'll use much less than a tank of petrol.

It can take a bit of courage to turn-off the main roads, but in the Goldfields it's the minor roads and tracks that will often be the most rewarding. Unless there has been heavy rain most gravel tracks in the region are easily negotiable in normal cars (but if you've rented a car, check the fine print). Once again, visitor information centres can provide up-to-date information if you are in doubt about conditions on a specific track. As a general rule, however, if there is a choice between a main road and a quieter alternative, take the quiet alternative!

Good maps to the region are hard to find, but it is an (almost!) universal rule that Goldfields tracks will lead you to a goldfield or a town or both. Otherwise the track will not have been built in the first place. Although you can quickly feel as if you are a million miles from civilisation, the distances between neighbouring farms and towns are generally small, so if worst comes to worst you might have to back-track or ask for directions.

There are plenty of attractive cafés scattered around the region, but pack a picnic (local wine and produce of course), because if you do get off the beaten track you will certainly find amazing places where you will want to stop and linger.

TRAIN

Two main lines service the Goldfields: Melbourne to Ararat (fast to Ballarat), and Melbourne to Swan Hill/Echuca (fast to Bendigo). Both lines are an important part of the region's heritage: the countryside is superb, the stations are historic, and some of the bridges you travel over and tunnels you travel through are significant heritage engineering feats.

The Melbourne to Ararat line links Melbourne, Bacchus Marsh, Ballan, Ballarat, Beaufort and Ararat. The Melbourne to Swan Hill/Echuca line links Melbourne, Gisborne, Macedon, Woodend, Kyneton, Malmsbury, Castlemaine, Bendigo – and on.

It's possible to carry bicycles on trains. For more information see www.viclink. com.au and www.vline.com.au

HERITAGE TRAINS

Daylesford Spa Country Railway

The old railway station is at the southern entrance to the town and an active preservation society runs the Daylesford Spa Country Railway. The railway has collected rail motors of the Victorian railways system from the early years of the 20th century. Trains run every Sunday on a preserved section of track through the Wombat State Forest as far as Bullarto (16 km). For information see www. chtr.org.au

Victorian Goldfields Railway

The Victorian Goldfields Railway runs vintage trains and rolling stock between Castlemaine and Maldon via Muckleford. Trains are hauled by steam locomotives, except on days of total fire ban. They run from the normal town railway stations. It's best to check the timetables, but generally they run on weekends and Wednesdays. For information phone 5475 2966, or visit www.vgr.com.au

Bus

Buses link all the major regional towns. Local visitor information centres can provide up-to-date information and relevant contact details. Most buses carry bicycles, although the fine print of their conditions often say 'if space allows'.

Bicycle

The Goldfields provide ideal cycling conditions. There are some fantastic hill and bush tracks for mountain bikers, and there are great options for tourers. Some of the most interesting sites in the region are down dirt roads, so the ideal machine may well be some sort of hybrid.

Most of the time, the terrain is relatively flat and with a bit of planning tourers can avoid steep climbs (if they want!). Traffic on most main roads is relatively light, but there are almost always even quieter alternative routes. It's possible to put together some creative routes using a combination of bikes, cars, trains and buses. Trains and buses do carry bicycles.

The ideal time to tour the region is in spring and autumn, when it's not too hot and not too cold. With the exception of the Daylesford and Ballarat regions most of the Goldfields lie north of the Great Divide, so the climate is generally considerably drier and warmer than that on the Victorian coast. Winter days can often be sunny, but nights can be very cold and frosty.

Local visitor information centres often have brochures about local cycle tours; in particular there are some great trails around Dunolly and Bendigo, and sections of the Great Dividing Trail are also suitable for mountain bikers.

Walking

Whatever transport you use to get to the Goldfields, walking is the best way to get around. Even the bigger cities are not so big that you can't see most of the sights best by foot. Start at the local visitor information centre; they nearly all have town maps with suggested walks.

The Box-Ironbark forests, which can look nondescript at 100 km per hour, reveal their riches when you walk through them – sometimes you actually need to be on hands and knees to really appreciate the tiny plants on the ground.

The network of national and state parks all have walks, ranging from short loops from picnic grounds to overnight expeditions, like the 18 km Pyrenees Walking Track, in the Pyrenees State Forest, or the 21 km Beeripmo Walk in the Mt Buangor State Park.

The premier walk, The Great Dividing Trail, traverses the Goldfields from Bendigo to Daylesford where it splits into two legs: from Daylesford to Ballarat and from Daylesford to Bacchus Marsh. Smaller sections are also designed to be accessible to day-trippers.

The Great Dividing Trail

The Great Dividing Trail combines goldrush heritage and natural beauty. It's a pathway to the Goldfields' hidden treasures: from deep gorges and fern-lined rivers through Box-Ironbark forests to the relics and cultural heritage of the goldrushes. It takes in an amazing variety of bushscapes and townscapes. And it is not just about walking, as many sections of the trail are appropriate for mountain bike riding.

The community-managed trail has four major tracks, each of which is in turn separated into thematic walks that can be easily accessed for day or shorter walks. Many sections of the trail are suitable for the casual stroller and there are also some very rugged parts, such as the Lerderderg section, that will test the most experienced walker. Green painted posts with a distinctive logo mark the trail.

The trail passes through, and can therefore be accessed from, large centres like Bendigo, Castlemaine, Daylesford, Bacchus Marsh, Creswick and Ballarat, and smaller villages such as Hepburn Springs, Chewton, Fryerstown, Vaughan Springs, Blackwood, Mollongghip and Dean.

Walking even a short section is a great way to get a feel for the country, to relax, and to experience (at least in part) the exhilarating freedom known by the diggers. William Ottey, a Castlemaine pioneer, described his digging days as 'a kind of Gipsy existence with a charm that is indescribable; one cannot call it *gambling* or a *lottery* or *speculation*'. It is 'a life of freedom, which can almost be felt; no rates, no shackles of any kind, in fact, it is a feeling that can only be understood by those who have experienced it'.

The trail is unlike any other in Australia for it offers what many have described as a 'European' experience: walkers are always close to heritage features, cities and villages, accommodation, services, cultural events, and good food and wine. Most trail users are content to do short sections of the trail at a time, making use of the cities and towns with their boutique B&Bs and cafés as bases and finishing the day with a great wine, a fine meal and a cosy bed.

Another difference between the Great Dividing Trail and other long distance walking tracks is its accessibility. The three major trailheads (Ballarat, Bendigo and Bacchus Marsh) are all within 1½ hours drive or train trip from Melbourne and all other entry points – such as Castlemaine, Daylesford, Creswick and Buninyong – can be reached by railway or main roads.

Highlights

The Great Dividing Trail has an impressive range of highlights, which include:

- the remains of Andersons logging tramway in the Wombat State Forest
- the Goldfields' best known volcano, Mt Franklin
- the two famous mineral springs – Hepburn and Vaughan/Glenluce
- the upper reaches of the River Loddon
- the original goldrush landscapes such as Sebastopol and Sailors gullies near Vaughan with their array of digger holes and hut sites
- Victoria's best preserved quartz mining landscapes near Fryerstown – Spring Gully and Eureka Reef
- the Garfield mine near Chewton, with the ruins of one of the world's largest water wheels
- abandoned mining villages such as Cornish Town (near Fryerstown) and the Welsh Village (near Chewton)
- one of Victoria's largest 19th century water supply schemes – the Coliban system

The Four Main Sections

FEDERATION TRACK – BALLARAT TO DAYLESFORD (72 KM)

There are entry points at the Ballarat Railway Station, the Creswick Information Centre and at Lake Daylesford.

There are four sections:

- Creswick Miners' Walk from Ballarat to Creswick (23 km)
- WG Spence Walk from Creswick to Mollongghip (20 km)
- Anderson's Tramway Walk from Mollongghip to Wombat Station (8 km)
- Wombat Forest Walk from Wombat Station to Daylesford (19 km).

Day/family walks on the Federation Track include:

- a walk from Ballarat to the edge of the Creswick State Forest
- any walk starting at Slaty Creek or St Georges Lake
- open country walking between Dean and Mollongghip
- a forest walk between Mollongghip and Wombat Station
- any walk in either direction from Sailors Falls
- a walk from Daylesford toward Sailors Falls

LERDERDERG TRACK – BACCHUS MARSH TO DAYLESFORD (71 KM)

Entry points for the Lerderderg Track are at Lake Daylesford – on the embankment adjacent to the spillway – and at the Blackwood General Store.

Day/family walks on the Lerderderg Track include:

- walks in the vicinity of Blackwood towards Garden of St Erth or down the Lerderderg River valley

- walks in either direction starting from O'Briens Crossing

DRY DIGGINGS TRACK – DAYLESFORD TO CASTLEMAINE (61 KM)

There are entry points at Lake Daylesford and at Castlemaine.

The three sections are:

- Castlemaine to Vaughan (17 km)

- Vaughan to Mt Franklin (23 km)

- Mt Franklin to Hepburn Springs (21 km)

Day/family walks on the Dry Diggings Track include:

- any walk between Vaughan and Fryerstown or Fryerstown and Castlemaine and Spring Gully

- a walk in either direction from Hepburn Springs

- a walk in either direction from Tipperary Springs or the Blow Hole

- a walk up Mt Franklin

LEANGANOOK TRACK – CASTLEMAINE TO BENDIGO (58 KM)

There are entry points at the Bendigo Railway Station at Castlemaine on the corner of Forest and Wheeler Sts.

There are four sections:

- Heritage Park Walk from Castlemaine to the Calder Fwy (14.5 km)

- Leanganook/Mt Alexander Summit Walk from Calder Fwy to Sutton Grange (11.5 km)

- Coliban Water Walk from Sutton Grange Rd to Sandhurst Rd (20 km)

- Bendigo Goldfields Walk from Sandhurst Reservoir to Bendigo (11 km)

Day/family walks on the Leanganook Track include:

- walks in the vicinity of Mt Alexander summit

- walks in the vicinity of Chewton

- walks along the existing Bendigo Bushland Trail towards One Tree Hill

More Information

Individual map/brochures are available for each of the four main walks; in Melbourne they can be bought at Information Victoria; in the Goldfields they're available from visitor information centres. For more information see the association's website at www.gdt.org.au

ROAD MILEAGE CHART

Ararat	- Avoca	63 km	45 min
	- Beaufort	44 km	30 min
	- Stawell	30 km	20 min
Avoca	- Ararat	63 km	45 min
	- Bung Bong	8 km	5 min
	- Maryborough	26 km	15 min
	- Moonambel	19 km	15 min
	- Stawell	81 km	55 min
	- Talbot	26 km	15 min
Ballarat	- Creswick	16 km	15 min
	- Daylesford	45 km	30 min
	- Melbourne	110 km	70 min
	- Smythesdale	19 km	15 min
Beaufort	- Ararat	44 km	30 min
	- Linton	50 km	35 min
Bendigo	- Castlemaine	38 km	30 min
	- Inglewood	44 km	30 min
	- Maldon	40 km	30 min
	- Tarnagulla	45 km	30 min
Blackwood	- Melbourne	84 km	50 min
	- Trentham	13 km	10 min
Bung Bong	- Avoca	8 km	5 min
	- Talbot	18 km	10 min
Castlemaine	- Bendigo	38 km	30 min
	- Daylesford	35 km	25 min
	- Maldon	20 km	15 min
	- Taradale	16 km	15 min
Clunes	- Creswick	18 km	10 min
	- Daylesford	52 km	40 min
	- Talbot	18 km	10 min
Creswick	- Ballarat	16 km	15 min
	- Clunes	18 km	10 min
Daylesford	- Ballarat	45 km	30 min
	- Castlemaine	35 km	25 min
	- Clunes	52 km	40 min
	- Hepburn Springs	4 km	5 min
	- Trentham	28 km	15 min
Dunolly	- Inglewood	45 km	30 min
	- St Arnaud	60 km	35 min
	- Tarnagulla	16 km	10 min
Halls Gap	- Stawell	25 km	20 min
Hepburn Springs	- Daylesford	4 km	5 min
	- Newstead	26 km	20 min

Inglewood	- Bendigo	44 km	30 min
	- Dunolly	45 km	30 min
Kyneton	- Macedon	25 km	15 min
	- Taradale	19 km	10 min
Linton	- Beaufort	50 km	35 min
	- Smythesdale	13 km	10 min
Macedon	- Kyneton	25 km	15 min
	- Melbourne	60 km	40 min
	- Taradale	50 km	35 min
Maldon	- Bendigo	40 km	30 min
	- Castlemaine	20 km	15 min
	- Newstead	14 km	10 min
Maryborough	- Avoca	26 km	15 min
	- Dunolly	22 km	15 min
	- Newstead	30 km	20 min
	- Talbot	14 km	10 min
Melbourne	- Ballarat	110 km	70 min
	- Blackwood	84 km	50 min
	- Macedon	60 km	40 min
Moonambel	- Avoca	19 km	15 min
	- St Arnaud	49 km	30 min
	- Stawell	62 km	40 min
Newstead	- Hepburn Springs	26 km	20 min
	- Maldon	14 km	10 min
	- Maryborough	30 km	20 min
Smythesdale	- Ballarat	19 km	15 min
	- Linton	13 km	10 min
St Arnaud	- Dunolly	60 km	35 min
	- Moonambel	49 km	30 min
Stawell	- Ararat	30 km	20 min
	- Avoca	81 km	55 min
	- Halls Gap	25 km	20 min
	- Moonambel	62 km	40 min
Talbot	- Avoca	26 km	15 min
	- Bung Bong	18 km	10 min
	- Clunes	18 km	10 min
	- Maryborough	14 km	10 min
Taradale	- Castlemaine	16 km	15 min
	- Castlemaine	24 km	15 min
	- Kyneton	19 km	10 min
	- Macedon	50 km	35 min
Tarnagulla	- Bendigo	45 km	30 min
	- Dunolly	16 km	10 min
Trentham	- Blackwood	13 km	10 min
	- Daylesford	28 km	15 min

ACCOMMODATION

The Goldfields region has a broad range of accommodation for all budgets and tastes. The choice ranges from historic and boutique B&Bs, 19th century miners' cottages, and grand old hotels with ornate Victorian architecture to motels and caravan parks (most caravan parks have comfortable on-site cabins). A number of working farms also offer accommodation and a chance to participate in country life. Start with recommendations from the local visitor information centres and where possible cross-reference with internet research.

FOOD

The Goldfields region has a reputation for culinary excellence, with a number of local eating establishments building a national reputation. Local produce is often a feature of their menus. As the area develops, more and more chefs are opting for a country lifestyle, so the range of good dining experiences continues to expand.

Superb local products are now on offer in the region. Many farmers are growing organic produce ranging from herbs and vegetables to grains and fruit.

In summer look for succulent pick-you-own berries, currants, cherries, and stone fruit. In autumn you'll find apples, sweet chestnuts and wild mushrooms. In spring you might spot morel mushrooms and the first new season baby vegetables, while sacks full of freshly dug potatoes appear throughout the year.

High quality olives and olive oils (sometimes infused with lemon or herbs) are also available.

Maldon Verandah, KS

WINE

MACEDON RANGES WINE REGION

The Macedon Ranges Wine Region is the coolest grape growing climate in mainland Australia and has a reputation for fine sparkling wines. Its altitude of about 500 metres and the resulting cool climate suits Chardonnay and Pinot Noir, much of which is used in the production of sparkling wine.

BENDIGO WINE REGION

The red soils and warm temperatures of the Bendigo Wine Region produce award-winning Shiraz and Cabernet wines – all deep red, ripe and boasting a distinct minty edge to the berry fruit. As with other Victorian regions, the early vineyards in Bendigo were destroyed by phylloxera and it was not until the late 1960s that a new vineyard was planted by local identity, Stuart Anderson. Many have followed his example.

HEATHCOTE WINE REGION

The Heathcote Wine Region is a little cooler than Bendigo, but this area is still dominated by Shiraz. The conjunction of climate and soil produces world-class Shiraz with a unique character. It is deep, rich and velvety with plum and sweetly spicy fruit, and fine tannins giving texture and sustaining length.

PYRENEES WINE REGION

The Pyrenees Wine Region has a variety of micro climates, so there is a great deal of variation. The biggest plantings are Cabernet Sauvignon and Shiraz, with high quality dry red wines resulting. Chardonnay also does well, as does Sauvignon Blanc in the cooler climates.

GRAMPIANS WINE REGION

Nowhere in Victoria is the link between gold and wine more pronounced, and nowhere are the legacies of the 19th century miners more impressive, than the underground cellars at Seppelt's Great Western. The Nursery Block at Best's includes grape varieties so rare that several have defied all attempts at identification and are, in all probability, the sole surviving examples in the world. The region is extremely varied, encompassing the hillsides of Mount Langi Ghiran and Mount Ararat in the east, and the more open countryside around Great Western. This is primarily a red wine area, producing wines that possess a combination of elegance and power, and an exceptional capacity to age.

GOLDMINING

Although the Goldfields have been turned upside down in the search for gold, prospectors continue to find nuggets and gold. The old miner's saying 'Gold is where you find it' remains true, but research is important. Local detector shops, bookshops and visitor information centres in the Goldfields often carry specialist fossicking guidebooks and maps.

There are two alternative approaches: low tech (panning) and high tech (metal detectors). The first is cheap, the second can be expensive. Purchasing a top line metal detector can set you back thousands of dollars, but it is also possible to hire them.

METAL DETECTORS

Look at the edges of old diggings for side gullies and slopes where the diggings become very shallow. If an area has been 'surfaced', meaning all the gold bearing gravel and dirt was taken away to be puddled for its gold content, all that will remain will be bare clay – but this is a good starting point for detecting.

Following a gully uphill may bring you to the quartz reef from where the gold was shed. Search all sides sloping away from any such reef. Or the gully could lead to a 'Hard Hill'. These were created by ancient river systems and are made up of water-worn pebbles in a conglomerate rock. Gold trapped in the conglomerate will have been released down the hillside; search all sides of the hill. Or the diggings in the gully might peter out. In this case look along the slopes above the diggings for those nuggets left behind as the gully deepened and changed course.

PANNING

Since gold is heavier that other metals it is normally found at the bottom of river silt on the inner edge of a bend in a stream, the point where stream widens, in a crevice or against a stone bar in the stream bed, or in the lee of a stone or stump protruding into the stream. Slate country is better than sandstone country because the slate tends to trap the gold particles. Avoid areas where no quartz reefs can be seen and granite areas in general.

Fill your pan with gravel and sand to 5 cm from the top. Immerse it in the stream and give it several vigorous half turns to bring large stones and pebbles to the top so they can be discarded.

Give the pan several more half turns then start rocking the pan alternatively towards and away from you. Keep the pan just beneath the water so the lighter material will wash over the edge. Frequently return the pan to the horizontal and give several half turns to bring the coarser material to the surface.

When most of the material has been washed away or discarded, immerse the dish gently and rotate until roughly two to three spoonfuls of sediment remain. Then gently pour out all but half a cup of water. Rotate the pan so the water swirls over the concentrate – if you have been successful black sands and gold will form the remaining particles.

PERMIT

A Miner's Right is required if you intend to prospect for minerals on Crown Land or private land. If you do wish to prospect on private land it is essential to get permission; without permission you are considered to be trespassing.

All adults searching for minerals, gold or gems require a miner's right, whether they are using a metal detector, hand tools or pans.

It is illegal to remove or damage trees and shrubs, and any relics or artefacts, including bricks and building stone) cannot be moved or taken.

Miner's Rights are $26.20 and are available from Department of Primary Industries at:

Ballarat, 402-406 Mair St, tel: 5336 6856

Bendigo, Cnr Midland Highway & Taylor Sts, Espsom, tel: 5430 4555

Maryborough, 126 High St, tel: 5461 1055

Melbourne, Level 8, 240 Victoria Pde, East Melbourne, tel: 9412 5103

Prospecting is generally not permitted in national or state parks. There are exceptions, in limited areas, in the following parks:

Castlemaine Diggings National Heritage Park

Greater Bendigo National Park

Heathcote-Graytown National Park

Kooyoora State Park

Paddy's Ranges State Park

St Arnaud Range National Park

Steiglitz Historic Park

Parks Victoria (tel 13 19 63) will provide full details of where you can prospect. There are local Parks Victoria offices at Ballarat (tel 5336 6817), Bendigo (tel 5444 6620), Castlemaine (tel 5472 1110) and Maryborough (tel 5461 1055).

Waterloo Diggings, JM

INDEX

This book belongs to

Maude

AUTUMN
PUBLISHING

Published in 2021
First published in the UK by Autumn Publishing
An imprint of Igloo Books Ltd
Cottage Farm, NN6 0BJ, UK
Owned by Bonnier Books
Sveavägen 56, Stockholm, Sweden
www.autumnpublishing.co.uk

1121 001
2 4 6 8 10 9 7 5 3 1
ISBN 978-1-80108-236-5

Wildlife consultancy by David Winnard

Illustrated by Amelia Herbertson
Written by Suzanne Fossey

Designed by Chris Stanley
Edited by Suzanne Fossey

The publisher would like to thank Elizabeth Wise for the index.

Printed and manufactured in China.

How to Find a

Dragon

AUTUMN PUBLISHING

Contents

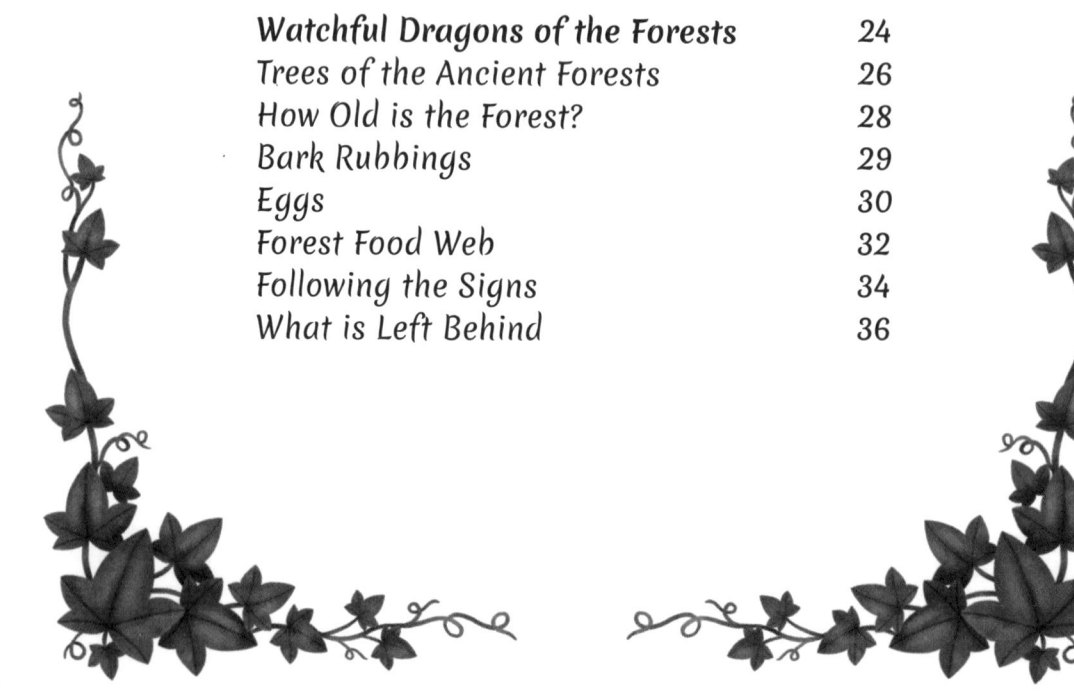

Wandering Wild

Most people think that dragons belong in myths and legends, and only exist between the pages of books. What they don't know is that dragons are as real as we are. They're just much better at hiding.

Dragon Wanderers devote themselves to searching for dragons, and protecting the wild places that they call home. They know that the roar of the wind whipping through hills and valleys hides the cries of dragons calling out to each other. A Wanderer has learned that the crashing of the waves against the shore disguises the thud of giant wing beats as dragons soar over the cliffs. They understand that the shrill calls of circling birds often conceal the cries of hungry baby dragons in their hidden nests.

There are signs of these grand and magnificent creatures all around us, waiting to be found...

... you just need to learn to listen carefully.

Wander wise...

Dragons are creatures of the Wilds.
They and their homes should be respected.
All Wanderers need to understand and
follow these important rules:

- Different Wilds have different dangers.
Make sure someone knows where you
are and what you're doing.

- Leave the Wilds as you find them.
Never leave behind anything that
doesn't belong.

- Don't eat anything you find in the
Wilds without checking with an adult first.

- Take only materials that are common
and plentiful. Leave most of the wild
in the Wilds for others to enjoy.

- Don't trespass on private property
or take anything without permission.

Only those who respect the Wilds will be
welcomed back.

... and Wander prepared

Wanderers understand the exciting but unpredictable nature of the Wilds they seek to protect. They are always prepared for every eventuality.

The call of the Wilds could take you far from home. Take water and snacks with you, as the dragons will not share.

Check the weather before you leave and wear the right clothing. The Wilds are as unpredictable as the dragons they hide.

Carry a notebook and pencil with you. You might be called upon to share your knowledge with other Wanderers one day.

Wanderers know that even the smallest or most distant things can be vital clues. Binoculars and a magnifying glass can be helpful tools.

Hidden Dragons of the Mountains

Mountain dragons spend most of the year high on snowy peaks. But when winter comes, bringing cold winds and a blanket of snow to the northern Wilds, the dragons fly down to the lower slopes and foothills. Their thick, icy blue scales, shimmering as they fly, protect them from the chill of the freezing winter air.

The dragons roam the skies, protecting all the creatures who live around the mountains. Their powerful wings clear away patches of snow, so deer and rabbits can nibble the grass underneath. Their scorching breath melts the ice on frozen ponds, so water birds can drink and find food.

When summer comes and the ground thaws, these majestic creatures head back into the mountains until the cold comes again.

Up Hills and Down Mountains

True Wanderers know that hills and mountains take different shapes according to how they were formed. By learning about the different types, a Wanderer knows where dragons are most likely to rest, hunt and play.

Tor

A tor is a rock formation on top of a hill. Tors are usually less than 5 metres (16 ft) high, so are not big enough to hide a mountain dragon. They are created by wind and rain wearing away the top of the hill, leaving only the hardest rocks behind.

Mound

A mound is a small hill made by people, rather than by nature. Larger mounds may have been forts built to defend villages long ago. Smaller ones sometimes hide the burial places of long-dead kings and queens.

Mountain

Mountains are the highest places on earth, towering above the surrounding land. The air gets colder higher up, so the tops of mountains are often too cold for trees to grow, and are the first places to be covered in snow. Some are so high that they are permanently topped with snow and ice. Mountain dragons can be spotted circling the peaks, calling out to the animals below.

Cliff

A cliff is a steep wall of rock. They are most commonly formed by water eating away at the bottom of a hillside, causing the slopes above to collapse. Look out for sunny ledges halfway up, where mountain dragons like to bask on warm days.

Drumlin

A drumlin is a long hill that looks like a half-buried egg. They were formed by glaciers scraping away the rock. The steep side of the hill faced towards the flow of ice, while the gentle side faced away. Playful young mountain dragons can sometimes be glimpsed sliding down the slopes on their stomachs.

Birds of Prey

Birds of prey, or raptors, are meat-eating birds that use their strong talons and hooked beaks to catch prey. Like dragons, these birds have adapted to be excellent airborne hunters.

Merlin

Merlins use their strong wings to fly fast and low to the ground. They turn quickly while flying and can catch prey in mid-air. Merlins will often venture into towns and cities looking for food. They eat insects, bats, small reptiles and other birds.

Harrier

A harrier's wingspan is almost three times the length of their bodies. They fly very slowly and then drop very quickly down onto prey. Special feathers around a harrier's ears help them pinpoint where a sound is coming from. This is helpful for hunting small birds, insects and frogs in the dark.

Barn owl

The heart-shaped face of a barn owl helps to direct sound straight into the their ears. Their hearing is so sharp that they can find a mouse in total darkness. Their wing feathers don't rustle, so they can fly silently when hunting. For the barn owl, the element of surprise is more important than speed.

Peregrine falcon

The peregrine falcon has stiff, pointed wings and a light, sleek body. When diving to attack their prey, they are faster than any other animal on Earth: about three times the speed of a cheetah. Their hooked beaks have a special notch which is used to sever the spinal cord of the peregrine's prey, killing it instantly.

Osprey

Because the osprey eats mostly fish, they usually nest near lakes. This powerful raptor dives down and snatches fish from the surface of the water. Their feathers are dense and oily, which stops them getting waterlogged. Ospreys also have reversible toes and barbed pads on their feet to help them hold slippery fish.

Golden eagle

Golden eagles have massive wingspans: wider than a grown human is tall. With these large wings, golden eagles can lift prey as heavy as themselves. They have excellent eyesight and bony ridges around their eyes to stop sunshine from blinding them during flight. Golden eagles will eat large prey, including deer, sheep and snakes. This eagle can swallow prey whole and then later regurgitate indigestible bits, like bones.

Plants of the Moors

Many beautiful and fascinating plants are unique to the boggy marshes of the moors. Plants here thrive on cold, wet soil and chilly air, much like the mountain dragons.

Heather

This short, shrubby plant has small pink flowers that appear in the middle of summer and last until late November. Heather is able to survive in freezing temperatures. Flattened heather patches are usually the result of a mountain dragon with an itchy back.

Cottongrass

The fluffy heads of cottongrass look like cotton wool balls on sticks, but the white tufts actually contain seeds. Cottongrass needs a cold climate to thrive, and is most likely to be seen near water or bogs. The white heads have been used as wound dressing, pillow stuffing, paper and nappies.

Sphagnum moss

Sphagnum moss lives in watery habitats like wetlands and bogs. It spreads out into colourful, spongy carpets that can soak up more than eight times their own weight in water. The stems and leaves can hold water for a long time after the soil around them has dried out.

Bilberry

The blue-black bilberry looks similar to a blueberry, but is purple inside rather than white. In the summer, the low bushes grow small, pink, bell-shaped flowers, and by August the bushes are covered in berries. These sweet, tangy berries can be made into jams, pies and sauces.

Sundew

These strange looking plants hide a secret: the red tendrils are tipped with droplets. Unwary insects get stuck to the "dew", then tangled in the red tendrils. The whole leaf then curls around the insect and breaks it down into plant food.

Scaly spiral mushrooms

These extremely rare, rainbow coloured mushrooms are only found in the middle of dense patches of boggy grass. To a human these mushrooms smell like rotting fruit, but to a dragon they are delicious. The scales of the mushroom help to clean dragons' teeth as they chew.

Inside a Wing

By moving its wings, an animal can produce lift and get off the ground. There are four kind of animals that have wings that are used for flight: birds, bats, insects and dragons.

Bird

The bones in a bird's wing are hollow and incredibly light. Each wing has a shoulder, elbow and wrist joint which allows the bird to fold their wings away neatly when they land.

The size and shape of a wing changes how the bird flies. A swift has pointed wings so it can dart about, and the albatross has long wings so it can glide easily. The hummingbird can beat its small, strong wings around 90 times per second.

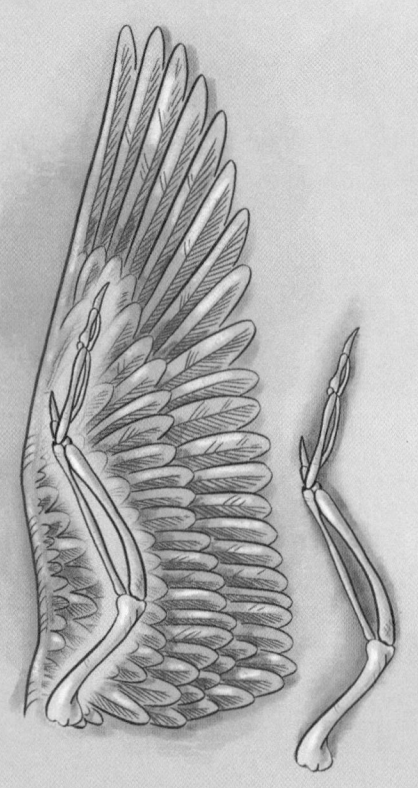

Bat

The bat has bony fingers that form the shape of its wings. The fingers move and curl so that they can bend their wings. Their bones are very thin and extremely long which allows the bat to weave through the air, twisting and turning in a way that birds can't.

Stretched across the wing between the fingers is a double layer of thin skin. A small clawed thumb extends out of the top of the wing, which the bat uses for climbing.

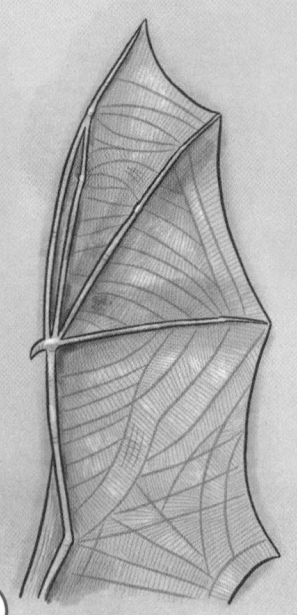

Honey Bee

The honey bee has four wings: two large forewings and two smaller hindwings. The forewing and hindwing hook together to form a larger wing.

Each wing is made up of three layers: a thin membrane on top, a layer of veins and nerves in the middle, and another thin membrane on the bottom.

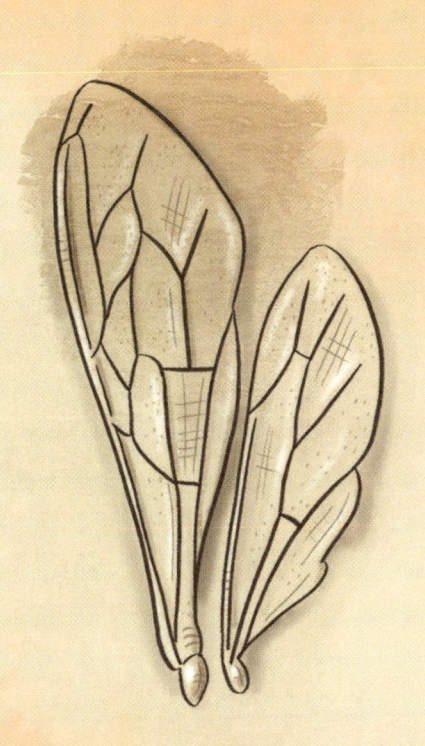

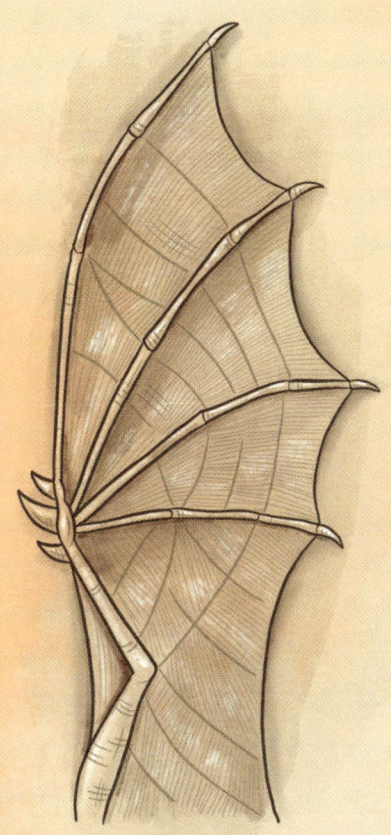

Dragon

The bones in a dragon's wing are hollow and light, which allows them to lift their massive bodies off the ground. The bones resemble fingers, like a bat's, and can bend in flight so they can swoop down on their prey.

The dragon's wing is made up of three layers: a thick layer of protective scales on top, a layer of veins and arteries in the middle, and a protective membrane on the underside of the wing.

Geological Marvels

The world around you is constantly changing. Even solid rock changes very slowly, as water, ice and wind act over millions of years. These rock formations are as old as the dragons who use them as ancient meeting places.

Basalt columns

These astonishing hexagonal columns are made of a black rock called basalt. They form when lava from a volcano cools and shrinks, cracking into six-sided columns. In the middle of the night, when no-one is around, coastal dragons meet on these rocks.

Pedestal rocks

These strange formations may look like lots of stones balanced in a pile, but they are in fact one piece of rock joined to the ground. They form where hard and soft types of rock mix together. The softer parts are worn away by wind and water, leaving the harder rock sticking out in strange, gravity-defying shapes.

Boulder fields

These strange expanses of broken rock are sometimes left behind by huge ice sheets called glaciers. As the glacier melts, water trickles down into the rock underneath. When the water freezes again, the ice pushes against the top layer of rock, cracking it into rugged boulders.

Sea arches

These arches form along coastlines where parts of the cliff jut out into the sea. As the water pounds against the rock, it starts to wear away. Eventually, the waves break through, creating an arch. Sometimes, a young mountain dragon learning to fly will crash into the arch, causing it to collapse.

Limestone pavements

Limestone is a soft, grey rock that sometimes forms flat "pavements" covered in thin grooves and channels. The flat surface was left by glaciers, heavy ice sheets that scraped away the top of the limestone. The channels are left by rainwater trickling across the rock over many thousands of years.

Karst caves

Karst caves are famous for plunging steeply far below the ground, with magnificent caverns and waterfalls. They form in areas where soft rocks, such as limestone, are carved away by rain and river water. Mountain dragons can occasionally be spotted swimming together in pools of the largest caverns. Wanderers, however, should never explore underground without a guide and safety equipment.

A Wanderer's Staff

Trekking up hills and mountains can be hard work. To help, many Wanderers carry a staff made from things found in the Wilds they protect. Here are some suggestions for ways to make your own staff.

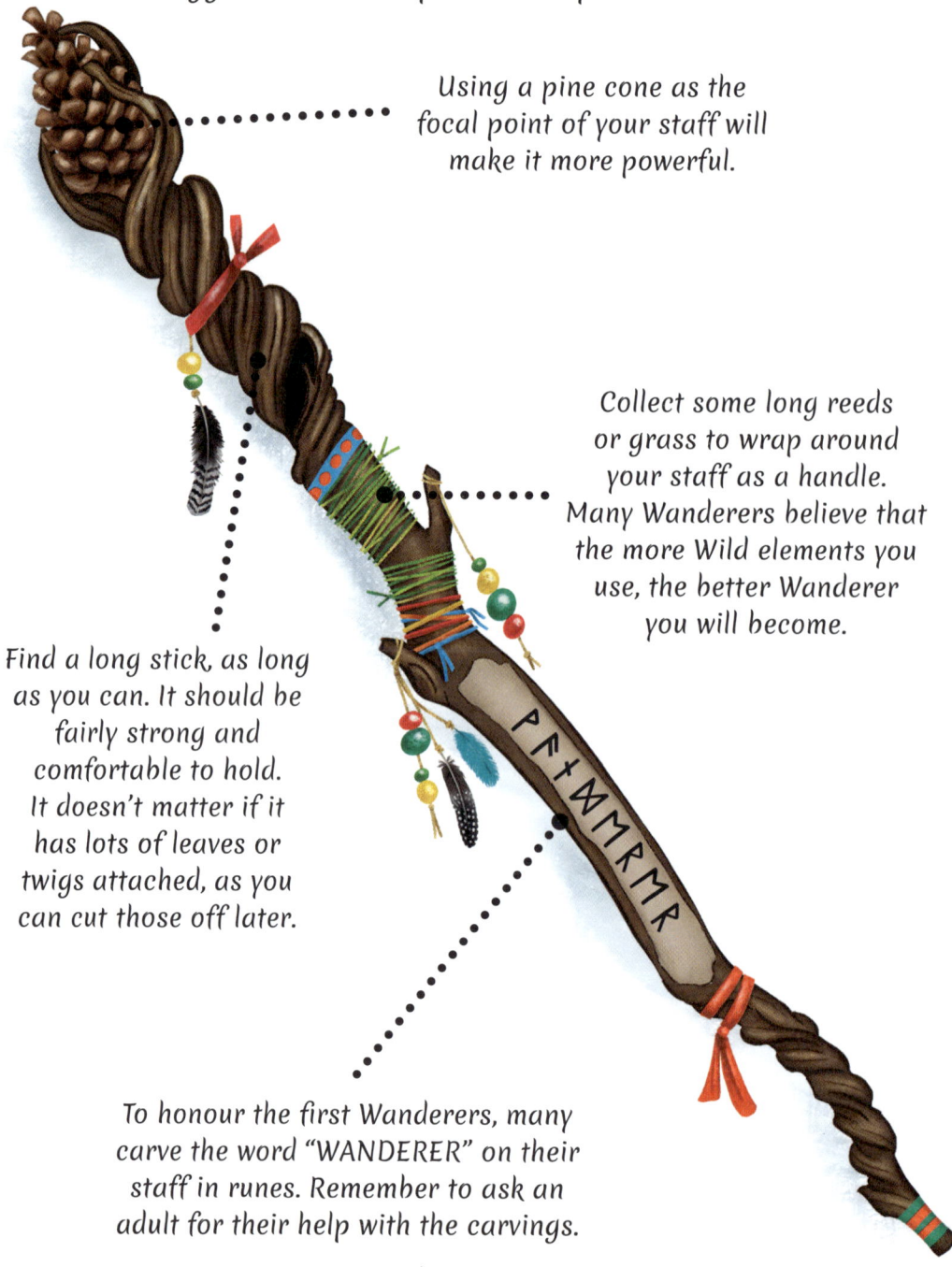

Using a pine cone as the focal point of your staff will make it more powerful.

Collect some long reeds or grass to wrap around your staff as a handle. Many Wanderers believe that the more Wild elements you use, the better Wanderer you will become.

Find a long stick, as long as you can. It should be fairly strong and comfortable to hold. It doesn't matter if it has lots of leaves or twigs attached, as you can cut those off later.

To honour the first Wanderers, many carve the word "WANDERER" on their staff in runes. Remember to ask an adult for their help with the carvings.

Elder Furthark is an alphabet of runes, ancient symbols designed to be easy to carve into stone or wood. Only the Wanderers know that it was once used to communicate with dragons.

Here is "WANDERER" in Elder Futhark to get you started:

WANDERER

Watchful Dragons of the Forests

In spring, when the lush, green forest is once again bursting with life, the forest dragons are at their most active. Moving unseen between the branches, they use their powerful claws to climb, then launch themselves from the very tops of the trees, diving and gliding through the air.

As spring arrives, animals slowly awaken from hibernation, emerging from their burrows and the flowers unfurl their petals, bursting into bloom. The sound of birdsong fills the chilly morning air and mixes with cries of the forest dragons, as they swoop through the trees, protecting all that live there.

Trees of the Ancient Forests

Knowing how to identify the trees in the areas around you is an important skill for a Wanderer, as the forest dragons favour certain trees in particular. Use the leaves and the bark to guide you.

Aspen

Recognisable by their smooth, white bark, aspen trees grow in cold places that have cool summers. When it's windy, the tree looks like its trembling. This is because the stalk that holds each leaf is flat, allowing the wind to flow over the leaf without tearing it off the branch.

Ash

These tall trees have pale brown and grey bark that becomes more grooved as they age. In summer, the oval leaves move in the direction of sunlight, and sometimes the whole tree may lean towards the sun. In winter, it grows black, velvety leaf buds.

Elm

The bark of the elm tree is grey-brown, and the twigs are covered with very fine reddish hair. The leaves, which are rough to the touch, have an asymmetric bottom: the two sides of the leaf don't quite match up.

Weeping Willow

This drooping tree is found in damp ground, near rivers and streams. The long leaves start off hairy all over. However, the hair underneath each leaf becomes white and velvety, and most of the hair on top of the leaf falls away. Sometimes a forest dragon can been found relaxing in the shade underneath the long branches.

Goat Willow

The goat willow is shorter than many forest trees, but it can live up to 300 years. The twigs are hairy at first but become smooth as the branch grows older. They also look orange in the sun. The undersides of the leaves are very hairy and have a pointed tip which bends to one side. The furry catkins, which are the tree's flowers, look like white cat paws.

Beech

Beech trees can grow very tall, making these trees the favoured hiding places of the forest dragons. Their bark is smooth and grey, with horizontal markings. The leaves are oval and pointed, with a wavy edge. The beech nuts, which grow in autumn, are filled with nutrients, perfect for a young forest dragon.

How Old is the Forest?

Forest dragons perch on older trees that are strong enough to support their weight. To have a better chance at spotting one use this Wanderer's trick to try and find the oldest tree in forest.

You will need:
• tape measure • pen • some paper

1 Choose a tree that you think looks old and big enough to support a dragon's weight. Try and identify the tree by looking at the information on the previous page. Make a note in your journal of the type of tree and its location.

2 Use a tape measure or piece of string to measure around the trunk at approximately 1 metre from the ground (about your head height). Measure to the nearest centimetre. This is the tree's girth or circumference.

Every 2.5cm of girth represents one year's growth. So, to estimate the age of a living tree, divide the girth by 2.5. For example, a tree with a girth of 40cm will be sixteen years old.

3 Write down the measurement in your journal and move on to the next tree. You don't need to do the whole forest, but try and measure between 6-10 trees to get a better idea of where a forest dragon might be.

Bark Rubbings

If you want to do a more detailed study of the forest around you, make a rubbing of the tree's bark after you measure it and then stick the research in your Wanderer's journal.

You will need:
· white paper · a crayon · sticky tape or glue

1 Place the paper against the trunk of a tree. You might want to ask a companion to hold it in place.

2 Rub the long edge of the crayon over the paper to make a print of the bark pattern. You might need to press hard if the bark isn't very bumpy.

3 Using what you've learned, find some different trees and repeat steps 1 and 2. Write the name of the tree somewhere on each piece of paper so you don't forget which one is which.

4 When you get home, put your bark rubbings into your journal. You can use the whole piece of paper, or cut a small square from each and stick those in.

Eggs

Many kinds of forest animals lay eggs. Learning to identify the different types is a good way to discover which animals are living around you. Never touch eggs that are still in their nests. Not only is it illegal, but it will really annoy the forest dragons.

Thrush

Birds like the thrush, that build their nests in trees and shrubs, usually have blue or greenish eggs. After the chick has hatched in the nest, the parents push the empty shell down onto the ground.

Large white butterfly

These bright yellow, skittle shaped eggs are laid on the underside of cabbage leaves. They can be found in clusters of 40 to 100, all neatly grouped together. The eggs take one or two weeks to hatch, depending on the temperature. The warmer it is, the quicker they hatch.

Shield bug

The female shield bug can lay up to 200 eggs at a time, in batches of 25 to 30. The eggs are laid in small, hexagonal groups on the underside of leaves. When the eggs are first laid, they are yellow-green, but they start to turn pink-grey as they get closer to hatching.

Snake

These eggs can be found in groups of anywhere between 6 to 40. They are creamy white and have a rubbery, soft skin. The eggs are usually laid in summer, in rotting plant matter or under logs, where they stay until the babies hatch during early autumn.

Slug

Slugs lay their eggs on moist, dark ground, covered by fallen leaves or branches to protect them from the cold. The eggs look like pearly white jelly sacks and they are coated with a slimy substance that is slightly gummy.

Forest dragon

The eggs of forest dragons are made of overlapping layers of green and yellow crystal. When the baby dragon breaks through the shell, it shatters into hundreds of little shards. The forest dragons are incredibly protective of all the shards, so if you find a piece you should feel incredibly lucky.

Forest Food Web

Every plant and animal needs energy to live. Plants get their energy from sunlight, and animals get energy by eating plants or other animals. A food web is a map of what different animals eat. In the forest, where lots of different animals and plants rely on each other for energy. The forest dragon is at the very top of the food web.

Forest dragon

Hawk

Owl

Red fox

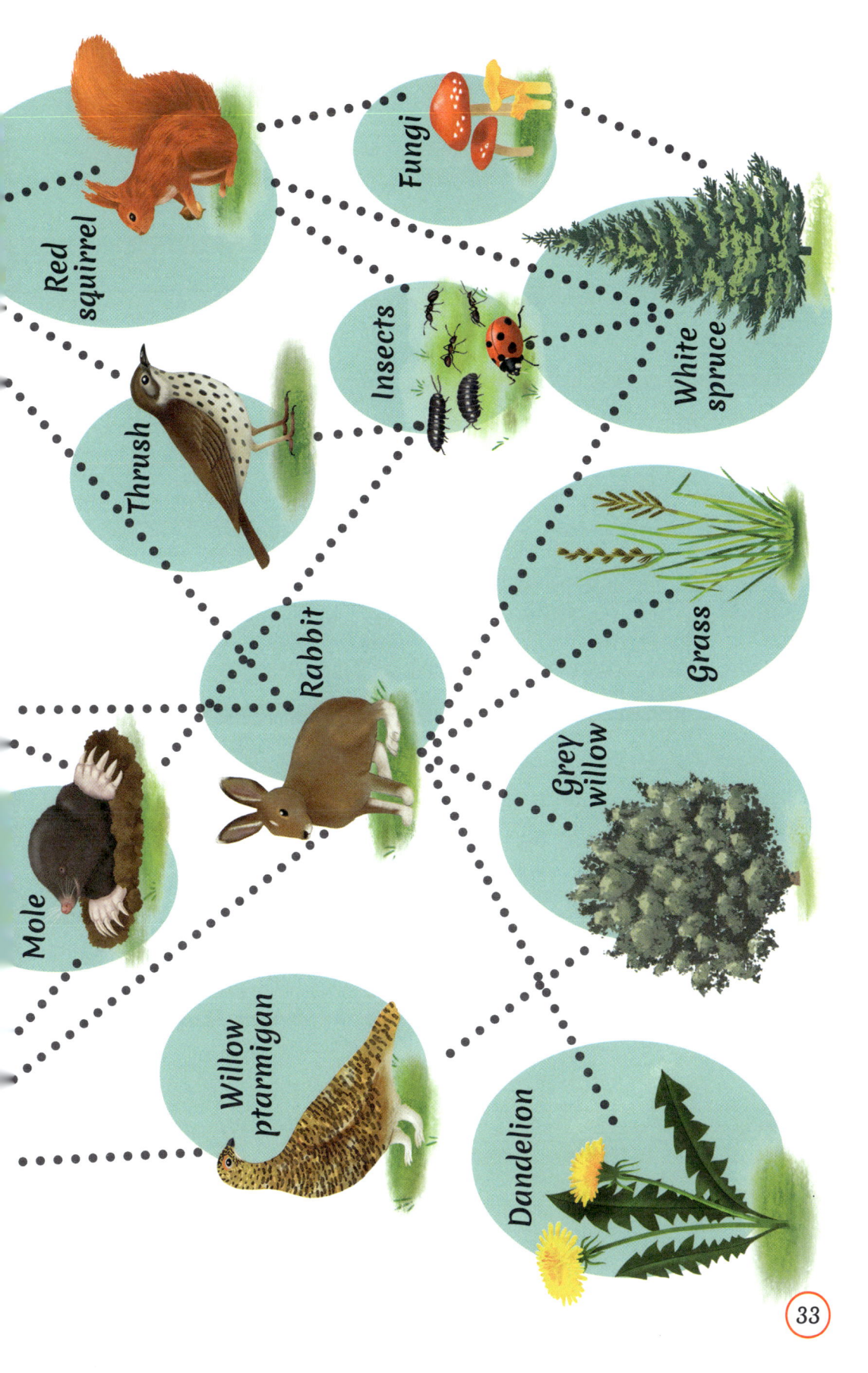

Red squirrel

Fungi

White spruce

Insects

Thrush

Grass

Rabbit

Grey willow

Mole

Willow ptarmigan

Dandelion

Following the Signs

Sometimes it's hard to get close to the animals of the forest. By examining their tracks, a skilled Wanderer can identify animals in an area without disturbing them. Treasure the skill and respect the Wilds.

Hare

Because of the bounding way the hare runs by swinging their hindlegs forward, prints of their long back feet are usually at the front, with the prints of their smaller forefeet between them.

Squirrel

Squirrels have five narrow, widely spread toes. Squirrel tracks almost always begin or end by a tree. As they are quite light on their feet, it might be hard to see their tracks, so look for pine cones that have been stripped to get at their seeds.

Mice

The prints of mice are tiny and hard to see. However, nibbled nuts and seeds are a sign that mice are around. When eating hazelnuts, they leave tooth marks on the surface of the nut and around the edge of the hole.

34

Beaver

Beavers cut down trees with their teeth. They use the branches to build dams across streams, with a home in the middle called a lodge. One beaver can cut down up to 300 trees each year.

Weasel

The paw prints of the weasel are very small and easy to miss. You should be able to see five toes spread out widely.

Forest dragon

Like squirrel tracks, the prints of forest dragons usually start and end near a tree. You should be able to see two pads on their three front-facing toes, with a thin claw at each end. They have a rear facing toe for balance, which also has two pads but no claw.

35

What is Left Behind

As well as eggs and tracks, you can identify woodland animals by the clues they leave behind. Do not take these things from the Wild, as another animal might be able to make use of them.

Crow feather

A crow's feather is long and black, with a blue, green or purplish sheen. The position of the shaft in the middle of the feather tells you what part of the crow it came from. If the shaft is in the centre, it's a tail feather. If the shaft is over to one side, it came from the wing.

Owl pellet

Owls eat small mammals such as shrews and mice, which they swallow whole. As they digest, they squish the fur and bones together into pellets, which the owl spits up. You can find out what the owl has been eating by looking at the bones inside the pellet.

Deer antler

Antlers are made of a single piece of bone. They have a soft, velvet covering that protects the antler when it first starts to grow in spring. At the end of the winter, the antlers naturally fall off and they will regrow the next spring.

Snake skin

The shed skin of a snake looks like a long, see-through tube. Snakes often shed their skin soon after they come out of hibernation, so that they have more room to grow. The old skin starts to come away around the snake's head, and the snake peels off the rest by rubbing its body against rough surfaces such as tree bark.

Snail shell

Snail shells can vary in colour. Some are dark brown or grey, but most are a light brown with yellowish stripes. The shell has between three to five spirals, and the shells of older snails have a lip at the edge of the opening. When snails die (or are eaten), their hard shells are often left intact.

Forest dragon scales

Forest dragons shed their shiny green and gold scales around four times a year. Sometimes, as the dragon climbs up the trunk of the tree and rubs against the bark, loose scales get caught in the grooves of the tree trunk.

Mischievous Dragons of the Grasslands

As the wildflower meadows turn green and full of life, watch for rabbits hopping through the grass and listen for the chattering of voles as they dig their tunnels just below the ground. Hear the buzz of dragonflies as they whizz past on their way to reservoirs and marshes, and keep an eye out for the families of foxes moving across the open fields.

On the moors, where the bees and butterflies flutter around in the sunshine, and the colourful flowers bob their heads in the warm, summer breeze, you might see a young grassland dragon swooping down over a lake or valley as it learns to fly. These golden dragons are notoriously playful, and delight in hide-and-seek, so only the most sharp-eyed Wanderers will spot them.

Insects of the Meadows

Meadows are lively places, bursting with colour and the hum of insects. Next time you are passing by, kneel down and have a closer look at the plants and animals. Can you spot some of these insects?

Ladybird

The ladybird has red wings and black spots. The bright red colour of ladybird is a warning to predators that it tastes horrible. During winter, they hibernate in hollow plant stems, emerging in spring. Long ago, some people used to eat ladybirds as they believed they would cure toothache.

Dragonfly

Found near ponds and lakes, the dragonfly has four, strong wings and a long, colourful body. They are powerful and fast fliers, and can capture their prey in mid-air. They eat other insects, including midges, mosquitoes, butterflies, moths and smaller dragonflies.

Crab spider

Unlike other spiders, the crab spider doesn't spin a web to trap prey. Instead, it disguises itself by changing colour to match its background and lies in wait on flowers and leaves for unsuspecting bees and moths to come by. It then traps them between its crab-like front legs.

Grasshopper

The green grasshopper has three pairs of legs and two pairs of wings. When found by predators, the grasshopper will try and scare them by flashing its brightly coloured wings and launching itself into the air. Male grasshoppers sing by rubbing their wings against their legs, creating a loud churring noise.

Tick

Ticks are small, brown, spider-like parasites that survive by feeding on the blood of mammals and birds. Each of their eight legs is tipped with a pair of claws, which they use to cling onto an animal's skin. They like to make their home at the edge of woodlands and open meadows, and will lay their eggs under the fallen leaves. They bite humans as well as animals, and carry nasty diseases, so cover up well in long grass where ticks are common.

Earthworm

The long, reddish earthworm burrows into the ground during the day and emerges at night to feed on fallen leaves and dead plants. As they burrow, they mix up the soil by eating and then pooping it back out. An earthworm can eat up to one third of its body weight in a day. They are covered in tiny hairs, which help them to grip the soil and move.

Dragon Kites

The wide open spaces of grasslands are perfect for flying kites. In times gone by, Wanderers flew specially shaped kites to attract grassland dragons. Here is how to make your own.

You will need:
• a strip of paper (approximately 60cm long and 8cm wide) • a large piece of card • a pencil • scissors • string • ribbon • glue or sticky tape • decorating materials (pens, pencils, paint, eco-glitter)

1 The long strip of paper is the body of the dragon. Fold the paper backwards and forwards to make a long zigzag shape.

2 Decorate the body of the dragon using pens, pencils or paint. Playful grassland dragons are attracted to shiny things, but if you use glitter, make sure it is biodegradable and eco-friendly.

3 On a piece of card, draw and decorate the head of your dragon. Use your experience as a Wanderer to decide what it should look like. Once you have finished, cut it out and stick it to one end of the zigzag body.

4 Cut some lengths of ribbon and use glue or sticky tape to attach them to the end of the body to make the dragon's tail. Add some to the dragon's mouth, to look like flames or smoke.

5 Use some sticky tape or glue to attach a long loop of string near the front of your kite, just underneath the head.

6 Once it is dry, your dragon kite is ready to go. On a breezy day, hold on tight to the string and let the wind lift the dragon into the air. Run towards the wind if your dragon needs a little extra lift. As you are flying your kite, keep an eye out for a grassland dragon, who will be curious about the colourful, unknown dragon in the sky.

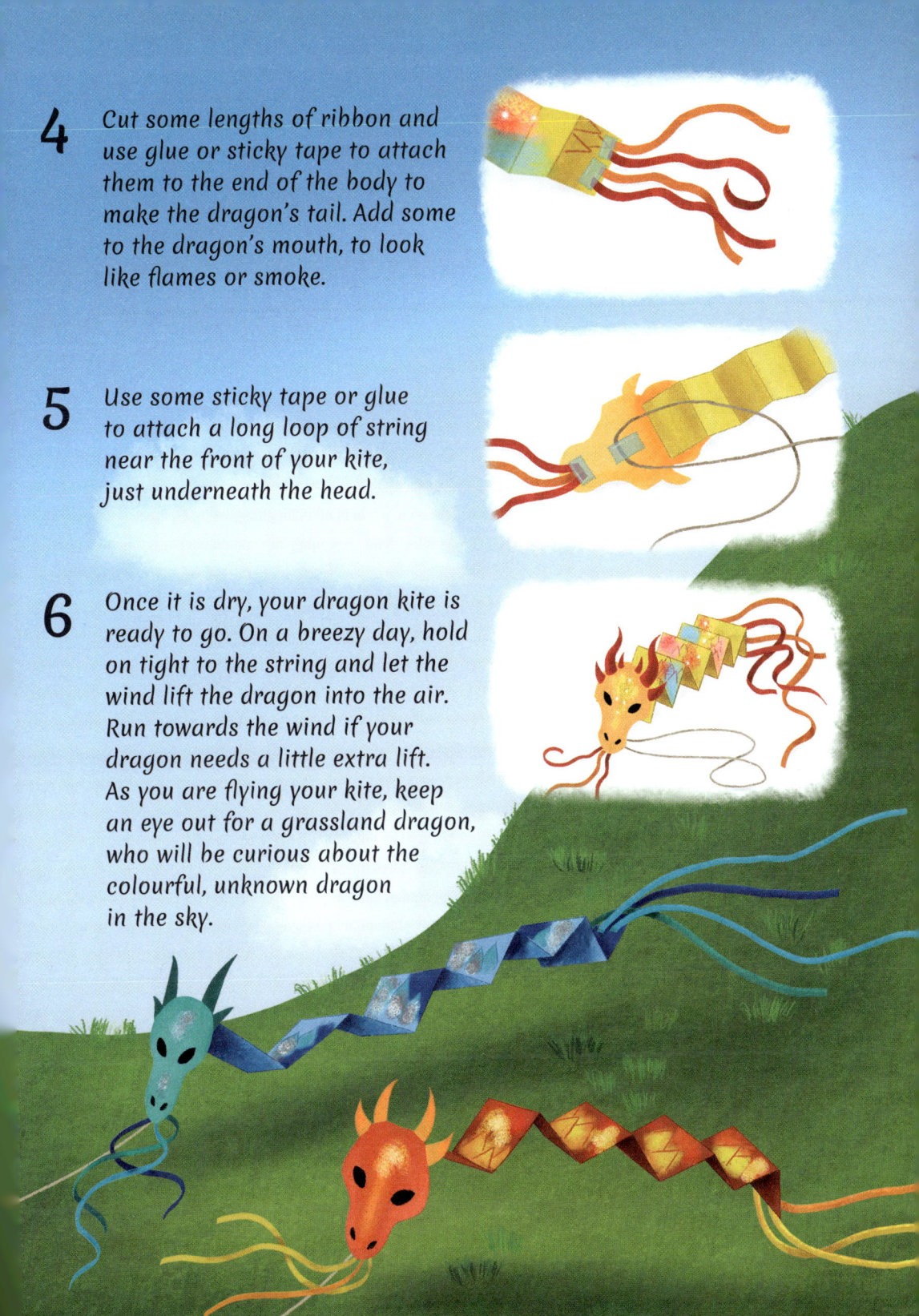

Grassland Flowers

The flowers of the open grasslands are colourful and distinctive. They support lots of different types of wildlife, including bees, butterflies, mice and grassland dragons.

Dandelion

The dandelion's bright yellow head is made up of lots of little flowers called florets, and its leaves are long with jagged edges. The flower closes up during the night. Young grassland dragons often suffer from aching wings. Dandelion sap soothes them, so the dragons can sometimes be seen rolling around in dandelion patches.

Cornflower

Like the dandelion, the head of the cornflower is made up of tiny blue florets. The outer florets are star-shaped, and there are smaller, purple ones in the middle. This flower is also known as Dragon's Delight, as it tastes like super sweet, juicy berries to a grassland dragon.

Red clover

Found in all types of grassy areas, the pinky-red flowers of the red clover are rounded in shape. The leaves are grouped into threes and each leaf has a white V-shaped marking on it. Wood mice like to collect the leaves for their nests.

Lupin

The flowers of the lupin plant are usually blue-purple, but you might also see them in shades of white, red, pink and yellow. The leaves are covered in short, silvery hairs. These plants can grow up to 1.5m tall. They are a favourite of grassland dragons, who swoop down low, snatch up long stems, and carry them away to eat.

Musk thistle

The reddish purple musk thistle flower has stiff, sharp spines. Its leaves are a shiny green colour, with white veins. The heavy flower heads can drag the plant down and make them look like they're nodding which is why they're sometimes called the Nodding Thistle.

Rippling wingspur

This dark red flower is a tasty snack for a young grassland dragon who has spent the morning flying. It hides amongst groups of cornflowers, but dragons are taught by their parents to seek them out. The wing-shaped petals are not only delicious but also contain magical properties: they can boost the energy of a tired grassland dragon.

Flight Patterns

At a distance, it can be hard to work out if you're seeing
a bird or a dragon. One way to tell the difference is by looking
at their flight patterns (the paths they take through the sky).

Direct Flight: Birds like ducks, herons, shorebirds and blackbirds fly
in a straight and level path by continuously flapping their wings.

Soaring: Large birds, such as kites and eagles, can travel long distances by
soaring. Their long wings catch currents of air, so they drift upwards
without having to flap.

Glide: Swifts and swallows glide in order to travel across long
distances without tiring themselves out by flapping.

Undulating Flight: Woodpeckers and finches fly in an up-and-down, roller-coaster pattern. They flap their wings to fly upwards, then glide as they descend.

V Formation: Migrating birds often fly in a V pattern, each bird slightly higher than the bird in front. The bird at the very front of the V has the hardest job leading the flock, so other birds take turns to help save energy.

Flap and Glide: Grassland dragons, ravens and owls will take a break from constantly flapping their wings by gliding through the air. This allows them to save energy for when they have to chase after prey.

Animals of the Grasslands

All throughout the year, the grasslands are bustling with life. The tall, unspoilt grass and shrubs provide a perfect hiding place for many types of wildlife, so you'll have to look carefully.

Rabbit

With its grey-brown fur and short, fluffy white tail, the rabbit may be hard to spot, so look for long ears poking out of the grass. They prefer meadows that have lots of grass and herbs for them to eat. They shelter in burrows deep underground.

Lizard

Lizards can be spotted basking in the warmth of the sun. Their tails can be twice as long as their bodies. If a lizard feels threatened, it will shed its still-moving tail to distract the attacker. It will regrow the tail later, but it is usually shorter.

Mole

Moles have dark brown fur and pink snouts. They live in underground tunnels, which they dig with their powerful front paws. Their favourite food is earthworm. Their small eyes are only able to see light and dark, but they have good hearing. When digging tunnels, they push soil upwards, creating molehills.

Starling

The dark feathers of the starling have a metallic purple-green sheen, and are speckled white. Their beaks are dark in winter, but turn yellow in spring. The starling prefers to live in open meadows and grasslands, where there are lots of insects and fruit for them to eat.

Red fox

The bright, rusty-red fur of the fox is very distinctive. They have white fur under their chins and bellies, and on the tips of their tails. Red foxes are usually seen in pairs or small family groups. They mostly hunt in the early morning before sunrise and in the late evening.

Vole

Voles have grey-brown fur and short, velvety tails. They have small, hairy ears and small noses. They make burrows under the grass and move around through a network of tunnels. Because they are so small, they're often hard to see, but you might hear them squeaking, chattering and rustling as they move through the grass.

Taking the Grassland Home

You can bring a bit of the Wild into your garden, and provide for visiting wildlife by collecting wildflower seeds. Wanderers are protectors of the Wilds and some flowers are very rare. Only choose flowers there are lots of and only ever take one of each type, leaving the rest for the Wilds to enjoy.

You will need:
• a newspaper • scissors • coloured paper • food processor
• wildflower seeds • silicone mould

1 Feel the flower head carefully. If the seeds fall away easily, it means they're ready to harvest. For poppies, you will hear the seeds rattling loosely inside when they're ready.

2 Cut off the flower head very carefully with scissors.

3 Spread the flower heads out on newspaper and leave in a cool, airy place so the seeds can dry out. When the heads feel dry and crumbly, they are ready for the next step.

Be careful not to handle plants like poison ivy or giant hogweed - they can cause a rash.

4 Tear your coloured paper into small pieces (2cm by 2cm is ideal). Two sheets of coloured paper makes about four seed bombs.

5 Soak the pieces of paper in a bowl of water for 20 minutes. Pour away most of the water and blend the squidgy paper in the food processor. Ask an adult to help with this part.

6 Once your paper is blended to a paste, squeeze out any excess water. Press some paper into the bottom of your silicone mould, just enough to cover the bottom.

7 Add a pinch of the wildflower seeds to the mould. Try to keep the seeds in the middle, away from the edges.

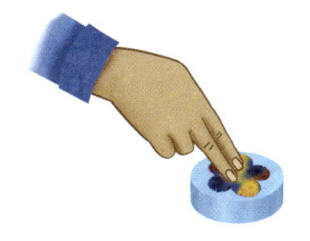

8 Add more paper on top of the seeds and press down to seal the edges. Let your seed shapes dry overnight.

9 Once they are dry, gently pop out your wildflower seed shapes from the moulds. Head out into the garden with an adult and bury your seeds in a flower patch or large pot.

Fearless Dragons of the Coasts

As the seabirds fly off to spend the winter in warmer climates, leaving the cliffs and seastacks empty, the coastal dragons come out to play. When the autumn sun starts to sink into the horizon and the nights become chillier, the coastal dragons grow bolder and more daring.

The dragons clamber up the edges of the rocky cliffs, and then leap off, diving down to the rocky waters below. Their wings brush the tops of waves as they glide just above the water. When play ends, they curl up in sheltered, rocky caves along the water's edge and sleep.

Coastal Plants

These are fascinating and unusual plants, and like the dragons that protect them, they thrive in wind-swept, salty coastal areas.

Bladderwrack

This olive-brown seaweed typically grows on rocky shores. The branches have round air-filled "bladders" which allow the plant to float upright when underwater. After it has washed up on the beach, it provides shelter for all kinds of shore creatures.

Gutweed

This fast-growing seaweed is found in rock pools, muddy puddles and on sandy beaches. Like the bladderwrack, it has small bubbles of air trapped inside its branches. Gutweed releases spores that not only turn pools of water green, but make coastal dragons sneeze. A lot. If you find a patch, listen carefully and you might hear one.

Sea bindweed

The sea bindwind has kidney-shaped dark green leaves and a trumpet-shaped pale pink flower with white stripes. This hardy vine grows on sand dunes, cliff-top paths and in salty gravel patches where few other plants can survive.

Viper's bugloss

This upright, spiky plant can be found on coastal cliffs, sand dunes and shingle beaches. It has hairy stems, narrow leaves and funnel-shaped flowers that start off pink and later turn vivid blue. It is very popular with bees and insects, who enjoy its blue pollen.

Marram grass

The spiky tufts of the marram grass are often found growing on sand dunes. The leaves are rolled up so they can trap moisture and not dry out in the sand. Marram grass roots form a large, underground blanket that traps sand and stops the dunes from being blown or washed away except in the biggest storms.

Sunstar dragon weed

If you're patient, you can see this weed creeping over rocks. If left undisturbed, it will cover a large boulder within two hours. The yellow star-shaped flower grows five black 'claws' that contain a foul-smelling liquid. It is poisonous to humans but the coastal dragon, who can smell it from a distance, will fight each other for the chance to eat it.

Mussel

The two halves of a mussel's shell are long and form a wedge shape. The outside of the shell is dark blue or black, but the inside is silvery and shimmery. Mussels eat microscopic creatures that float in seawater.

Octopus

With its big head, large eyes and warty skin, you'd expect the octopus to be easy to spot. However, it can hide in plain sight by instantly matching the colour of its surroundings. Predators like sharks will swim past without noticing it.

Dog Whelk

The shell of the dog whelk is small and pointed, and can be found in a variety of colours. They feed on other small sea creatures using their toothed tongue to bore a hole the shell of their prey.

Rock Pool Dwellers

As the tide goes out over a rocky shore, it leaves behind pools of salt water full of sea life. Rock pools are like miniature worlds full of colourful creatures, from wriggling octopuses to scuttling crabs. Here are some that you might spot.

Shrimp

Shrimp can be found in shallow rock pools, hiding in cracks or under stones. They grow to around 2-5cm, or about the length of your thumb. They eat tiny floating animals, catching them with their front limbs.

Hermit Crab

Hermit crabs make their homes inside abandoned seashells. As the crab grows, it swaps its home for bigger and bigger shells. Small hermit crabs are shy and hide in their shell if they feel threatened, but bigger ones run away backwards with their claws raised to defend themselves.

Shells

Some sea creatures grow hard shells to protect their soft bodies from sharp rocks and hungry predators. When the animals die, their shells can be found washed up along the shore.

Sea glass

Most of these smooth, colourful treasures are pieces of glass that the sea has worn smooth. But the luckiest Wanderer may find a coastal dragon scale lying among them.

A Dragon's Eye

This type of stone has a naturally occurring hole through it. Legend says these stones belong to the Wanderers of old. They would press the stone to the side of a cliff and look through it to see if there were dragons sleeping in hidden caves underground.

Treasures of the Sea

The sea hides many mysterious things, but on rare occasions it shares its secrets. If you keep your eyes peeled and happen to be in the right place at the right time, you might find some treasure along the shoreline.

Fossils

Fossils are the outlines of animals or plants from millions of years ago, that have been preserved in layers of rock. They can be as tiny as a grain of sand or as huge as a dragon's leg bone. You can sometimes see them in rocks that the sea has worn away.

Teeth

Both sharks and baby coastal dragons shed their teeth, and they can often be found on the beach. A Wanderer who has found a coastal dragon's baby tooth will wear it on a cord around their neck as a sign of their skill to other Wanderers.

Looking for Rocks

Nature's treasures are just beneath our feet, waiting to be discovered. You can find interesting rocks in almost any natural environment, as long as you know what to look for.

Quartz

Quartz is a hard, white crystal. There are lots of different types and you might have seen some in jewellery. You can often find it left behind where softer rocks have been worn away, for example in the beds of streams. It also occurs in gravel, as white, slightly see-through lumps in among the other stones.

Sandstone

This stripy orange rock is made when layers of different coloured sand gets buried underground. Over thousands of years, the grains of sand are pressed together to form a rock. The best place to find striped sandstone is where ancient rivers and oceans once flowed.

Slate

Slate is a hard, dark grey rock made from volcanic ash that has been pressed together over millions of years. When struck in a particular way, slate splits into thin, flat sheets. You might have seen slate tiles on the roofs of houses.

Mica Crystals

If you look at a rock and see tiny sparkles or flashes of light, then you may well be seeing mica crystals. They form in thin, slightly see-through sheets that can be peeled off. While it is mostly white, you might see pink, purple, yellow or red crystals, too.

Flint

In areas with light, chalky soil, look for grey or black stones that have pointy edges. Flint breaks into sharp-edged pieces, which is why it was used to make arrowheads in the Stone Age. Flint arrowheads are triangular with a sharp point. If you're really lucky, you might spot one when exploring the Wilds.

Mysterious Dragons of the Dusk

Wanderers will often speak in hushed whispers about the dusk dragons. These beautiful creatures haven't been seen in centuries, but dedicated Wanderers continue to look for signs that they still live among us.

As the sun sinks into the distance and twilight settles across the land, keep looking to the skies. Watch for a shadow across the moon, or a single star burning more brightly and fiercely than the rest.

It may be that the dusk dragons are finally returning.

In a Split Second...

As autumn moves into winter, two coastal dragons meet.
They show off to each other, diving and swooping through
the clouds. Their dance causes swirling air currents, creating
a thunderstorm. As lightning strikes the sand, it leaves behind
a very rare memento of that meeting: a fulgurite,
or petrified lightning.

When lightning hits
the sand it does something
amazing: it makes a sculpture of itself.
Damp sand conducts electricity, and the lightning is so
hot that when it strikes, it melts the sand into a delicate glass tube.

Fulgurites don't look like the transparent glass in your windows.
They can occur in lots of colours, from black or sandy brown,
to green or translucent white, depending on the type of sand.
The largest excavated fulgurite was 5m (17ft) long, but it is more
common to find smaller pieces because they are very delicate and
break easily.

Stories in the Stars

Since the earliest times, people have tried to find patterns and stories in the stars. When someone found a pattern, they grouped it together into a constellation and gave it a name. There are 88 constellations that all fit together in the sky like a twinkling puzzle.

Draco is a constellation that circles the North Pole, where it can be seen all year round. Its name is Latin for dragon and Wanderers believe it to be the last dusk dragon.

Centaurus, the centaur, is a bright constellation in the southern sky. Alpha Centauri, the nearest star to our solar system, is the star at the end in the front leg of this constellation.

The **Phoenix** constellation can be seen in the southern skies. The Phoenix, along with the constellations Grus, (crane), Pavo (peacock) and Tucana (toucan), are known as the Southern Birds.

Ursa Minor, "Little Bear" in Latin, is an ancient constellation in the northern sky. It is very important for navigation, because Polaris, the brightest star in this constellation, is the North Star.

Flight of the Last Dusk Dragon

Wanderer legend tells that, when the last dusk dragon flew above the earth, it scattered its scales across the night time sky, creating the stars. Wanderers continue to look for dusk dragons but the only information we have comes from pictures in the stars.

This constellation is **Egan**, the light dragon. Wanderers believe that Egan made the sun rise every day. It can be seen for a split second as the sun dips below the horizon each day.

Dalinda, the sea serpent constellation, was the protector of the oceans. It can only be seen when standing on a 200m sea cliff at exactly 11.59pm. It lights up the sky for just 30 seconds each night.

Tyson and Tatsuya were brother and sister dusk dragons, who couldn't stop fighting. Their fight spilled out into the sky where they became entwined forever. This constellation can only be seen after a meteor shower during a snowy day in December.

The Legends of the Polar Lights

In the far north and south of this world, when the conditions are right, lucky Wanderers may see an astonishing display. Billowing curtains of multicoloured light stream across the night sky in a phenomenon known as the aurora, or polar lights. This incredible sight is the result of streams of tiny particles speeding from the Sun and striking Earth's atmosphere.

However, many cultures around the world have different explanations for this beautiful light show.

In tales told by the Inuit in Alaska, the lights are the souls of animals, like salmon and deer, travelling to heaven.

Some Chinese legends say that the lights are the result of a celestial battle between good and evil dragons who breathe fire across the sky.

Finnish myths tell of a sly Arctic fox that would run so quickly across the snow that sparks from its tail flew into the night sky, creating the lights. The Finnish word for the lights is "revontulet", which means "fox fire".

In the stories of Menominee people of northeastern USA, the lights are the torches of great giants living in the North.

According to the traditions of Aboriginal communities in parts of Australia, the lights are the blood shed by warriors fighting a great battle, or spirits of the dead rising to the heavens.

The Vikings believed that the lights were the reflection of the armour of female warriors called the Valkyries. Others believed that they were seeing the Bifröst Bridge, a giant rainbow bridge that connected Asgard, realm of the gods, to Midgard, the human world.

Wander Free

Your journey has taken you up hills to hear the piercing cry of young mountain dragons. You have made a path through dense trees to discover watchful forest dragons. You travelled through the wild flower glades of the valleys to find the playful grassland dragons. You have trekked across shingle beaches to watch fearless coastal dragons diving off the cliffs. And you have looked to the skies in search of the rare dusk dragons.

You now have all the skills you need to go out into the Wilds and protect the magnificent creatures that call them home.

You are a true Dragon Wanderer.

A Wanderer's Notebook

Use these pages to keep a record of the things you find on your explorations. You could write down all you hear, take a rubbing of fossils on the beach or stick a fallen feather onto the page. If you are lucky and keep very calm, you may even be able to sketch a sleeping dragon.

Wing

the wing has strong bones and thin skins. the rein ha is sm to that lists up in the wind.

Dragons are avery good at with and hide are rede

I've finly found a forist dragon!

Why Dose a forist dragon egg shater.

This is a Forist dragon
↓

its because if it dozent shator the dragons mother will not bno that its hache but the dha mother is allwas cladio

good ya mink

The Flaming Fire Dragon

flaming horns;

flaming body when asleep terns red hot

all dragons have to have a ruby they all are connecting to the dragons

flaming tale

Gisland Dragons

Dragon rubys

Have you

Index

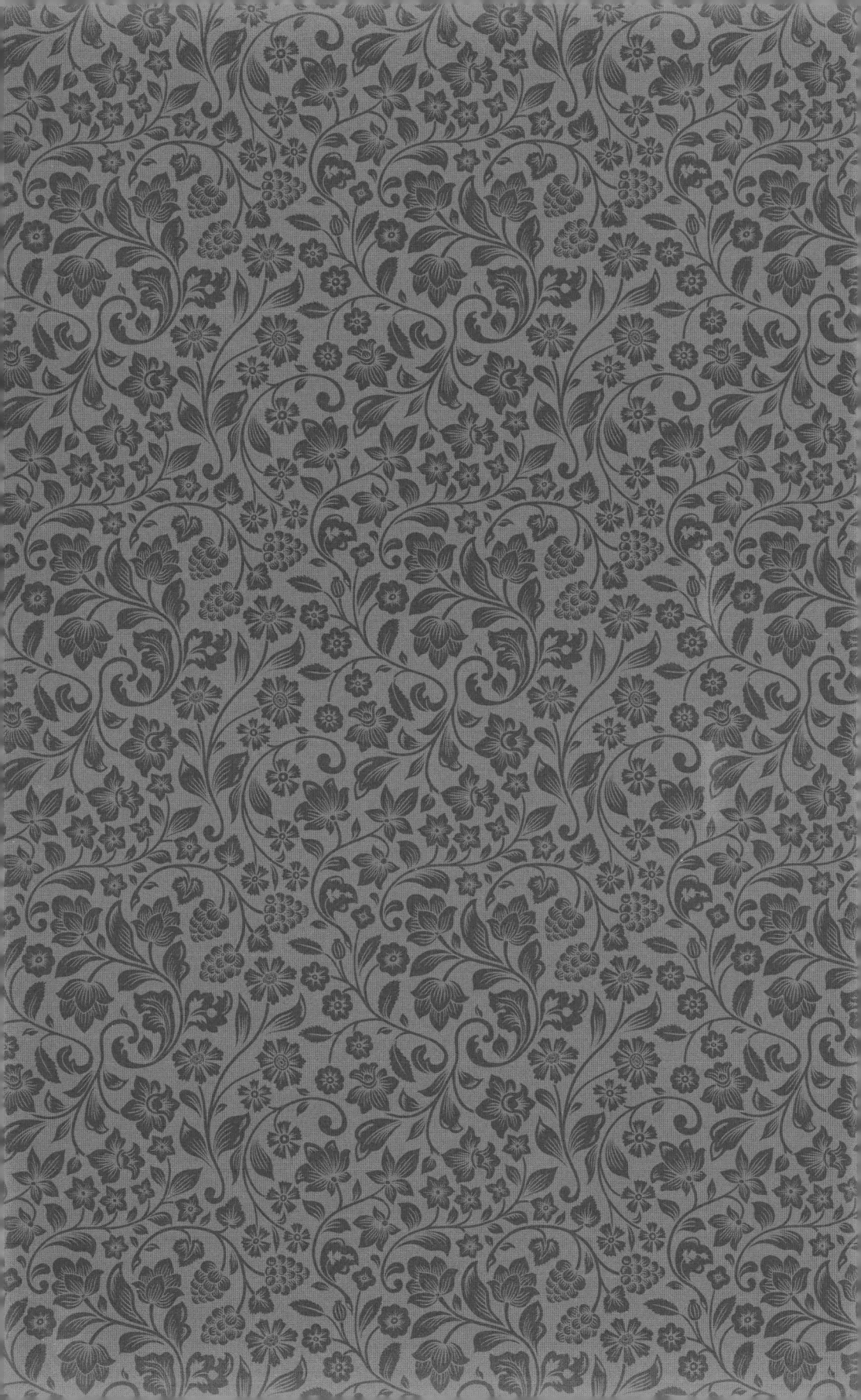

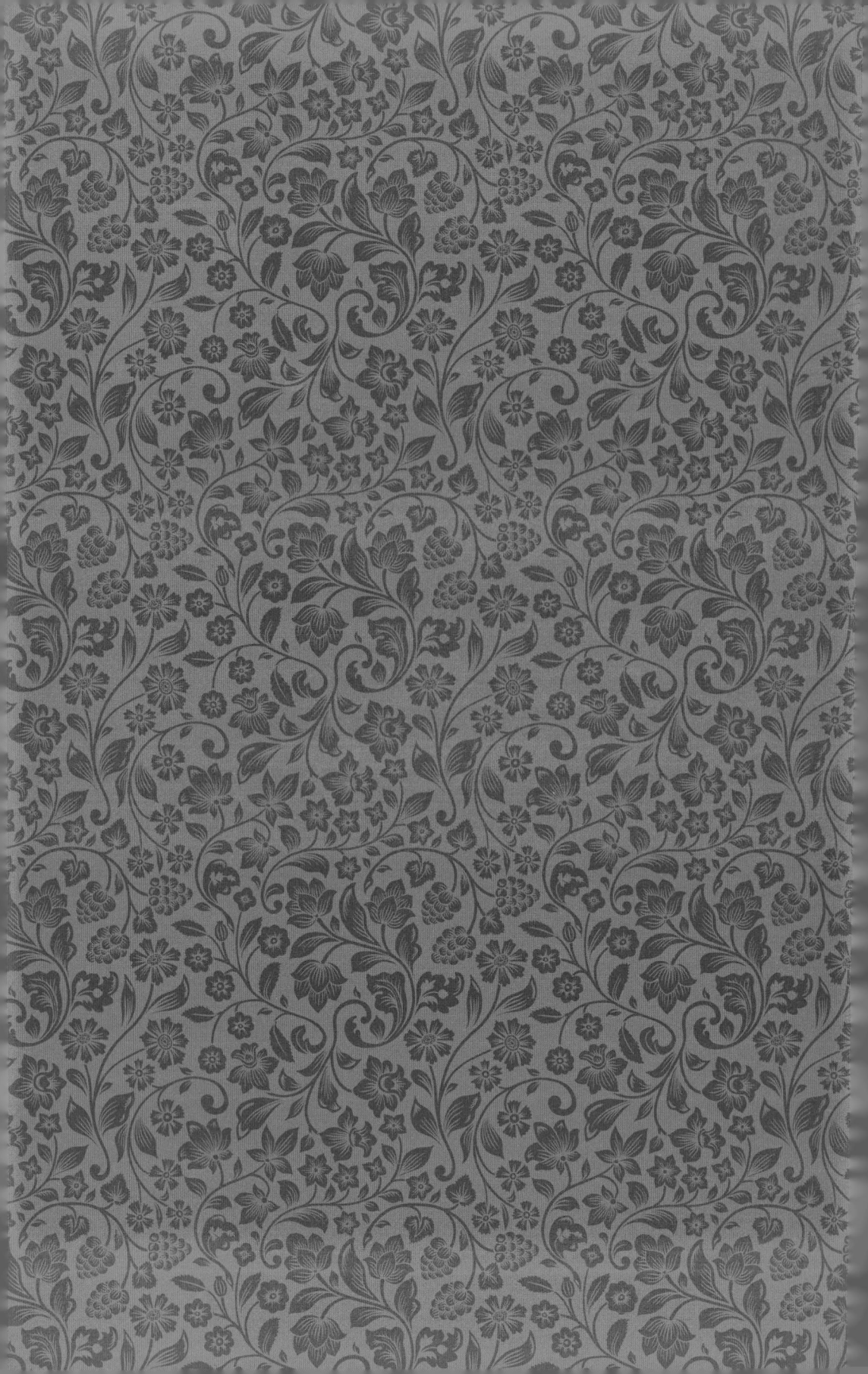